THE SEARCH FOR OTHER WORLDS

Previous Proceedings
in the Series of Annual Astrophysics Conferences
in College Park, Maryland

Year		Short Title	Publisher	ISBN
Thirteenth	2002	The Emergence of Cosmic Structure	AIP Conference Proceedings 666	0-7354-0128-4
Twelfth	2001		*"unpublished"*	
Eleventh	2000	Young Supernova	AIP Conference Proceedings 565	0-7354-0001-6
Tenth	1999	Cosmic Explosions	AIP Conference Proceedings 522	1-56396-943-2
Ninth	1998	After the Dark Ages	AIP Conference Proceedings 470	1-56396-855-X
Eighth	1997	Accretion Processes	AIP Conference Proceedings 431	1-56396-767-7
Seventh	1996	Star Formation	AIP Conference Proceedings 393	1-56396-678-6

Other Related Titles from AIP Conference Proceedings

689 Neutrinos, Flavor Physics, and Precision Cosmology: Fourth Tropical Workshop on Particle Physics and Cosmology
Edited by José F. Nieves and Raymond R. Volkas, October 2003, 0-7354-0160-8

686 The Astrophysics of Gravitational Wave Sources
Edited by Joan M. Centrella, October 2003, 0-7354-0157-8

672 Short Distance Behavior of Fundamental Interactions: 31st Coral Gables Conference on High Energy Physics and Cosmology
Edited by Behram N. Kursunoglu, Metin Camcigil, Stephan L. Mintz, and Arnold Perlmutter, June 2003, 0-7354-0139-X

668 Cosmology and Gravitation: Xth Brazilian School of Cosmology and Gravitation; 25th Anniversary (1977-2002)
Edited by Mario Novello and Santiago E. Perez Bergliaffa, May 2003, 0-7354-0131-4

655 Particle Physics and Cosmology: Third Tropical Workshop on Particle Physics and Cosmology – Neutrinos, Branes, and Cosmology
Edited by José F. Nieves and Chung N. Leung, February 2003, 0-7354-0112-8

624 Cosmology and Elementary Particle Physics: Coral Gables Conference on Cosmology and Elementary Particle Physics
Edited by B. N. Kursunoglu, S. L. Mintz, and A. Perlmutter, July 2002, 0-7354-0073-3

To learn more about these titles, or the AIP Conference Proceedings Series, please visit the webpage
http://proceedings/aip.org/proceedings

THE SEARCH FOR OTHER WORLDS

Fourteenth Astrophysics Conference

College Park, Maryland   13-14 October 2003

EDITORS

Stephen S. Holt
Franklin W. Olin College of Engineering
Needham, Massachusetts

Drake Deming
NASA/Goddard Space Flight Center
Greenbelt, Maryland

SPONSORING ORGANIZATIONS
NASA/Goddard Space Flight Center
University of Maryland

Melville, New York, 2004
AIP CONFERENCE PROCEEDINGS ■ VOLUME 713

Editors:

Stephen S. Holt
Franklin W. Olin College of Engineering
1735 Great Plain Avenue
Needham, MA 02492
USA

E-mail: steve.holt@olin.edu

Drake Deming
NASA/Goddard Space Flight Center
Code 693
Greenbelt, MD 20771
USA

E-mail: ddeming@pop600.gsfc.nasa.gov

L.C. Catalog Card No. 2004106926
ISBN 0-7354-0190-X
ISSN 0094-243X
Printed in the United States of America

CONTENTS

DUSTY DISKS AND UNSEEN PLANETS

GIANT PLANETS

TERRESTRIAL PLANETS

DYNAMICS AND DYNAMICAL EVOLUTION OF PLANETARY SYSTEMS

PROGRAMS AND TECHNIQUES FOR DETECTION AND
CHARACTERIZATION OF EXTRASOLAR PLANETS

Preface

This is the fourteenth installment of the current series of annual October Astrophysics Conferences in Maryland. These conferences are organized by astrophysicists at the Goddard Space Flight Center and the University of Maryland. The topic for each conference is selected with the help of an International Advisory Committee, the current membership of which is:

Marek Abramowicz, Göteborg	*Jim Peebles,* Princeton
Roger Blandford, Pasadena	*Sir Martin Rees,* Cambridge, UK
Claude Canizares, Cambridge	*Vera Rubin,* Washington
Arnon Dar, Haifa	*Joseph Silk,* Oxford
Alan Dressler, Pasadena	*David Spergel,* Princeton
Guenther Hasinger, Munich	*Rashid Sunyaev,* Moscow
Steve Holt, Needham	*Yasuo Tanaka,* Tokyo
Bob Kirshner, Cambridge	*Scott Tremaine,* Princeton
Dick McCray, Boulder	*Simon White,* Munich

In the spirit of this series, where we attempt to identify "hot" topics with broad appeal to both observers and theoreticians, *"The Search for Other Worlds"* clearly qualifies as an appropriate conference theme. The programme for this year's conference was developed by a Scientific Organizing Committee comprised of:

Adam Burrows	*Steve Holt*
Paul Butler	*Lee Mundy*
Mark Clampin	*Sara Seeger*
Drake Deming	*Caroline Terquem*

After a brief welcoming address, the conference opened with what is now a traditional tandem of introductory invited presentations. The first of these was an review of the history of planetary searches by *Virginia Trimble*, and the second, delivered by *David Spergel*, defined the outstanding issues to which the conference attendees were obliged to address their attention. The programme of non-paralleled sessions then proceeded through the next two days. Each session was devoted to a specific topic with two or three invited talks and an extensive discussion period that included one-minute, one-viewgraph opportunities for each conference attendee to advertise his/her poster paper.

The banquet at the conclusion of the second day of each of these conferences always features a distinguished speaker. This year our speaker was *Peter Goldreich*, who shared some of his insights and experiences in a career that was devoted to more than just planetary studies.

The banqueteers were also regaled by the *Chromatics*, a marvelous singing group with a core of astrophysicists from the Goddard Space Flight Center who feature songs that educate as well as entertain. If we're lucky, we'll be able to make the *Chromatics* a regular feature of future conference banquets.

As usual, *John Trasco* and *Susan Lehr* made sure that all the logistics were handled flawlessly. Thanks also to *Drake Deming*, the co-editor of these Proceedings, and to all of the attendees who contributed to the success of the meeting.

Steve Holt
January 2004

Introduction

The Quest for Other Worlds, 350 BCE to 1995 CE

Virginia Trimble

Physics Department, University of California, Irvine, CA 92697
and
Department of Astronomy, University of Maryland, College Park, MD 20770

Abstract. The concept of aperoi kosmoi (multiplicity or plenitude of worlds) has had at least four meanings at various times, two capable of yielding images and two not. We explore here the several meanings and their connections with modern astronomical ideas, focusing finally on the one in the minds of the organizers of the present conference. There seem to be at least two dozen ways of potentially detecting or limiting the incidence of "other worlds," in the sense of potentially habitable planets orbiting other stars, a handful of which have led to detections, false alarms, or both.

INTRODUCTION

In the Fourth Century BCE, Epicurus taught that there are an infinite number of worlds both like and unlike ours, and Aristotle taught that there is only one. Neither hypothesis can currently be falsified. Both had in mind entire universes, with a center to which everything should fall, each observable only by the inhabitants thereof (else how should they know which center to fall toward), and Aristotle's objection was a logical one focused on the impossibility of multiple centers (Dick 1982). When theologians Roger Bacon and Thomas Acquinas reconsidered Greek ideas around 1250 and rejected plurality, this was the concept they had in mind. For all of them, a "world" would be centered on an "earth."

The modern descendent of Epicurian other worlds would seem to be the universes of self-reproducing inflation and the like, in which we may be merely one of many (or infinitely many) four-dimensional space-times, embedded in higher dimension space and apparently without the possibility of communication among them (Rees 1997, 1999). This concept cannot reasonably be expected to result in pretty pictures of the other worlds or universes, though one might be able to imagine some of their properties by deducing the consequences of different values for the constants of nature from the ones in our universe.

The second pluralistic concept is the possibility of worlds succeeding worlds in temporal succession. This was considered positively by Origen in the 3rd century CE, somewhat negatively by Buridan, of whom more shortly, and again positively by Oresme in 1277. So far, nobody mentioned has gotten into serious trouble by discussing these ideas. Modern descendants are of two sorts. First, an oscillating (big bang - big crunch) universe, in which the same matter and energy participate repeatedly. This is not possible within the framework of general relativity (or any other metric theory of gravity within

CP713, *The Search for Other Worlds: Fourteenth Astrophysics Conference*,
edited by S. S. Holt and D. Deming

which trapped surfaces form) as long as all the stuff has positive energy (as does all the stuff we know about, even the dark energy with its negative pressure). Second, the ekpyrotic and brane cosmologies, in which higher-dimension branes collide and bounce repeatedly. Essentially all of the stuff in each cycle belongs to that cycle, so that we cannot, in effect, communicate with those before or after, but some entropy perhaps carries over so that an infinite series is not possible. Again, no pictures.

The third, illustratable, concept is that of the moon, sun, and perhaps planets of our own system being habitable and inhabited worlds. This belongs almost entirely to the post-telescope era, but appears third because Lucian of Samosata wrote in the Second Century CE a *Vera Historia* of a voyage to the moon, intended primarily to explore terrestrial foolishness (Dick, 1982, Guthke 1990).

Fourth is the idea of other systems like ours and potentially detectable. Initially these were earth-centered and regarded favorably from about 1330 within the Catholic Church tradition. Proponents reasonably well known (e.g. Dick 1982) include John Buridan (ca. 1295-1358) of Paris, remembered for his parabolic ass, who starved half way between two equally desirable piles of hay, unable to make a decision between them, and William of Occam or Okham (ca. 1280-1347), who was probably bearded, despite his razor. So obscure as to have missed Dick's eagle eye was a certain Bradwardine, also educated at Oxford, who wrote around 1330 in favor of plurality (but without taking a stand on whether there were inhabitants or salvation elsewhere), and, remaining unpublished until many years later, was elected Archbishop of Canterbury in 1348, shortly before succumbing to the plague (Leff 1984). Their worlds were earth-centered. After Copernicus came Giordano Bruno and Thomas Digges advocating planets in orbit around other stars shortly before 1600. Digges died peacefully in 1595, Bruno less so in 1600. For what it is worth, the pioneering author (and translator of Fontenelle) Aphra Behn was perhaps the first woman to deny Copernican plurality in 1686, and Fanny Burney, the author of *Evalina*, the first to affirm it, in 1786. Still a century later, Maria Mitchell, the first American woman astronomer, declared that more data were needed.

Finally, if our goal is really other earths, then we must ask what it takes to make an earth-like planet. A solid body, presumably, (so you will have someplace to put your car keys), but with air and water. These require both the presence of volatiles and location within the fuzzily-defined habitable zone of a star not too different from the sun. Third, perhaps, tides, to permit tide pools. Is the sun enough, or would you also like a moon? The latter also comes in handy in helping to stabilize the rotation axis of its parent planet. Though Mars at the moment is only a bit more oblique than Earth, it is said to be capable of lying on its side from time to time.

Mantle convection and plate tectonics are useful for preserving a mix of land and water, and also in due course for producing ores and fossil fuels. Required are at least radioactive heating or melting of the core from kinetic energy of accumulation and (apparently, from a comparison with Venus) either rapid rotation or liquid water to cool descending slabs, or both. Magnetic fields for protection from solar energetic particles again require a molten core and rotation (compare both Venus and Mars). You might want a Jupiter (large mass, further out) to clear out debris and protect from excessive bombardment. But the ultimate earth-like signature would be the presence of chemically-based life, potential evidence for which was discussed by Seager during the conference.

WORLDS WITHIN THE SOLAR SYSTEM

Here we encounter the first false alarms. Galileo described the mountains and craters he had seen on the moon as implying the moon might be earth-like. It took Kepler, rushing in within weeks of seeing a preprint of *Sidereus Nuncius* to declare the round craters to be circular embankments in which the lunar inhabitants had dug their cave dwellings. Indeed he had written in a proto-science-fiction mode of an inhabited moon beginning still earlier, though his *Somnium* was published only posthumously in 1634. Kepler's moon had a moist atmosphere (hence the smooth appearance of the limb, despite the mountains and craters), and he credited observational evidence for rain there to his teacher Michael Maestlin.

By 1650, the idea of the moon and sun as inhabited worlds was common. But our favorite narrator is Cyrano de Bergerac (no, and nose, he was not just an invention of Edmund Rostand), who wrote in 1687, with accompanying wood cut, of a voyage to these other worlds, powered by a giant dew drop (he rode inside) that, like smaller dewdrops, rose at dawn. He was perhaps the last spacecraft designer to use his own invention.

The 1686 *Entretiens sur la pluralite des mondes* of Bernard le Bovier de Fontenelle was perhaps the most influential work in this tradition, helped by his living 71 years beyond its publication (50 of them as perpetual secretary of the Academie des Sciences) and by translations into English and other languages. His illustrations are so romantic that the long eyelashes suggest a gal in the moon rather than a man.

The last main-stream astronomer in this lunatic tradition was William Herschel, who began sketching the moon in 1776 (Crowe 1986). He has understood that the atmosphere of the moon must be much sparser than that of earth (or the limb would not look so sharp), but he remained convinced of the presence of inhabitants, centered around the circular structures and drew what he described as giant forests in their vicinity. I have seen Herschel's lunacy described as a deliberate hoax, but, though he never published the forest sketches, he took them seriously enough to write to Neville Maskelyne, the Astronomer Royal.

A century ago, the best-known inhabited world was Mars, with the canals, oases, seasonal agriculture, and all the rest of Giovanni Schiaparelli (1835-1910), Percival Lowell (1855-1916), and, less famously, William Pickering and others, though the opposition was nearly equally famous. Some exciting science fiction arose from these canals (*War of the Worlds*, Edgar Rice Burroughs, C. S. Lewis's *Out of the Silent Planet*, etc.) but they fell out of fashion, and disappeared forever in the cratered surface revealed by the first orbiter images. The human predisposition for pattern recognition had been carried to excess. Less well known and still more illusory were Lowell's Venusian canals. At least one of the drawings appears to be the pattern of blood vessels at the back of an eye, reflected in the lens of a telescope stopped very far down to accommodate a very bright object.

Incidentally, the very recent Galileo spacecraft images of cracks in the ice on Europa do not, without some advance instruction, look all that different from the Martian and Venusian drawings. At least no one seems to think them artificial, and they are clearest in the best images, not the most fleeting.

WORLDS OUTSIDE THE SOLAR SYSTEM

The parallel track of thinking about plurality elsewhere is marked mostly by images that look increasingly like modern images of stars and galaxies. Bruno's 1591 diagram is simply spheres of influence with letters marking their centers and the points where a test particle might have difficulty deciding which way to fall. Apparently he had in mind much larger separations between (planet-centered) worlds than could be shown on even folio paper. Descartes tesselated the universe into vortices in 1644, and Henry Regius in 1654 enhanced a Copernican diagram by putting at least one planetary orbit around each star in the "sphere of fixed stars," still apparently finite. Digges's stars extended explicitly to infinity (thought he too had the small-paper problem in drawing this), and only the caption tells you that they have their own planets.

A century later, Thomas Wright who appears most often in modern books for having drawn the Milky Way as a slab rather than a sphere, provides a close up, in which the spheres of influence of the sun, Sirius, and Rigel are all the same size (with others lurking in the background), and separated by distances not too much larger than their sizes, so that communication might not be impossible. Each system has a few circular planet orbits and some highly eccentric comets. An 1821 edition of Fontenelle once again has the solar system bigger than any of the others, as well as in the middle. Indeed the plurality issue remains somewhat entwined with the question of the location of the center of the Milky Way (often then thought of as the entire universe) until the work of Harlow Shapley between 1918 and 1922 moved us firmly to the outskirts of a very large galaxy (shown soon after by Edwin Hubble to be merely one of many).

It is perhaps not unexpected that Alfred R. Wallace (Darwin's pacer) might think of using a central location to claim that the earth was uniquely habitable and inhabited. Darwin, however, was a pluralist, and Arthur Eddington, who drew a very similar Milky Way at much the same time (1912 to Wallace's 1903) could envisage a central star cluster and ring around it without insisting upon anthropocentric implications. Their ring, incidentally, was more or less Gould's belt of OB stars, not the sort of ring or arc recently found and attributed to a torn up companion galaxy.

A last barrier to plentitude was the Chamberlin-Moulton hypothesis for the formation of the planetary system. Through most of the 18th and 19th centuries, most scholars had imagined something like the Kant-Laplace nebular hypothesis, in which the planets (etc.) arose from a disk of residual material around a star as it formed. This would make planetary systems common. Indeed there was a school of thought that identified the spiral nebulae as such systems in formation. A serious physical objection to this picture was that it would seem to imply that the sun should contain its fair share of the angular momentum of the solar system, instead of 99.9% of the mass, but only 1% of the angular momentum. Much of this disparity we now blame on the gradual slowing of solar rotation by the solar wind, much more powerful in the past.

But to Chamberlin and Moulton (and later Jeans, Jeffreys, Lyttleton, and others), it meant that the material to form the planets must have been dragged out of the sun by a close stellar encounter. If so, then there might be only one or a few systems in the whole Milky Way, since such encounters are rare (it's a long way between stars).

The tidal encounter hypothesis began to go out of fashion in the 1940s, because it is hard to make the hot stuff condense as required. Using modern data, we know that the presence of deuterium in the planets means their substance can never have been inside the sun, where all the deuterium long ago fused to helium. But the prediction of planetary rarity did not actually keep people from looking in the early years of the 20th century or, indeed, from announcing that they had found planets orbiting other stars.

RECENT HISTORY - SEARCH AND DETECTION STRATEGIES

The table shows the most complete list I have been able to compile of methods whose success or promise has been claimed in the (more or less) refereed literature. A few were actually added in the course of the conference (evaporated comets, X-ray flashes from planetary collisions). References are given for only a few of the less well known, and the meanings of the codes F, T, ?, and * are briefly explained again at the foot of the table. Less succinctly (and the opinions are not necessarily those of the editors, all participants, or a majority of the total community), the F's are false alarms. That is, someone published an announcement of one or more planets found in the tabulated way that later turned out to be something else (often though not always underestimated errors). This does not preclude real planets having been found by the same method later. The T's I regard as observations truly indicative of planetary mass objects in orbits around solar-type (mostly) stars. The ?'s indicate real astronomical phenomena for which other explanations are possible. In some cases (Mira variables) I think the others are much more probable. In other cases (gaps in disks around young stellar objects, much discussed at the conference) planets may well be the most likely explanation.

Table: Search and Detection Methods

1. Direct imaging (F)
2. Periodic residual in proper motion (F, T)
*3. Periodic variation in pulsar period (F, T)
*4. Periodic variation in radial velocity (F, T)
*5. Transits (T)
6. Perturbation of line profiles by planetary atmosphere (?)
*7. Blips in microlensing light curves (T)
8. Gaps in YSO accretion disks (?)
9. Warps in YSO accretion disks (?)
10. Collimation of bipolar ejecta (?)
11. Pollution of host atmosphere (F)
12. Distortion of times of CBS eclipses (gravitational effect, not transit, F?)
13. Nova-like outburst of V838 Mon as AGB swallowing planets (?, R1)
14. X-ray or gamma ray bursts as collisions (?, R2)
15. OH, H_2O in evolved carbon star (?, R3)
16. Embedded planet in RG atmosphere mimics Mira pulsation (?, R4)
17. Masers in planetary atmospheres, vs. host stars (?)

18. Zodiacal light
19. Independent confirmation of panspermia
20. Success of SETI
21. Incredible luck with Pioneers or Voyagers
22. GRBs as exhaust from spaceships: 4-d lines of events with 511 keV features (R5)
23. Arrival of LGMs
24. Something even more outlandish

F: False alarms   T: Real detections
*: Discoveries   ?: Phenomenon exists; planetary connection uncertain

R1: Retter & Marom (2003)
R2: Zhang & Siggurdsson (2003)
R3: Ford et al. (2003)
R4: Berlioz-Arthaud (2003)
R5: Harris (1990)

The *'s mark techniques that have led to discovery of one or more planets where none was previously known. Notice that not all T's are *'s (and indeed not all snarks are boojums). Notice that "pollution of host atmosphere" is marked F. Clearly some planets do migrate to destruction and this will increase their (former) hosts' surface metallicity. But, as emphasized by Fischer and others at the conference, this is not the dominant effect. Host stars were relatively metal-rich to begin with.

The story of the successes really belongs to the rest of the conference, so we begin here with the others. The title of King of the False Alarms (indeed arguably King of Frauds) surely belongs to Thomas Jefferson Jackson See (1866-1962), who spent the last 60 years of his life in scientific exile (largely on Mare Island) but received a fairly nice tribute in *Sky and Telescope*. In 1897, he announced (in *Atlantic Monthly*!) that he had actually imaged planets around some nearby stars, though he never said which ones, and, curiously, his intent seems to have been to demonstrate the uniqueness of our earth and solar system. This claim never reached the astronomical literature, but his discovery of a planet orbiting the secondary star of the visual binary 70 0ph did (See 1896). So acrimonious, however, became his disputes with doubters that the pages of AJ and other standard journals were soon closed to him. He argued for the rapid rotation of Venus (true anyhow for its atmosphere) and against the Chamberlin-Moulton hypothesis (false, though not for his reasons) and has had mild cult status, like Velikovsky, at times.

Curiously 70 0ph B was one of the two stars (the other was 61 Cyg) for which Kaj Strand recorded a false-alarm planet in 1943. His announcement, like that of See, was based on small deviations in proper motion from the expected orbit. Best known in this balliwick is Peter van den Kamp, who gave proper motion planets to Barnard's star and Lalande 21185 in 1944 and upped Barnard's star's allotment to two in 1963. Most of us have heard at least a folk-tale version of the problem with the Sproul Observatory discoveries, which was that the telescope had been disassembled in 1949 and reassembled with some slight difference in the alignment which then mimicked periodic residuals.

George Gatewood of Alleghany Observatory and Heinrich Eichhorn of the University of South Florida were instrumental in showing that the Sproul results could not be duplicated for either star. Curiously, Gatewood (1996) later put forward another very tentative planet for Lalande 21185 which has also not been confirmed.

Even the two radial velocity methods, the pride of our discipline, have had their junior moments. Campbell et al. (1988) reported periodic velocity residuals with periods of a few years and amplitudes of 10's of m/sec for more than half the solar-type stars examined, including ε Eridani, object of the first SETI search. Some combination of variable star activity and under-estimated error bars seem to have been responsible. Implied masses were a few jupiters.

Most recent and so best remembered is the timing residual for psr 1829-10 reported by Bailes et al. (1991). Dozens of models were in press before the authors announced that the six-month period was due to a planet all right, but the planet was earth, whose orbital eccentricity had not been correctly removed from the data. Those models then acquired the status of predictions and were ready and waiting when Wolszczan and Frail (1992) announced the first two (of three) terrestrial mass planets orbiting the millisecond pulsar 1257+12. Lightning has not really struck again. The only other strong pulsar-planet candidate belongs to a binary pulsar in a globular cluster.

But we are always ready and waiting for the next good chance. As early as 1920, Scientific American published a grid-type drawing to be radioed in dots and dashes to some promising life-host planet as soon as one should be discovered. The image resembles the plaque actually sent with the Pioneer spacecraft, except Carl and Linda were better looking.

On the side of the successful methods, only pulsar timing and distortions of MACHO light curves (methods 3 and 7) can currently reveal masses as small as that of the Earth, though transits and proper motions measured from space someday should. As for discoveries within our own solar system, Uranus was found by direct imaging, Neptune (and to a certain extent Pluto) from proper motion residuals, and Vulcan from transits, of which several were reported soon after Leverrier announced the anomalous advance of the perihelion of Mercury. Motion of the earth was actually discovered as aberration of starlight, a special relativistic effect, but later from annual variations of radial velocities.

ACKNOWLEDGMENTS

I am grateful as always to Steve Holt and his SOC for the invitation to provide this review and to Susan Lehr for transforming a typed(!!) original into keystrokes. Ideas not elsewhere credited have been lifted bodily from Dick (1982, 1998), Crowe (1986), Guthke (1990), Whewall (1853), and Fontenelle (1990). Some are even now being recycled for undergraduate courses on life in the universe.

REFERENCES

1. Bailes, M. et al. 1991, Nature, 352, 311.
2. Berlioz-Arthaud, P. 2003, A&A, 397, 943.

3. Campbell, B. et al. 1988, ApJ, 331, 902.
4. Crowe, M. J. 1986, The Extraterrestrial Life Debate, 1750-1900, CUP.
5. Dick, S. J. 1982, Purality of Worlds (Democritus to Kant) CUP.
6. Dick, S. J. 1982, The Biological Universe, Cambridge University Press.
7. Fontenelle, B. (Trans. H. A. Hargreaves) 1990, Conversations on the Plurality of Worlds, Univ. of California Press.
8. Guthke, K. S. 1990, The Last Frontier: Imaging Other Worlds from the Copernican Revolution to Modern Science Fiction (translation H. Atkins), Cornell UP.
9. Harris, M. J. 1990, J. Brit, Interplan. Assoc., 43, 551.
10. Leff, G. 1984, Bradwardine and the Pelagians, Cambridge UP.
11. Rees, M. J. 1997, Before the Beginning: Our Universe and Others, Cambridge UP.
12. Rees, M. J. 1999, Just Six Numbers, Basic Books.
13. Retter, A. & Marom, 2003, MNRAS, 345, L25.
14. See, T. J. J. 1896, AJ, 16, 17.
15. Whewell, W. 1853, Of the Plurality of Worlds (ed. M. Ruse, 2001), U. Chicago Press.
16. Wolszczan, A. & Frail, D. A. 1992, Nature, 355, 145.
17. Zhang, B. & Siggurdsson, S. 2003, ApJ, 596, L95.

Statistics of Planetary Systems

Detection and Characterization
of Extrasolar Planets

Chris McCarthy[*][†], Paul Butler[†], Debra Fischer[**][‡], Geoff Marcy[‡] and
Steve Vogt[§]

[*]San Francisco State University, CA
[†]Carnegie Institution of Washington, DC
[**]San Francisco State University,CA
[‡]University of California, Berkeley, CA
[§]University of California, Santa Cruz, CA

Abstract. Doppler monitoring of a sample of 1330 stars with a precision of 3 m s^{-1} is discussed. Seventy five planets residing in 65 systems have been discovered within this sample, a detection rate of 5%. Eight multiple systems are found in the sample, and roughly half of the systems with one known planet appear to harbor additional companion(s). The incidence of "hot jupiters", orbiting within 0.1 AU is 1.0 %. There appears to be a "pile up" of planets at P=3 days. Planets with even shorter periods could easily be detected but none was found indicating they are exceedingly rare.

INTRODUCTION

Precise Radial Velocity (PRV) observations of 120 stars began in 1987 at Lick Observatory, with a typical precision of 10 m s^{-1} , which improved to 3 m s^{-1} in 1995. By 1998, observations of additional stars had begun at Keck and the Anglo Australian Observatories, and the Lick target list was expanded, so that in total 1330 stars have been studied for at least 5 years, rendering giant planets in orbits smaller than 3 AU detectable.

Active PRV research programs are underway at other observatories as well. The result of all such studies have been summarized recently [1]. Surveys such as these provide us with the end state of the planet formation and evolution process, allowing theoretical studies of these processes to be constrained.

In the present paper we will attempt to draw statistical conclusions from a known sample of extrasolar planets. For this purpose we will consider only the 1330 stars studied at the three observatories mentioned above since all three programs have an identical analysis algorithm [2] and similar errors of $\sim$ 3 m s^{-1} . Also, several other exciting planet detection topics, such as transits and direct imaging, which are discussed elsewhere in this volume, will not be reviewed here.

SURVEY SAMPLE

With the exception of stars on the original Lick program, the vast majority (1200) of stars under study were drawn from the HIPPARCOS catalog subject to the following criteria.

CP713, *The Search for Other Worlds: Fourteenth Astrophysics Conference,*
edited by S. S. Holt and D. Deming

Spectral types later than F8V (B-V > 0.55) were chosen because earlier types preclude precise radial velocity measurements. Binaries with separations less than 2 arcseconds were excluded in order to distinguish the spectra of the two components. Young, active stars (age > 2 billion years) were removed using chromospheric activity and rotation period as age diagnostics [5] Giants were excluded. No F and G stars fainter than V = 10 were selected, while for K and M stars the cutoff was V = 11. The vast majority of all FGKM stars brighter than V = 8 are on our survey.

RESULTS

Hot Jupiters

Of the 1330 stars studied to date, 13 (1%) harbor a "hot jupiter" orbiting within 0.1 AU. Such planets are easy to detect and it is very unlikely any more will be found in this sample of stars nor in in the samples similarly studied at other observatories. While constituting only $\sim 20\%$ of the presently known extrasolar planets, hot jupiters nevertheless present an exciting prospect for understanding exoplanets:

1. Hot jupiters present the only realistic hope of finding more transiting planets. Already the transiting planet around HD 209458 has provided a wealth of information on planet density and composition [6].

2. Stars known to harbor one planet have a greater likelyhood of harboring a second planet [7]. Hence any future hot jupiters detected can serve as potential tracers of multiple planet systems.

3. Hot jupiters (HJ's) can also serve as a probe of the internal structure of giant planets. The rate at which a planet's orbit circularizes due to tidal interactions with the star depends upon the internal structure of the planet, in particular the presence or absence of a solid core. While the few known HJ's have eccentricities consistent with zero, discovery of additional HJ's is required to better assess the composition of giant planets and hence constrain formation scenarios.

4. Hot jupiters provide a laboratory to study the planet formation process. No planets on our surveys were found with orbital periods shorter than 2.99 days. This apparent "pile up" suggests that a process which disrupts the planet formation mechanism is operating near most stars. While our data constrain the frequency of exceedingly close (P < 2.5 d) HJ's around nearby field stars to < 0.1%, a planet with just such a period has been reported around a star in the galactic bulge.
As discussed by Marcy et al. [1] the discovery of photometric and radial velocity variability in the bulge star OGLE-TR-56 consistent with a 1.4 M_{JUP} planet companion in a 1.2 day orbit. This discovery [3] [4] presents a conundrum requiring better knowledge of HJ's to solve, to wit: "Why would the first planet ever discovered by its transit have an unprecedented orbital period, 1.2 d, and be found before any of the dozens of normal 3.5d orbital periods that must also exist in the OGLE field?" [1]

To better address all of the above issues, we have organized an international consortium (D. Fischer, PI) to survey the 2000 stars not currently being monitored by radial velocity groups, with the specific intention of discovering of order 1 dozen more hot jupiters.

Multiple Planets

To date eight multiple planet systems have been identified [1] two of which (Ups And [8] and 55 ρ^1 Cancri, [9]) have three planets. Such systems provide a host of opportunities for the study of orbital physics. Multiple planets are often found in resonances such as the 2:1 resonance exhibited by the two planets orbiting Gl876. It has been shown [10] [11] have shown that models which incorporate planet-planet interactions achieve a better fit to the doppler data than simple two-Keplerian models. Since interacting models depend upon the true masses of the planets, such systems, when adequately modeled with sufficient data, proffer the opportunity to resolve the $\sin i$ ambiguity that otherwise affects individual doppler planet detections.

Metallicities

Fischer & Valenti (in prep.) have performed LTE spectral synthesis of all stars currently surveyed by our group for extrasolar planets. The frequency of planetary companions is a strong function of host star metallicity (e.g., [12]). This clean result, derived from the actual stars being surveyed, supports the conclusions of earlier work [13] which made the less direct comparison between planet bearing stars and field stars.

Statistical Summary

Doppler survey are least sensitive to the lowest mass planets, with our current threshold being somewhere below 1 saturn mass. It is therefore remarkable that the most populated bin in a mass histogram of detected planets (see e.g., http://www.exoplanets.org for a recent histogram) is the 0-1 M_{JUP} bin, which reflects the greatest bias against detecting planets. The paucity of larger mass planetary companions and brown dwarfs is clear, and likely indicative of a different formation mechanism. The fact that the histogram is rising for the lowest detectable masses might indicate a large population of lower mass planets to be discovered only after observational sensitivities improve. The lack of knowledge of inclination angle (i) for most known planets has little bearing on this statistical conclusion [14].

ACKNOWLEDGMENTS

The Doppler surveys discussed here were carried out by Chris Tinney, Hugh Jones, Brad Carter, Alan Penny, Jason Wright, Mel Munoz, and the authors. We thank the Lick, Keck and Anglo-Australian Observatories for their outstanding personnel and research facilities.

REFERENCES

1. Marcy, G., Butler, R.P., Fischer, D., & Vogt, S. "Extrasolar Planets, Today and Tommorow", ASP Conference Series 2004, eds. J.-P. Beaulieu, A. Lecavelier, & C. Terquem, in press.
2. Butler, R.P., Marcy, G., Williams, E., McCarthy, C., Dosanjh, P. & Vogt, S.S. 1996, PASP, 108, 500
3. Konacki, M., Torres, G., Jha, S., & Sasselov, D. 2003, Nature, 421, 507
4. Torres, G., Konacki, M., Sasselov, D. & Jha, S. 2003, ApJ Lett., submitted.
5. Tinney, C., McCarthy, C., Jones, H.R.A., Butler, R.P., Carter, B., Marcy, G., & Penny, A. 2002, MNRAS, 332, 759
6. Charbonneau, D. 2003, in *Scientific Frontiers in Research on Extrasolar Planets*, ASP Conference Series, Vol. 294, eds. D. Deming & S. Seager, p. 449
7. Fischer, D., Marcy, G., Butler, R., Vogt, S. Frink, S., & Apps ,K. 2001, ApJ, 551, 1107.
8. Butler, R.P., Marcy, G., Fischer, D., Brown, T., Contos, A., Korzennik, S., Nisenson, P., & Noyes, R., 1999, ApJ, 526, 916.
9. Marcy, G., Butler, R.P., Fischer, D., Laughlin, G., Vogt, S., Henry, G., & Pourbaix, D. 2002, ApJ, 581, 1375
10. Laughlin, G. & Chambers, J. 2001, ApJ, 551, L109
11. Rivera, E. & Lissauer, J. 2001, ApJ, 558, 392
12. Fischer, D., & Valenti, J. 2003, in *Scientific Frontiers in Research on Extrasolar Planets*, ASP Conf. Series Vol. 294, eds. D. Deming & S. Seager, p. 117.
13. Gonzalez, G. 2000, ASP Conf. Ser. 219, 523
14. Jorissen, A., Mayor, M., & Udry, S. 2001, A&A, 379, 992

What is Unknown About the Statistics of Extrasolar Planetary Systems?

Hugh R.A. Jones

Astrophysics Research Institute, Liverpool John Moores University, Twelve Quays House, Egerton Wharf, Birkenhead CH41 1LD, UK

Abstract. While the study of extrasolar planetary systems has made astonishing progress during the last decade we are very much at the beginning of this new topic. Our study so far has only been sensitive to gas giant extrasolar planets. The 110 or so that we have discovered probably represent the relatively close-in subset of the population of gas giant planets. Here we consider some of the questions that arise from the study of the properties of this subset. In particular we focus on the semimajor axis of extrasolar planets. Values of semimajor axis are a key parameter because observationally they are well determined and theoretically they are a key test of planet formation and migration. While many features of the semimajor axis distributions are well produced by models, features such as the apparent drop in eccentricities and metallicities towards large values of semimajor axis are not yet reproduced by simulations. A full understanding of the statistics of extrasolar planets awaits the discovery and study of a much broader range and larger sample of planets as well as continued intensive work on numerical simulations to produce more physical models of formation, migration and interaction.

INTRODUCTION

Ten years ago it would have been the stuff of crazy wild dreams to imagine being asked to give a talk about "What is unknown about the statistics of extrasolar planets". My allocated title implies that this field is already well developed. The discovery of the 100th extrasolar planet suggests that the sample is large enough to make reasonably broad inferences about the population of planets as a whole. While I will argue that this is not true it is fun to consider a moment longer what astonishing progress has been made. In 1994 Geoff Marcy gave a somewhat despairing talk on his and Paul Butler's searches for brown dwarfs[12]. While this was an excellent much needed conference that probably organised the community for the era of discovery, the bottom line of this conference titled "The bottom of the main sequence and beyond" was that there were no brown dwarfs and certainly no planets. We conference participants passionately believed in the exciting future to come, though I don't think any of us would have predicted the field blossoming so quickly. Back then the idea of more than 100 extrasolar planets, as well as the current plans for Darwin/TPF, would have seemed utterly preposterous.

The confirmation of extrasolar planets by radial velocity, transit and astrometry has provided the motivation and justification for the next generation of ground (CELT, OWL) and space-based missions (MOST, SIRTF, COROT, Kepler, JWST, GAIA, SIM, Darwin/TPF). Nonetheless there are very severe biases in the discoveries made so far. While in many fields a sample of 100 objects might suffice to satisfy our curiosity

CP713, *The Search for Other Worlds: Fourteenth Astrophysics Conference,*
edited by S. S. Holt and D. Deming
© 2004 American Institute of Physics 0-7354-0190-X/04/$22.00

the key difference is that extrasolar planets represent the possibilty to investigate life elsewhere and the next frontier in human exploration. The human race desires and thrives on discovery and exploration. So far we have experience of one planet and have a strong desire (and perhaps necessity) to explore other planets. The scientific and cultural interest in this field mean that, based on the bibliography of "The Extrasolar Planets Encyclopaedia"[18], the field is now producing more than 500 papers per year. However, none of the known extrasolar planets are equivalent to any of the planets in our Solar System, thus it seems apt to quote caution from Alexander Pope in his Essay on Criticism: "A little knowledge is a dangerous thing: drink deep, or taste not the Pierian spring; there shallow draughts intoxicate the brain, and drinking largely sobers us again."

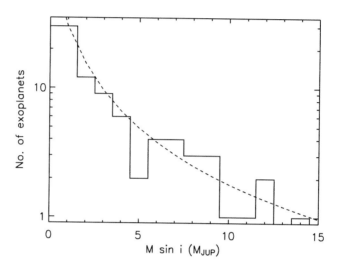

FIGURE 1. Number versus companion mass for all radial velocity extrasolar planets as given by [13]. The dashed line depicts the number of extrasolar planets proportional to Mass to the power -1.6 (not corrected for sensitivity function) and is normalised to fit through the mass bin of 1-2 M_{JUP} sini i assumed to be reasonably complete for the sample of known extrasolar planets.

WHAT IS AN EXTRASOLAR PLANET?

With any new field nomenclature is important, in our case particularly crucial, since the information we have about extrasolar planets is all indirect. While we have confidence in the various indirect techniques that have been used we still await the detection of photons directly from an extrasolar planet. The word planet means 'wanderer' in Greek, thus planets move on the sky relative to the fixed background stars. This definition needs updating since with modern equipment even quite distant galaxies can be seen to move. Discussions about the status of Pluto and 'free-floating planets/brown dwarfs'[16] describe the issues. According to the Oxford English Dictionary (http://oed.com) the old astronomical meaning of planet 'A heavenly body distinguished from the fixed stars by having an apparent motion of its own among them; each planet, according to the Ptolemaic system, being carried round the earth by the rotation of the particular sphere or orb in which it was placed.' has evolved to 'The name given to each of the heavenly bodies that revolve in approximately circular orbits round the sun (primary planets), and to those that revolve round these (secondary planets or satellites)'. More specific definitions are provided elsewhere, e.g., http://dictionary.com: 'A non-luminous celestial body larger than an asteroid or comet, illuminated by light from a star, such as the sun, around which it revolves.' The IAU working group on extrasolar planets provides a half page working definition of extrasolar planet and a working list of candidate extrasolar planets (www.ciw.edu/boss/IAU/div3/wgesp).

Planetary mass objects may have already been imaged in young star forming regions (e.g. Tamura et al.[22]). Apart from the masses of these objects being very dependent on poorly constrained theoretical models, the 'free-floating' nature of these objects means they fall outside the currently accepted notion of extrasolar planet. An important strand in most definitions is the concept that to be a 'planet' an object must be in orbit around a 'star'. This proximity to a much brighter object as well as their relative faintness makes planets so difficult to find. Discovery would be simplified were it possible to directly image extrasolar planets. The best opportunity so far is perhaps the controversial planet around Epsilon Eridani. At only 3.2 pc, it should soon become feasible to image this object although with a separation of 1 arcsecond and a magnitude difference of 15 (a factor of 1,000,000 in brightness). While this observation is at the limit of current technology, a number of inovative techniques seem promising (e.g., [4]).

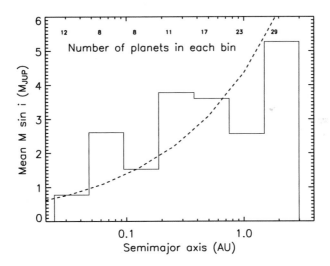

FIGURE 2. The mean mass (M_{JUP} sin i) of extrasolar planets plotted with semimajor axis. The dotted line represents a crude approximation of the sensitivity function of the Doppler technique (semimajor axis to the power 0.5).

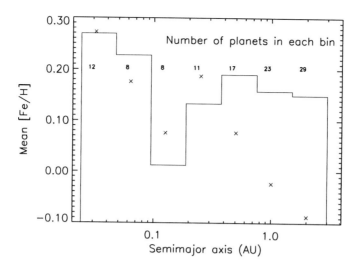

FIGURE 3. Mean spectroscopic metallicities of the primaries of extrasolar planets plotted as a function of semimajor axis. The crosses represent the low mass (0.1–1.1 M_{JUP} sin i) third of detected extrasolar planets.

BROWN-DWARF DESERT / PLANET JUNGLE AROUND SOLAR-TYPE STARS

1995 was a watershed year, not just for extrasolar planets but also for brown dwarf research. A conference on cool stars in Florence saw the announcement of the extrasolar planet 51 Peg b as well as the brown dwarf Gl229B. Brown dwarfs bridge the gap between stars and planets. Too small and cool to be a star and sustain thermonuclear hydrogen burning but yet too massive to be a planet. Whilst a steady refinement of radial velocity searches means they are spectacularly sensitive to extrasolar planets, such searches are actually far more sensitive to the presence of brown dwarfs which have hardly been found. For many years it was expected that planets would be found by extending radial velocity searches of brown dwarfs to lower masses. However, while stellar radial velocity companions are relatively abundant, there is a relative deficit of companions from around 100 to 10 M_{JUP}, approximately the brown dwarf regime. However, once sensitivity to below 10 M_{JUP} is achieved, Fig. 1 indicates how detections rapidly increase. This sharply rising detection rate at low masses is found against a sensitivity function for finding planets that falls in proportion to mass.

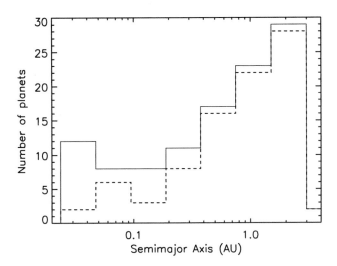

FIGURE 4. The number of extrasolar planets discovered with semimajor axis: solid line (all extrasolar planets), dashed line (those with masses above 0.8 M_{JUP} sin i).

WHY ARE EXTRASOLAR PLANETS SO DIFFERENT FROM SOLAR SYSTEM PLANETS?

While the discovery of 51 Peg b was a landmark, as with the earlier discovery of planets around pulsars it was met with skepticism in part because of the difficulties of the measurement but as much because 51 Peg b seemed to have nothing in common with our Jupiter (apart from mass). Its orbit seemed to require radical new ideas about the formation of planets; though in fact all that was required was the rediscovery of the robust theoretical concept of inward planetary migration driven by tidal interactions with the protoplanetary disk[8]. Thus while such a large mass planet could not form in the glare of radiation from its Sun, it was entirely plausible that it had migrated into position through the disk of material around 51 Peg[10]. Although 51 Peg b-like objects dominated the early discoveries, other types of planets are much more common. Of the 110 or so extrasolar planets that have been discovered the 51 Peg b-like planets (3-5 day orbital periods) represent a class of planets circling about 1% of stars[14]. The 51 Peg b class were found first because the radial velocity technique is biased towards relatively heavy planets close to their stars, as they produce the largest signal. This makes these planets easiest to find. As more planets are discovered other types of biases in our understanding of extrasolar planets resulting from our experimental sensitivity will start to reveal themselves. Fig. 2 shows how the average mass of extrasolar planets increases with semimajor axis in accordance with the changing sensitivity of the radial velocity technique. This suggests that the long-term stability of the radial velocity searches is excellent providing the mass function of extrasolar planets is approximately constant with semimajor axis[1]. Fig. 2 is thus very encouraging for the potential detection of Solar System like extrasolar planets.

The extrasolar planets discussed and plotted in this paper all come from the compilation of Marcy et al. [13] and were all detected by the radial velocity technique. However, they are not discovered from a single well documented and quantified methodology. The compilation relies on a number of different ongoing surveys operating with different samples, sensitivities, instruments, scheduling, strategies and referencing techniques. Cumming et al.[2] have thoroughly investigated the observational biases inherent in the Lick and Keck surveys but has yet to report findings for the bulk of detected extrasolar planets. So far none of the surveys have the 3 m/s precision over 15 years necessary to detect Jupiter and thus do not yet constrain the frequency of Solar Systems analogs. The relatively large number and fraction of planets that we have discovered before achieving sensitivity to our own Solar System together with the wide range of parameters discovered suggests that planetary formation and survival are robust. This seems to be borne out by theoretical work, which indicates that planet formation is an 'easy come, easy go' business, with many planets created and many destroyed, and with an important minority – including our own Jupiter – surviving[23]. Simulations, e.g. [1], suggest that gas giant planets will form at around 5 au. Over the next decade radial velocity searches should be well placed to constain such predictions. So far the compilation of extrasolar planets offers little constraint on extrasolar planets:

- around O, B, A and M stars,
- beyond 4 au,
- less massive than Saturn,
- in regions outside the Solar Neighbourhood (e.g. clusters, bulge, halo),
- in binary systems,
- in multiple systems.

WHAT IS THE EXTRASOLAR PLANETARY MASS FUNCTION?

Fig. 1 shows the mass function for extrasolar planets is rising fairly steeply towards lower masses. The simulations of Tabachnik & Tremaine[21] and Zucker & Mazeh[24] using around 60 planets favour a flatish mass distribution. The Bayesian approach[3] promises to incorporate a detailed knowledge of detection sensitivities. So far, the relatively small numbers of objects as well as the selection biases preclude much confidence in a particular value of the mass function. Nontheless, it is clear that we find an increasing number of objects towards lower masses which is consistent with our expectations based on our Solar System as well as simulations of planet formation.

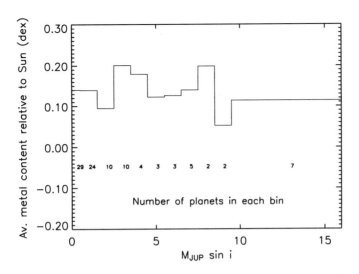

FIGURE 5. Mean spectroscopic metallicities of the primaries of extrasolar planets plotted as a function of extrasolar planet masses M_{JUP} sin i.

EVIDENCE FOR MIGRATION - SEMIMAJOR AXIS AND METALLICITY DISTRIBUTIONS?

The distribution of extrasolar planets with semimajor axis in Fig. 4 suggests that extrasolar planets show key differences with semimajor axis. The solid line may suggest two separate features in the extrasolar planet semimajor axis distribution. A peak of short-period extrasolar planets is seen in the 51 Peg-type objects, then a dearth, followed by an smooth rise in the number of extrasolar planets toward longer periods[9]. However, except for a peak at 3 days [14], the short-period peak does not appear to be so significant when a completeness correction is made. The dotted line represents all extrasolar planets with masses greater than 0.8 M_{JUP} sin i and suggests that the short-period peak may well be a selection effect. The rise in the number of extrasolar planets towards larger semimajor axes is becoming more apparent as more are discovered and is well reproduced by extrasolar planet migration scenarios which envisage planets migrating inwards[1] as well as outwards[15].

An important characteristic of a star is its metallicity. Gonzalez[5] found extrasolar planet host stars to be metal-rich. This conclusion has been confirmed by many authors with different samples, methodologies and spectral synthesis codes (e.g., [7], [19]). Fig. 3 shows just how metal-rich the extrasolar planet primaries are. Only a single bin is around solar metallicity. All other bins are at least 0.1 dex above the solar; whereas the Sun and other solar type dwarf stars in the solar neighbourhood have an average metallicity of 0 or even slightly less[17]. The probability of detecting an extrasolar planet is proportional to its metallicity. By a metallicity of +0.3 dex the frequency of stars with

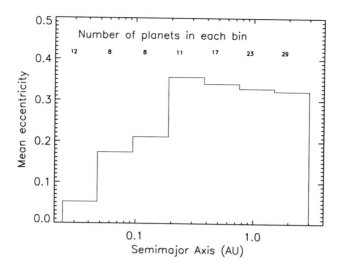

FIGURE 6. Eccentricity versus semimajor axis for extrasolar planets.

extrasolar planets is effectively 100%. This result, representing the only link between the presence of planets and a stellar photospheric feature is explained by the classical view that giant planets are formed by runaway accretion of gas on to a 'planetesimal' having up to 10 Earth masses. In such a case, we can expect that the higher the proportion of dust to gas in the primordial cloud (i.e. metals), and consequently in the resulting protoplanetary disc, the more rapidly and easily may planetesimals, and subsequently the observed giant planets be built.

So far, it seems that the higher metallicity of most planet-harbouring stars arises because high metallicity environments have a higher probability of planet formation and migration. The relatively low metal content of the Solar System may be consistent with the relative lack of migration[11]. Thus it appears that analogous to the early detections of 51 Peg b-type extrasolar planets we are finding a surfeit of extrasolar planets around metal-rich stars because they are easier to detect. The idea that migration is dependent on metallicity also seems to be borne out by the overall decrease in metallicity for increasing semimajor axis. Given that the best explanation for the close-in planets is migration and that migration is modelled to be mass dependent, it is interesting to see if there is any mass dependency. The crosses in Fig. 3, represent the low-mass extrasolar planets (less than 1.1 M_{JUP} sin i) and suggest the slight decline to long periods is contributed primarily by the low-mass third of the extrasolar planet sample. This result is as expected by migration theories which predict more migration of lower mass objects. One might expect to see a relationship between metallicity and mass though yet none is readily apparent, e.g. Fig. 5. Such results are very preliminary and need confirmation with more extrasolar planets and higher precision metallicities.

WHY DON'T EXTRASOLAR PLANETS HAVE CIRCULAR ORBITS LIKE OUR SOLAR SYSTEM?

Apart from the short-period extrasolar planets whose orbits are circularised by the tidal pull from their parent star, Fig. 6 shows the eccentricity of extrasolar planet orbits is much higher than in the Solar System. Eccentricities rise steeply out to semimajor axis values of around 0.2 au at which point a mean eccentricity of around 0.35 is reached. This mean eccentricity shows a slight decline out to several au. Fig. 7 suggests that metallicity does not play an important role in the determination of eccentricity. Fischer et al.[6] find that it is rather close to that observed for stars. According to our paradigm of planetary formation, a planet (formed in a disk) should keep a relatively circular (low eccentricity) orbit. In order to boost extrasolar planet eccentricities it is necessary to imagine interactions between multiple planets in a disk and between a planet and a disk of planetesimals and perhaps the influence of a distant stellar companion. In fact dynamical interactions between planets seem inevitable since even with the fairly poor sampling of known extrasolar planets, 10 multiple systems have already been found. Of these a number are in resonant orbits. Thus 'dynamical fullness' is probably important and suggests that interactions play a vital role in determining the properties of many extrasolar planets. These orbital complexities mean that to understand extrasolar planets more generally, it will be necessary to find all the main components in planetary systems. This will require using results from different techniques, particularly radial velocity and astrometric, to disentangle the various planetary components in orbit around nearby stars. Fischer et al. [6] suggest that selection effect may play a role in the high eccentricities observed among the single extrasolar planets discovered to date. Most known extrasolar planets reside within 3 au due to the limited duration (10 yr) of the Doppler surveys. Thus the planets detected to date represent a subset that ended up within 3 au. Giant planets within 3 au may systematically represent the survivors of scattering events in which the other planet was ejected while extracting energy from the surviving planet and throwing it inward. This would give rise to us systematically detecting the more massive, surviving planet residing in an orbit with a period less than 10 years.

CONCLUSIONS

Detecting the true distributions of extrasolar planets for the radial velocity technique will require a quantitative knowledge of detectability function for the radial velocity technique. Detectability can easily be corrected for the enhanced sensitivity of the radial velocity technique to large mass planets with short orbital periods. However, corrections for sampling, duration of observations, velocity jitter and differing sensitivity of different surveys are much more subtle. The move to the automation of radial velocities should help in the quantification and optimisation of radial velocity surveys. Overall our sample of extrasolar planets is still subject to biases which have not yet been quantified and not representative of planetary systems in general which we expect to include terrestrial planets and ocean planets as well as gas giants. Nonetheless, if knowledge in this field

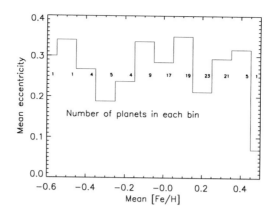

FIGURE 7. Mean eccentricities as a function of spectroscopic metallicities of the primaries of extrasolar planets.

can progress in the next decade as fast as the last, an incredible era of discovery awaits.

REFERENCES

1. Armitage P.J., Livio M., Lubow S.H., Pringle J.E., 2002 MNRAS, 334, 248
2. Cumming A., Marcy G.W., Butler R.P., 1999, ApJ, 526, 896
3. Cumming A., Marcy G.W., Butler R.P., Vogt S.S., 2003, ASP Conf. Ser. 294, p27, eds. Deming D., Seager S.
4. Danchi W.C., Deming D., Kuchner M.J., Seager S., 2003, ApJ, 597, 57L
5. Gonzalez G., 1997, MNRAS, 285, 403
6. Fischer D.A. et al., 2003, ApJ, 586, 1394
7. Fischer D.A., Valenti J.A., Marcy G.W., 2004, IAU 219, in press, ed. Dupree A.
8. Goldreich P., Tremaine S., 1980, ApJ, 241, 425
9. Jones et al., 2003, MNRAS, 341, 948
10. Lin D.N.C., Bodenhimer P., Richardson D.C., 1996, Nature, 380, 606
11. Lineweaver C.H., 2003, Icarus, 151, 307
12. Marcy G.W., Butler R.P., 1995, The Bottom of the Main Sequence and Beyond, p98, ed. Tinney, C., Springer-Verlag,
13. Marcy G.W., Butler R.P., Vogt S.S., Fischer D.A., et al., 2003, http://extrasolar planets.org
14. Marcy G.W., Butler R.P., Vogt S.S., Fischer D.A., 2004, ASP Conf Ser., in press, eds Beaulieu J-P., Lecavelier des Etangs A., Terquem C.
15. Masset F.S., Papaloizou J.C.B., 2003, ApJ, 588, 494
16. McCaughrean M.J. et al., 2001, Science 291, 1487
17. Reid I.N., 2002, PASP, 114, 306
18. Schneider J., 2003, http://www.obspm.fr/encycl/encycl.html
19. Santos N. et al. 2003, A&A, 398, 363
20. Santos N. et al. 2004, A&A, in press
21. Tabachnik S., Tremaine S., 2002, MNRAS, 335, 151
22. Tamura M., Itoh Y., Oasa Y., Nakajima T., 1998, Science, 282, 1095
23. Trilling D., Lunine J.I., Benz W., 2003, A&A, 394, 241
24. Zucker S., Mazeh T., 2002, ApJ, 568, 113L

Quantifying the Uncertainty in the Orbits of Extrasolar Planets with Markov Chain Monte Carlo

Eric B. Ford

Astronomy Department, 601 Campbell Hall, UC Berkeley, Berkeley, CA 94720-5275

Abstract. Precise radial velocity measurements have led to the discovery of ~ 100 extrasolar planetary systems. It is important to understand the uncertainties in the orbital elements that have been fit to these data. While detections of short-period planets can be rapidly refined, planets with long orbital periods will require decades of observations to constrain the orbital parameters precisely. Already, in some cases, very different orbital solutions provide similarly good fits, particularly for long-period and multiple planet systems. Thus, it will become increasingly important to quantify the uncertainties in orbital parameters, as future discoveries are likely to include many planets with long orbital periods and in multiple planet systems.

Markov chain Monte Carlo (MCMC) provides a computationally efficient way to quantify the uncertainties in orbital elements and to address *specific* questions directly from the observational data rather than relying on best-fit orbital solutions. MCMC simulations reveal that for some systems there are strong correlations between orbital parameters and/or significant non-Gaussianities in parameter distributions, even though the observational errors are Gaussian. Once these effects are considered the actual uncertainties in orbital elements can differ significantly from the published uncertainties. This has implications for the interpretation of the orbits of extrasolar planets.

METHODS TO ESTIMATE MODEL UNCERTAINTIES

In this paper we focus on estimating the uncertainies in model parameters, and neglect the important issue of how to identify the best fit model. When fitting a model to a set of observational data, there are several possible ways to estimate the uncertainties in the model parameters. Here we briefly review two methods commonly used in the literature and then focus our attention on the application of MCMC.

Fischer Matrix

Perhaps the most common method of estimating uncertainties in model parameters is based on the Fischer matrix. With this method the surface, $\chi^2(\vec{x})$, is approximated as a quadratic function of the model parameters ($\vec{x}$) near the best-fit model. We outline the procedure below.

1. Identify best-fit model: $\chi^2_{\min} = \chi^2(\vec{x}_{\min})$

2. Calculate $F_{ij} = \frac{\partial \chi^2}{\partial x_i \partial x_j}(\vec{x}_{\min})$

CP713, *The Search for Other Worlds: Fourteenth Astrophysics Conference,*
edited by S. S. Holt and D. Deming
© 2004 American Institute of Physics 0-7354-0190-X/04/$22.00

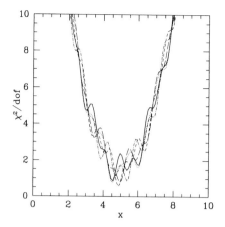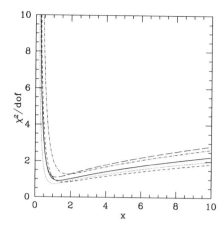

FIGURE 1. We plot χ^2 surfaces for two hypothetical sets of observational data (solid lines) as a function of one parameter, x, and for resampled data sets (dotted lines). *Left:* A Fischer matrix analysis significantly underestimates the uncertainty in x, since the χ^2 surface has narrow bumps near the minimum. Here, either resampling or MCMC would give accurate uncertainties. *Right:* A Fischer matrix analysis gives inaccurate uncertainties, since the χ^2 minimum is not well approximated by a quadratic. Resampling would underestimate the uncertainty, since the shift in the minima does not reflect the very gradual increase in χ^2 for large x. Here, only MCMC would accurately characterize the uncertainty in x.

3. Assume $\chi^2(\vec{x}) \simeq \chi^2_{\min} + \Sigma_{ij} F_{ij}(x_i - x_{\min,i})(x_j - x_{\min,j})$
4. Calculate confidence intervals for $\vec{x}$

While this method is easy and computationally very fast, it can give inaccurate results when the approximation for the $\chi^2(\vec{x})$ surface is not accurate. In particular, often $\chi^2(\vec{x})$ is a "rough" function. I.e., there are many local minima near the best-fit solution and the curvature of $\chi^2(\vec{x})$ at the global minima does not reflect the overall shape of $\chi^2(\vec{x})$ (See Fig. 1 (left)). The Fischer matrix can also give misleading uncertainties if the $\chi^2(\vec{x})$ surface is significanly asymmetric near the best-fit solution (see Fig. 1 (right)). In such cases, a Fischer matrix analysis may not accurately reflect the potential for a large tail in the distribution of possible parameter values.

Resampling

Another common method for estimating parameter uncertainties involves repeatedly finding the best-fit model to synthetic data sets which are generated from the actual observational data and the uncertainties in individual observations. There are several variations, e.g., drawing observations with or without replacement.

While resampling based techniques are much slower than calculating the Fischer matrix, resampling can accurately estimate parameter uncertainties, even for rough $\chi^2(\vec{x})$ surfaces (see Fig. 1 (left)). However, resampling may have trouble accurately estimating parameter uncertainties for parameters with heavy tails (see Fig. 1 (right)).

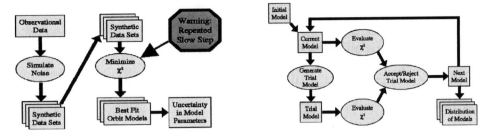

FIGURE 2. These flow charts illustrate the procedure for estimating parameter uncertainties using resampling (left) and MCMC (right). Resampling requires solving a difficult minimization problem for each set of resampled data, while MCMC only requires that χ^2 be calculated for each trial model. However, when using MCMC, one must be careful to using base inferences on Markov chains which have converged.

The main disadvantage is that a time consuming minimization problem must be solved for each synthetic data set. The repeated minimization problems are difficult, especially for high dimensional parameter spaces.

Markov Chain Monte Carlo

The MCMC technique has a close relationship to Bayesian inference. The Bayesian framework considers a joint probability distribution function for both the observed data ($\vec{d}$) and model parameters which can not be directly observed ($\vec{x}$). This joint probability, $p(\vec{d},\vec{x})$, can be expressed as the product of the probability of the observables given the model parameters, $p(\vec{d}|\vec{x})$, and a prior probability distribution function, $p(\vec{x})$, which is based on previous knowledge of the model parameters. Bayes's theorem allows one to compute a posterior probability density function, $p(\vec{x}|\vec{d})$, which reflects the knowledge gained by the observations $\vec{d}$. That is

$$p(\vec{x}|\vec{d}) = \frac{p(\vec{x},\vec{d})}{\int p(\vec{d},\vec{x})p(\vec{x})\,d\vec{x}} = \frac{p(\vec{x})p(\vec{d}|\vec{x})}{\int p(\vec{d},\vec{x})p(\vec{x})\,d\vec{x}}. \tag{1}$$

Unfortunately, the lower integral can be extremely difficult to compute, particularly when $\vec{x}$ has a large number of dimensions. MCMC offers a relatively efficient method of performing the necessary integrations.

Our application of the MCMC method is to generate a chain (i.e. sequence) of states (i.e. sets of parameter values, $\vec{x}_n$) which are sampled from the joint posterior probability distribution ($f(\vec{x}) = p(\vec{x}|\vec{d})$) for the model parameters ($\vec{x}$) conditioned on the observational data ($\vec{d}$). The Monte Carlo aspect of MCMC simulation refers to randomness in the generation of each subsequent state. The Markov property specifies that the probability distribution for determination of $\vec{x}_{n+1}$ can depend on $\vec{x}_n$, but not previous states. More information about the MCMC algorithm can be found in [1], and details of the application to radial velocity observations can be found in [2].

MCMC has the advantage that it results in the joint posterior distribution, even for "rough" and asymmetric $\chi^2(\vec{x})$ surfaces. Furthermore, each additional step in the Markov chain can be computed quickly, and the algorithm is efficient for large dimensional parameter spaces. However, in some cases (e.g., highly correlated parameters) the Markov chain may converge to the joint posterior distribution slowly. While it is impossible to prove that a given chain has converged, there are many statistical tests which can detect if a chain has not converged. When several such tests do not demonstrate a failure to converge, it suggests (but does not prove) that the chain has converged.

DISCUSSION

MCMC is well suited to quantifying the uncertainties in orbital parameters of extrasolar planets derived from radial velocity observations. Besides calculating probability distributions for orbital elements, MCMC also allows specific questions to be addressed directly from the observational data. For example, one could ask questions like, "What is the probability that the eccentricities of the short-period planets differ from zero?", "What is the probability distribution for the pericenter separation for HD 80606?", and "What is the probability that the planets around 47 UMa are in a particular resonance?" Additionally, MCMC can be used to help plan observations by predicting possible transit times or times when the additional radial velocity observations would be most useful.

In conclusion, MCMC methods provide a valuable tool for characterizing the uncertainties in estimated orbital parameters for extrasolar planets. Presently, several systems have large uncertainties and correlations in orbital parameters, particularly the long period planets and multiple planet systems. MCMC simulations reveal that resampling can significantly over- or underestimate the uncertainty in orbital parameters. In addition to quantifying the uncertainties of various orbital parameters, MCMC allows specific questions to be investigated directly from the radial velocity data itself bypassing fits to orbital parameters. As more long period and multiple planet systems are discovered, such methods could become increasingly valuable.

ACKNOWLEDGMENTS

The author would like to thank Gilbert Holder, Hiranya Peiris, David Spergel, and Scott Tremaine for valuable discussions. This research was supported in part by NASA grant NAG5-10456, the EPIcS SIM Key Project, and the Miller Institue for Basic Research.

REFERENCES

1. Gilks, W.R., Richardson, S., & Spiegelhalter, D.J. 1996, Markov Chain Monte Carlo in Practice. New York: Chapman & Hall/CRC.
2. Ford, 2003, submitted to AJ. (astro-ph/0305441)

A Search for Wide (Sub)Stellar Companions Around Extrasolar Planet Host Stars

M. Mugrauer*, R. Neuhäuser†, T. Mazeh**, M. Fernández‡ and
E. Guenther§

*AIU, Schillergässchen 2-3, 07745 Jena
†AIU, Schillergässchen 2-3, 07745 Jena, Germany
**Tel Aviv University, P.O. Box 39040, 69978 Tel Aviv, Israel
‡IAA, Apdo. Correos 3004, 18080 Granada, Spain
§TLS, Sternwarte 5, 07778 Tautenburg, Germany

Abstract. We present an overview of our ongoing systematic search for wide (sub)stellar companions around the stars known to host rad-vel planets. By using a relatively large field of view and going very deep, our survey can find all directly detectable stellar and massive brown dwarf companions (m$>$40 M_{Jup}) within a 1000 AU orbit.

INTRODUCTION

Circumstellar disks are discovered with sizes up to 1000 AU and also binary stars with the same comparable separations are known. Because the formation of stars and brown dwarfs seems to follow a similar scheme (fragmentation of large gas clouds) substellar objects may indeed reside in that distance around stars hosting rad-vel planets. Adaptive optics search programs to find very close companions around those stars already exist, but they leave out an interesting regime of objects, namely the wide companions because of a too small field of view (e.g., [3]).

As of October 2003, more than one hundred extra-solar planets were discovered. Many of those have extremely close orbits which could be explained by a migration process in the early history of the system. During this migration angular momentum is transferred from the inner part of the accretion disk to its outer border. A wide companion can cut off the disk and be a sink for the lost angular momentum. Furthermore theories predict that wide companions can induce rapid instability in disks which otherwise would be stable, hence they could have a strong influence on the planet formation and on the longtime evolution of planetary orbits.

Actually, some extrasolar planets were found to reside in binary stellar systems. Those few cases are intriguing, and might exhibit some statistically different features than the planets around single stars [5]. This could be a first hint about an interaction between the (sub)stellar wide companions and the extrasolar planets. Nevertheless, the whole sample of extrasolar planetary systems has not been surveyed completely for wide companions with sensitive IR cameras that are able to find faint low-mass companions. For this reason we have started in 2001 a systematic deep imaging of all the stars known to harbor planets, in order to look for faint companions in wide orbits. The companionship

CP713, *The Search for Other Worlds: Fourteenth Astrophysics Conference*,
edited by S. S. Holt and D. Deming
© 2004 American Institute of Physics 0-7354-0190-X/04/$22.00

of those faint objects can be established only by follow up observations which will detect common proper motion with the nearby star that host the planet.

To find all the companion-candidates around the planet hosting stars we secure deep IR images, obtained with the IR cameras SOFI at the 3.58 m NTT and UFTI at the 3.8 m UKIRT, with detection limit of $H \sim 19.5$ mag. To detect common proper motion we obtain two images about one year apart. We also make use of the 2 micron all sky survey (2MASS) images, which were taken several years before our exposures. However, the limit of 2MASS is $H \sim 15$ mag [2], and the proper motion of fainter objects needs to be measured only by our two images.

ASTROMETRY - AN EFFECTIVE WAY TO FIND COMPANIONS

Astrometry is a very effective tool to find unknown wide companions. All stars hosting planetary systems are bright and therefore are listed in the Hipparcos catalogue, hence their proper motions, which are relatively high (~ 200 mas per year), are known with an accuracy of a few milliarcsecs (mas). With the IR cameras — UFTI/UKIRT and SOFI/NTT, which have pixel scales of 91 mas and 144 mas, respectively, one year difference means a shift of the photocenter by 1.5 to 2 pixel. This is easily detectable. Only two observations are necessary to distinguish real companions from background stars. A real companion is bound to its host star and therefore they form common proper motion pair. Co-moving companions stand out with non varying separation, whereas the separation between background stars and the target star changes according to the well known proper motion of the target star. The orbital motion of the companion around its host star can be neglected, because the motion of wide companions with orbital radii larger than 100 AU is generally much smaller than the proper motion.

An example of the results of our astrometric search campaign is illustrated in Figs. 1 & 2. We observed the star HD 37124 (for planet data see [4]) with SOFI/NTT in Dec. 2001 and Dec. 2002. Total integration time was 10 min in H in both exposures. The SOFI/NTT field of view is 144 arcsecs, hence companions with maximum projected separation of 2200 AU can be detected. Close to the bright star the detector becomes non-linear and saturation occurs (see the white circle in Fig. 1 left and the dashed line in Fig. 1 right). In this region the detection of any companion is impossible. To avoid saturation at large parts of the detectors we choose as short integration time as possible. Many of those short integrated images have to be superimposed to reach a high signal to noise.

In the next step we have derived the detection limit of our observations by measuring the local noise in the images for a range of separations to the rad-vel planet host star. The detection sensitivity is increasing for larger separations due to a lower local photon noise. In the right panel of Fig. 1 we have plotted the detection limit for a range of angular distances to HD 37124. Only a small fraction of the whole SOFI/NTT field of view (147 arcsecs) is shown here. Beyond 24 arcsecs the image is background limited and the detection limit is constant, around 19.5 mag (see the dashed black circle in Fig. 1, left panel). In this region brown dwarf companions with $m > 40 M_{Jup}$ are detectable. All stellar companions ($m > 78 M_{Jup}$) can be detected outside ~ 3 arcsecs, with projected separation of ~ 100 AU. We use evolutionary models to derive the apparent magnitudes

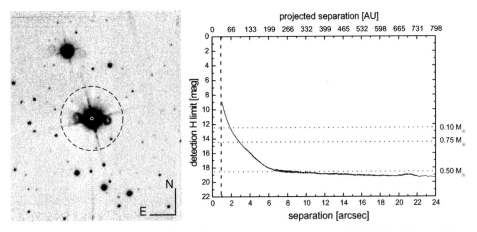

FIGURE 1. NTT H band image of the field of the rad-vel planet host star HD 37124 which was taken in Dec. 2002. The averaged FWHM is ~0.8 arcsec. Several companion-candidates are detected close to the rad-vel planet host star with magnitudes down to H = 19.5 mag. Inside 1 arcsec saturation occurs (see the white small circle in the left image and the dashed line in the right panel). The right panel shows the detection limit for a range of separations. The field of view of the derived detection limits is marked in the left image as dashed black circle. The upper x-axis shows the projected separation in AU. The right y-axis scales in companion mass. We used Baraffe et al. COND models [1] for the magnitude-mass transformation, assuming an age of 5 Gyrs for the system.

of objects with different masses at the distance of HD 37124 for an age of 5 Gyrs.

The proper motion of all detected objects around the star can be measured by using relative astrometry. Only the star itself has a non-negligible proper motion which is consistent with the Hipparcos proper motion data (see the white box in Fig. 2). All other detected objects are clearly not co-moving, hence they are all non-moving background stars. Due to the results of our relative astrometry and the derived detection limit we can conclude that there is no further stellar companion around HD 37124 between 100 and 2200 AU (3...73 arcsecs).

So far nearly all rad-vel planet host stars were observed in first epoch and second epoch follow-up observations are on the way. Several new wide companions were detected which we will be published soon.

ACKNOWLEDGMENTS

We would like to thank the technical staff of NTT and UKIRT for all their help and assistance in carrying out the observations. The United Kingdom Infrared Telescope (UKIRT) is operated by the Joint Astronomy Centre on behalf of the U.K. Particle Physics and Astronomy Research Council. This publication made use of data products from the Two Micron All Sky Survey, which is a joint project of the University of Massachusetts and the Infrared Processing and Analysis Center/California Institute of Technology, funded by the National Aeronautics and Space Administration and the

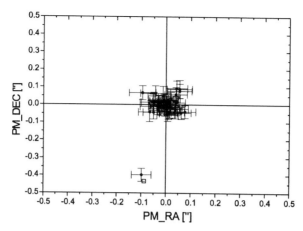

FIGURE 2. Measured proper motion of all objects in Fig. 1, obtained by comparing two SOFI/NTT H band images (relative astrometry) taken with an epoch difference of one year. The proper motion of HD 37124 is well known from Hipparcos measurements, and is presented here by the small squared box. All detected objects have proper motions negligible within the astrometry precision, hence they are none-moving background stars. Only HD 37124 itself is moving with the expected proper motion.

National Science Foundation. We have used the SIMBAD database, operated at CDS, Strasbourg, France. Finally we would like to thank Christopher Broeg, Andreas Seifahrt and Alexander Szameit for carrying out some of the observations.

REFERENCES

1. Baraffe, I., Chabrier, G., Barman, *A&A*, 2003, 402, 701
2. Cutri, R. M., Skrutskie, M. F., van Dyk, S., *yCat*, 2003, 2246, 0C
3. Patience, J., White, R. J., Ghez, A. M., *ApJ*, 2002, 581, 654P
4. Vogt, S. S., Marcy, G. W., Butler, R. P., *ApJ*, 2000, 536, 902V
5. Zucker., S. & Mazeh, T., *ApJ*, 2002, 568L, 113

Establishing the Environment
for Planetary Formation

Observations of "Gas-Rich" Disks

François Ménard

FOST, Laboratoire d'Astrophysique de Grenoble, 414 rue de la Piscine,
BP 53, 38041 Grenoble Cedex 9, France

Abstract. Accretion disks are pivotal elements in the formation and early evolution of solar-like stars. On top of supplying the raw material, their internal conditions also regulate the formation of planets. Their study therefore holds the key to solve the mystery of *the formation of our Solar System.* This chapter focuses on observational studies of circumstellar disks associated with pre-main sequence solar-like stars. The direct measurement of disk parameters poses an obvious challenge: at the distance of typical star forming regions (e.g., $\sim$140pc for Taurus), a planetary system like ours (with diameter $\simeq$ 50AU out to Pluto, but excluding the Kuiper belt) subtends only 0.35". Yet its surface brightness is low in comparison to the bright central star and high angular and high contrast imaging techniques are required if one hopes to resolve and measure these protoplanetary disks.

Fortunately, capable instruments providing 0.1" resolution or better and high contrast have been available for just about 10 years now. They are covering a large part of the electromagnetic spectrum, from the UV/Optical with HST and the near-infrared from ground-based adaptive optics systems, to the millimetric range with long-baseline radio interferometers. It is therefore not surprising that our knowledge of the structure of the disks surrounding low-mass stars has made a gigantic leap forward in the last decade. In the following pages I will attempt to give an overview of the structural and physical parameters of protoplanetary disks that can be estimated today from direct observations.

A VERY BRIEF HISTORICAL BACKGROUND

Compelling observational evidence that rotationally supported disks exist around solar-like young stars is only recent. In his pioneering observational study of YY Orionis stars, Walker [1] was first in 1972 to propose disk accretion as a possibility to explain the inverse P Cygni profiles and strong UV excess displayed by these objects. When Lynden-Bell & Pringle (hereafter LBP) suggested, in their insightful theoretical paper of 1974 [2], that viscous dissipation in a disk might account for the continuous excesses observed in T-Tauri stars (TTSs), there was basically no data available to support, or refute, this hypothesis. A decade went by before two satellites provided the observational means for testing LBP's model: IRAS (e.g.,[3]) in the infrared and IUE in the UV. Together with ground-based broad-band photometry, these facilities provided the data needed for an extensive study of the accretion disk model, mostly via fitting of spectral energy distributions (e.g., [4-6]). The disk idea finally reached consensual status shortly after, but that was on the basis of converging indirect evidence rather than on firm observational facts. The first direct circumstellar disk detections around T-Tauri stars date from less than ten years back, e.g., [7] at millimeter wavelength and [8] in the optical.

Since these discoveries, our understanding of the accretion / ejection process, and especially the pivotal role played by the disk, has evolved considerably. Most aspects of

CP713, *The Search for Other Worlds: Fourteenth Astrophysics Conference,*
edited by S. S. Holt and D. Deming
© 2004 American Institute of Physics 0-7354-0190-X/04/$22.00

the classical LBP disk model are being challenged on both observational and theoretical grounds. The idea that the disk is flat and homogeneous has given way to the more complex view of a warped, inhomogeneous disk in which tidal interactions (induced by close companions, large planetesimals, or by non-axisymmetric instabilities) play a major role. Direct images to support those claims are slowly becoming available (e.g., [9,10]).

In the following section I will present the current observational status of our knowledge of young gas-rich disk. In particular, I will focus on how, in details, some of the structural and physical parameters of the disk, relevant to the planet formation process, are extracted from the observations. The properties of more evolved, second generation disks will be presented in A. Weinberger's contribution, in this volume.

HIGH ANGULAR RESOLUTION OBSERVATIONS OF DISKS

In a broad sense, the observers benefit from two sources of information when trying to probe the disk structure directly. The first one is the *scattered* starlight the disk is sending back to the observer, the second one is the disk's (or dust's) own *thermal* emission. The scattering process occurs at the surface of an optically thick disk. At millimeter wavelengths the dust thermal emission is usually assumed optically thin, except at the very center. The two regimes therefore probe different volumes within the disks and both approaches complement each other nicely.

FIGURE 1. VLT/NACO K-band (2.2μm) image of the pre-main sequence binary star HK Tau. The separation between the two stars is 2.''4. The companion (to the South) is entirely nebulous and the central star remains undetected at $\lambda \leq 2.2\mu$m. The dark lane is interpreted as a disk seen edge-on. Image from [13].

There is little doubt now that disks exist around TTSs. The images of the edge-on systems HH 30 [8], HK Tau/c [11], and IRAS 04158+2805 [12,13], the rotating gas rings around the binaries GG Tau [7] and UY Aur [14], the silhouette disks in Orion [15], are as many proofs. But these images contain much more: they contain the traces of the disk structure itself. In the following paragraphs, I describe a few of these structural parameters and critically discuss how they are estimated.

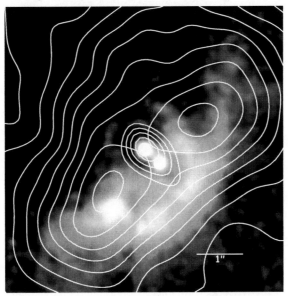

FIGURE 2. Overlay of the ^{13}CO $(2 \rightarrow 1)$ integrated area (large contours) and 1.3mm continuum emission (smaller contour, close to the center) over an Adaptive Optics J-band image (grey shade) obtained at CFHT of the T-Tauri star UY Aur. As expected if the dust and gas are well mixed radially, the ^{13}CO countours trace roughly the scattered light. See [16] for the AO image, and [14] for the radio data.

Outer Radius. Because the instruments have finite sensitivities, the outer radii measured are lower limits. Furthermore, a severe flaring of the disk, or large "puffing" of the inner edge of the disk [17] can also cast shadow on the outer parts, significantly reducing the amount of scattered light received. Nevertheless, the disks that have been imaged so far are in general *larger* than predicted by SED fitting, e.g., [18].

Whether the gaseous disk is actually larger than the dust disk in the radial direction remains an open issue. From existing mm-interferometric observations, one finds that either the dust disks are smaller or their emission drops below the detection limits faster (i.e., at 200-300AU) than the gas disks do (they often extend out to >850AU). However, in the case of GM Aur (a single star), UY Aur (see Fig. 2) and GG Tau – two binaries – the disks seen in scattered light extend over the same distance as the CO disks. This suggests that both techniques are probing the same physical structure and the grains (at least those responsible for scattering) are well mixed with the gas over large radial distances. The disk sizes measured today are in the range $50 \leq R_{out} \leq 2000$AU, larger than our own Solar system, and possibly similar to the size of the Kuiper Belt.

Inclination. The inclination of a resolved disk is usually derived from the axis ratio of either the scattered light, the dust thermal continuum, or the CO gas total emission map. The kinematic pattern can also be used to estimate the inclination in mm-line work. When disks are seen close to edge-on, models of the scattered light are used to estimate the inclination relying on the assumption that the nebula is uniform and symmetric and both lobes should have equal brightness for $i = 90^o$. This technique has been applied for HH 30 [8], HK Tau/c [11], HV Tau C [19], and IRAS 04158+2805 [13] for example, yielding inclination estimates with error bars of a few degrees only.

Density Profiles and Scale Height. For edge-on systems, the scale height H_0 can be measured directly from optical and infrared images. However, a large flaring or a thick envelope may forbid direct view of the central parts, reducing the radial coverage and forbidding an assessment of the radial behaviour of the scale height. A wavelength coverage as wide as possible is necessary to extract a more reliable estimate. For an arbitrary inclination, estimating the surface density law $\Sigma(r)$ from scattered light is difficult because the disk is optically thick. For edge-on systems, $\Sigma(r)$ and $H(r)$ are coupled (degenerate) and isolating each one is difficult. Coherent models are nevertheless useful as other constraints exist.

For example, the size of millimeter continuum sources detected in the course of a survey was modelled by Dutrey et al. [20] who concluded that the radial dependence of the surface density is probably shallower than was imagined in previous models using single dish (sub-)millimeter data alone. This suggestion of shallowness is confirmed by better, more representative models for GM Aur [21] and DM Tau [22] for example based on higher resolution observations and better fitting procedure.

The fact that the scale height of disks is roughly of order $h/r = 0.1$ implies that the disks surrounding T-Tauri stars are gas rich (i.e., the mean molecular weight is low). This is a necessary condition for the formation of giant planets. Older disks, in β Pictoris for example, are dominated by dust and are much flatter.

Rotation Curve. From molecular line observations, solid evidence for rotating disks have been found in a number of Classical T-Tauri stars now (e.g., [23] for LkCa 15; [24] for MWC 480, an Herbig Ae star; [25] for UZ Tau E; [22] for DM Tau; [21] for GM Aur). In most cases, the measured rotation profile is consistent with Keplerian rotation in the outer parts of the disks where data exist (outside 0."5 or so), with $V(r) \propto r^{-0.5}$ within the error bars.

An interesting application of the rotation curves is the possibility to estimate the stellar masses dynamically with an accuracy of 10%–20% [26].

Disk Mass. Disk mass estimates from scattered light models are severely affected by the large optical thickness of the disk and are mere lower limits, usually. An exception to that is found, once again, in the edge-on case. Then, the optically thick disk draws a dark lane in the equatorial plane, the *thickness* of which is directly related to the disk mass. For HH 30, the mass derived from such a model agrees reasonably well with the mass estimated from millimeter continuum flux.

Estimating disk masses from millimetric observations is not much easier (see, e.g., [18,27]). One usually assumes that the dust is optically thin throughout the disk and the continuum flux therefore a good tracer of the total mass. This assumption may fail close to the center. Also, the grain size distribution, the dust emission properties and the dust-to-gas ratio are quantities subject to uncertainties, up to a factor of 10. Similarly, mass estimates based on gas phase tracers can suffer from optical thickness and unknown depletion factor of the molecular species used. Usually, dust masses are considered more reliable because the dust opacity at 1mm is considered accurate to within a factor of a few (e.g., [28]) for a wide-range of grain sizes and shapes.

Assuming that a representative CTTS is 10^6 years old and that its accretion rate is $10^{-7} M_\odot$ yr^{-1}, then the accreted mass is $0.1 M_\odot$. This is more than most of the disk masses estimated so far. However, the accretion rate may be lower, $10^{-8} M_\odot$ yr^{-1}, and it has been shown to decline with time [29]. The mass accreted over the pre-main sequence lifetime then becomes comparable to the disk masses deduced from mm continuum emission (typically a few $0.01 M_\odot$), and there may be no need to "feed" the disk to sustain the accretion process in the Classical T-Tauri phase. This mass is of the order of what is needed in models of the minimum mass pre-solar nebula.

Disks have been resolved around stars spanning the spectral range M5 to early A/late B. So far, only indirect evidence exist for the presence of disks around brown dwarfs. They show excess infrared emission and sub-millimeter dust continuum emission suggestive of disks (see e.g., [30]), but none has been imaged directly at the time of writing.

THE EVOLUTION AND DISSIPATION OF ACCRETION DISKS

Although the evolution of an accretion disk is probably very fast on astronomical timescales, a few millions years at most!, it nevertheless remains exceedingly long with respect to observational astronomy's history. Today's astronomers are therefore faced with two choices when trying to elucidate the fate of accretion disks: either take a statistical approach by studying a large number of young targets to evaluate the disk fraction; or study young stars with resolved disks on a case by case basis to try and learn more about their properties and hope to find relevant clues to identify the dissipation mechanisms.

Disk dissipation timescales: the statistical approach

The statistical approach is extremely useful to assess the general fraction of disks and the typical age at which none can be found anymore. Hillenbrand & Meyer [31] compiled near-infrared data for a large number of stars located in numerous star forming regions. The presence of a near-infrared excess, well above the values expected from a photosphere alone, is generally accepted as a good tracer of the presence of an active and optically thick disk in accretion onto the central object. As expected, there is a large spread in the magnitude of the excess as a function of age for a given group of stars and a precise timescale for disk dissipation is difficult to evaluate. It is however a solid result that the fraction of stars with NIR excess is decreasing with age [32]. In young Star Forming Regions, i.e., those a few million years old or less, the disk fraction traced by NIR excess is high, 50%–60%. On the contrary, other more evolved open clusters where the population of low-mass stars is closer to the ZAMS show a NIR excess fraction that is practically null. So there is little doubt that between the early age of a few millions years (say less than 5Myr) and the later age of a few hundred million years, the NIR infrared excess becomes undetectable.

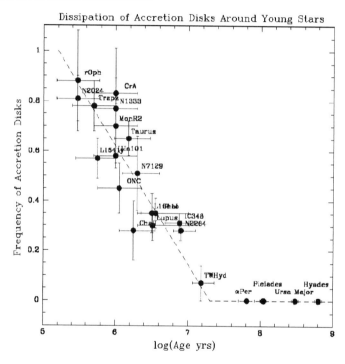

FIGURE 3. Evolution of the fraction of stars having hot inner disks (inferred from location in the J-H vs. H-K diagram) as a function of time in various star forming regions and stellar clusters. Image from [31].

The presence of NIR excess is probably telling us that the inner part of the disks are still present, hence probably that the disk is still opaque and accreting actively. However, the disappearance of the NIR excess barely means that the hottest dust in the disk has gone. But by no means does it certify that the gas and/or colder dust has also been removed from the disk.

Colder dust can also be probed by its thermal emission, but at far-infrared or millimeter wavelengths. Intuitively, and roughly speaking, this would probe the outer parts of the disks. Habing et al. [33] performed this experiment at 60μm with the Infrared

Satellite Observatory (ISO). They observed a number of stars spanning a wide range of ages, from pre-main sequence to old main sequence, i.e., from a few million- to many billion years old. Their results suggest that up to an age of 400Myr or so, about 60% of the observed sample had 60μm emission in excess of photospheric emission, suggesting the presence of a (cold) dust disk. For the older stars in their sample, i.e., those older than 400Myr, the fraction of objects with detectable 60μm excess drops dramatically to only 8% or so. The latter sample includes stars ranging from 400Myr up to about 10Gyr. A clear break is seen in their histogram at an age of about 400Myr. The straightforward conclusion is that by age 400Myr, most disks have also dissipated their cold dust components. Two dedicated Spitzer Legacy surveys targetting much larger samples will refine these numbers.

These global approaches are powerful to assess timescales. However, they remain relatively powerless to identify the exact process(es) leading to the disappearance of dust and gas in the accretion disks. The next generation of instruments, like Spitzer now flying, and ALMA, will help understand these processes.

Evolution of the disk content

Around pre-main sequence low-mass stars, a gas-to-dust ratio close to the one derived for the interstellar medium, i.e., a factor 100 in mass, is generally assumed in the disks. At a later age, in our Solar System for example, the gas is almost completely absent. In between, the situation is not clear yet!

What happens to their constituents as disks evolve? Common belief has it that planetesimals rapidly form and gravitationally settle to the disk mid-plane. There is also evidence in Vega-like stars, older stars of intermediate mass, that cometary-like bodies fall onto the central star (e.g. [34]). However, these "disks" appear almost devoid of gas, the gas/dust mass ratio being small, with both components (gas and dust) probably not original. A large number of these Vega-like stars have now been found, they represent nearly 15% of nearby A-F stars [35] and disks have been imaged around a few of them.

In this section we will focus on dust properties in the circumstellar environment of young (a few Myr old at most) and low-mass stars in the hope of catching a glimpse of the planetesimal formation process.

Evidence from scattered light images

Numerous disks around T-Tauri stars have now been imaged with HST in the optical and with ground-based adaptive optics in the near-infrared. Model fitting of these images suggest that the grain size distributions needed to reproduce the observations are roughly similar to the one found in the interstellar medium (e.g., [11,13,19]). Although their exact optical properties may differ slightly from the ISM ones, these grains are small and there is therefore no clear signs of grain growth based only on intensity images in the optical and near-infrared wavelength range. These grains are probably located at the surface of the disks however, away from the midplane. Deeper in the disk, closer to the midplane, there is evidence that larger grains are needed to explain the thermal emission detected at longer wavelengths. The results that small grains are present in large numbers, based on scattered light images, suggest nevertheless that the small grain population *has not (yet) been depleted significantly* at an age of a few Myr, although grain growth may have already started.

Interestingly, images of the scattered light at longer wavelenghts, i.e., from 3.5 up to 12 microns, are slowly becoming available with the improved sensitivity of mid-IR cameras attached to large aperture telescopes [36]. These offer the possibility to probe deeper into the optically thick accretion disk's surface. Also, they probe sligthly larger grains. Not surprisingly, preliminary results [36] suggest that the dust size distribution may be segregated vertically, with only the small grains being lifted to the surface, and the larger grains remaining deeper below the surface as their sizes increase.

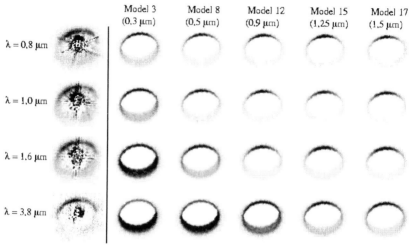

FIGURE 4. Scattered light images of the circumbinary ring of GG Tau from 0.8 to 3.8 microns (left column) compared with models of the scattered light where only the maximum grain size has been changed in the dust distribution. The maximum grain radius used is given between brackets for each model, ranging from 0.3 micron (left) to 1.5 microns (right). No single model fit all wavelengths. Image from [36].

Evidence from thermal emission measurements

At longer wavelengths, [20] fitted the observed flux at 0.8, 1.3 and 2.7mm for a number of T-Tauri stars. A power-law of the form $\kappa(\nu) \propto \nu^{-\beta}$ was assumed for the emissivity. They found that $\beta = 1$ provided a good fit to the data, a value much smaller than what is found in molecular clouds ($\beta = 2$). These results were obtained independently by several groups (e.g., [18],[37]) and they too suggest grain growth ([28],[38]).

CONCLUDING REMARKS

Recent and direct images of accretion disks around young solar-like stars have allowed us to measure some of the key structural parameters of these disks. Although the census is still far from complete, we are slowly unveiling important features like their masses, their density profiles, and even their radial and vertical temperatrure structures.

Nonetheless, much remains to be discovered and many of the next generation of instruments like Spitzer, Herschel and Alma to name only those three, will be well suited to push our understandnig forward. They will enable a more complete survey of the gas phase content of the disks, opening the way for more realistic chemistry models and a better understanding of the temperature structure throughout the disk. The study of grain growth, already initiated, will benefit from mid-IR spectroscopy, as will the characterisation of the chemical content of solid-state particles.

In a different direction, near-infrared long baseline interferometry will quickly allow us to estimate the geometrical properties of the very inner disk, first results are already available from the Keck and ESO/VLT interferometers. In the near future, aperture synthesis imaging should become available with these instruments. Optimistically, they will help us further understand the temperature structure and the exact density distribution in the inner regions, and possibly probe the important star-disk interaction zone, where winds/jets are launched from.

Supplemented by high contrast high resolution imaging, e.g., from ground based next generation AO systems, features like gaps and warps driven by orbiting giant planets may become detectable. Needless to say, the next few years should be fascinating.

I would like to warmly thank the organisers for inviting me to participate in this very stimulating conference. I also appreciate the patience of the editors. I acknowledge financial support form the Programme National de Physique Stellaire *(PNPS) of* CNRS/INSU, France.

REFERENCES

1. Walker, M.F. 1972, ApJ, 175, 8
2. Lynden-Bell, D. , Pringle, J.E. 1974, MNRAS, 168, 603
3. Rucinski, S.M. 1985, AJ, 90, 2321
4. Hartmann, L., Kenyon, S.J. 1987, ApJ, 312, 243
5. Bertout, C., Basri, G., Bouvier, J. 1988, ApJ, 330, 350
6. Basri, G., Bertout, C. 1989, ApJ, 341, 340
7. Dutrey, A., Guilloteau, S., Simon, M. 1994, A&A, 286, 149
8. Burrows, C.J., Stapelfeldt, K.R., Watson, A.M., et al. 1996, ApJ, 473, 437.
9. Mouillet, D., Lagrange, A.M., Augereau, J.C., Ménard, F. 2001, A&A, 372, L61.
10. Clampin, M., Krist, J.E., Ardila, D.R., et al. 2003, AJ, 128, 385
11. Stapelfeldt, K.R., Krist, J.E., Ménard, F., Bouvier, J., Padgett, D.L., Burrows, C.J. 1998, ApJ, 502, L65
12. Ménard, F., Dougados, C., Magnier, E., Cuillandre, J.C., Fahlman, G., Forveille, T., Lai, O., Manset, N., Martin, P., Veillet, C. 2001, AAS, 198, 4803.
13. Ménard, F., Dougados, C., Magnier, E., Cuillandre, J.C., Fahlman, G., Forveille, T., Lai, O., Manset, N., Martin, P., Veillet, C. 2004, ApJ, in press.
14. Duvert, G., Dutrey, A., Guilloteau, S., Ménard, F., Schuster, K., Prato, L., Simon, M. 1998, A&A, 332, 867

15. O'Dell, C.R., Wen, Z., Hu, X. 1993, ApJ, 410, 696
16. Close, L.M., Dutrey, A., Roddier, F. et al. 1998, ApJ, 499, 883
17. Dullemond, C.P., Dominik, C., Natta, A. 2001, ApJ, 560, 957
18. Beckwith, S.V.W., Sargent, A.I., Chini R.S., Güsten, R. 1990, AJ, 99, 924
19. Stapelfeldt, K.R., Ménard, F., Krist, J.E., et al. 2003, ApJ, 589, 410
20. Dutrey, A., Guilloteau, S., Duvert, G., Prato, L., Simon, M., Schuster, K., Ménard, F. 1996, A&A, 309, 493
21. Dutrey, A., Guilloteau, S., Prato, L., Simon, M., Duvert, G., Schuster, K., Ménard, F. 1998, A&A, 338, L63
22. Guilloteau, S., Dutrey, A. 1998, A&A, 339, 467
23. Duvert, G., Guilloteau, S., Ménard, F., Simon, M., Dutrey, A. 2000, A&A, 355, 165
24. Mannings, V., Koerner, D.W., Sargent, A.I. 1997, Nature, 388, 555
25. Jensen, E.A., Koerner, D.W., Mathieu, R.D. 1996, AJ, 111, 2431
26. Simon, M., Dutrey, A., Guiloteau, S. 2000, ApJ, 545, 1034
27. André, P., Montmerle, T. 1994, ApJ, 420, 837
28. Pollack, J.B., Hollenback, D., Beckwith, S., Simonelli, D., Roush, T., Fong, W. 1994, ApJ, 421, 615
29. Hartmann, L., Calvet, N., Gullbring, E., D'Alessio, P. 1998, ApJ, 495, 385
30. Pascucci, I., Apai, D., Henning, Th., Dullemond, C.P. 2003, ApJ, 590, L111
31. Hillenbrand, L., Meyer, M. 1999, AAS, 195, 02.09
32. Strom, S.E. 1995, RevMexAA (serie de Conferencias), 1, 317
33. Habing, H.J., Dominik, C., Jourdain de Muizon, M., Kessler, M.F., Laureijs, R.J., Leech, K., Metcalfe, L., Salama, A., Siebenmorgen, R., Trams, N. 1999, Nature, 401, 456.
34. Beust, H., Vidal-Madjar, A., Ferlet, R., Lagrange-Henri, A.M. 1990, A&A, 227, L13.
35. Lagrange, A.-M., Backman, D., Artymowicz, P. 2000, in *Protostars & Planets IV*, eds. V. Mannings, A. Boss & S. Russell (Tucson: Univ. of Arizona Press), p. 639
36. Duchêne, G., McCabe, C.E., Ghez, A. 2004, ApJ, in press. (astro-ph/0401560)
37. Mannings, V., Emerson, J.P. 1994, MNRAS, 267, 361
38. Miyake, K., Nakagawa, Y. 1993, Icarus, 106, 2

The Disk, Jet, and Environment of the Nearest Herbig Ae Star: HD 104237

C. A. Grady[*], B. Woodgate[†], Carlos A. O. Torres[**], Th. Henning, D. Apai,
J. Rodmann[‡], Hongchi Wang[§], B. Stecklum, H. Linz[¶], G. M. Williger[||],
A. Brown, E. Wilkinson, G. M. Harper[††] and G. J. Herczeg[††]

[*]*Eureka Scientific & GSFC, Code 681, NASA's GSFC, Greenbelt, MD 20771*
[†]*LASP, NASA's GSFC, Greenbelt, MD 20771, USA*
[**]*LNA-MCT, Itajubá, 37504-364 Brasil*
[‡]*MPIA, Heidelberg, D-69117, Germany*
[§]*Purple Mountain Observatory, Academica Sinica, Nanjing 210008, PR China and MPIA,
Heidelberg, D-69117, Germany*
[¶]*TLS Tautenburg, D-07778 Tautenburg, Germany*
[||]*The Johns Hopkins University, Baltimore, MD 21218*
[††]*University of Colorado, Boulder, CO 80309, USA*

Abstract. The environment of the nearest Herbig Ae star has been investigated through a program
of multi-wavelength, high contrast and high spatial resolution imagery, FUV through optical integrated light spectroscopy, and FUV spatially resolved spectroscopy. HD 104237 is the primary
of a 5 Myr old aggregate of at least 4 PMS stars, 2 of which in addition to HD 104237 have IR
excesses indicating the presence of dust disks. HD 104237 is actively accreting, and is driving a
bipolar outflow (HH 669) which can be traced 2.65" from the star and which is viewed at an inclination of $18^{+14°}_{-11}$. The counterjet can be traced no closer than 0.6" (79 AU) from the star, providing
a firm upper limit to the size of the disk. The absence of spatially extended H_2 emission, FUV reflection nebulosity, and mid-IR PAH emission features are all consistent with dust settling and the
presence of a geometrically shadowed disk. The combination of proximity, low reddening, and the
high density of disks in the HD 104237 association make this group of stars an ideal laboratory for
probing the comparative evolution of planetary systems.

INTRODUCTION

Understanding the evolution of planetary systems is essential both for understanding
the history of our Solar System, and for accounting for the diversity of the known exoplanetary systems. Nearby, associations of coeval stars are likely to play a crucial role
in extending our understanding of the evolution of protoplanetary systems, since they
can be securely dated, and are accessible to a variety of multi-wavelength observational
techniques. One such group of stars is associated with the optically brightest and nearest
Herbig Ae star, HD 104237 (DX Cha, d=116^{+8}_{-7} pc, [1]). Two T Tauri companions have
been identified within 15" of HD 104237 ([2][3]).

The combination of HST/STIS coronagraphic imagery and VLT/NACO high contrast,
high resolution near-IR imagery have revealed an additional late type companion to
HD 104237, some 1.3" from the star, and further demonstrate (Fig. 1) that, in addition to
HD 104237 itself, 2 of the companions have the red near-IR and mid-IR colors indicating

CP713, *The Search for Other Worlds: Fourteenth Astrophysics Conference*,
edited by S. S. Holt and D. Deming
© 2004 American Institute of Physics 0-7354-0190-X/04/$22.00

the presence of circumstellar dust [3]. Together with the Ae star disk, the discovery of disks around these two low mass stars brings the disk fraction to 75%.

Initial efforts to date HD 104237 resulted in age estimates of 1-2 Myr [1] to 2-4 Myr [2], in disagreement both with the estimated ages of the T Tauri companions, and other nearby young stars. The discrepancy is partially due to use of a range of spectral type and extinction estimates for HD 104237, both of which affect the star's location in the HR diagram. The STIS data indicate that the line of sight to HD 104237 is very lightly reddened, with E(B-V)=0.004 for R=3.1. The optical colors of HD 104237 and its UV spectrum, as measured by FUSE, STIS, GHRS, and IUE all indicate that the star is closer to A8Ve/F0Ve with a significant UV excess [3]. This new spectral type makes HD 104237 the brightest known Herbig Fe star, accounting for the UV emission line spectrum resembling a late type star [4], and resolves the age discrepancy, with both the T Tauri companions [5], and the Herbig star [6] at 5 Myr.

THE DISK AND JET OF HD 104237

The STIS spectra of HD 104237 reveal the presence of a bipolar microjet, discovered in Lyman α, with two jet knots within 1.15" of the star (Fig. 2). This is the second microjet discovered in association with an older Herbig Ae star. The combined proper motion and radial velocity of the inner knot enables us to derive $i=18^{+14o}_{-11}$ for the outflow. The counterjet can be traced to r$\geq$0.6" from the star, demonstrating that the disk has r$\leq$79 AU. If the nearest companion at 1.3" is coplanar with the disk, the true disk size may be closer to 0.4", making the HD 104237 disk the closest known early Solar System analog. The absence of spatially extended fluorescent H_2 emission, FUV reflection nebulosity, and mid-IR PAH emission features are all consistent with the dust disk being geometrically shadowed at and beyond terrestrial planet distances from the star. This shadowing is an expected consequence of grain growth in the disk [7][8], and is consistent with the 9.7μm silicate emission profile, which indicates average grain sizes in the 2 μm range [9]. The HD 104237 data indicate that this process, an expected precursor to large-scale planetesimal formation, is well-advanced by 5 Myr.

Despite the grain growth, the UV/FUV data indicate that the star is still accreting. It has a temporally variable UV/FUV excess which does not appear to simply be due to blending of fluorescent H_2 emission lines. The continuum is correlated with the FUV emission spectrum, particularly the infalling gas features seen in species ranging from Ly α through O VI. If interpreted in the context of funnelled accretion models ([10]), the continuum is due to the accretion shock and is consistent with a mass accretion rate of a few times 10^{-8} M$_\odot$ yr^{-1}.

A REMNANT ENVELOPE AT T=5 MYR

The presence of so many companions retaining disks suggests that star formation in the HD 104237 association was rather quiescent. Under these circumstances, one might expect traces of the natal molecular cloud to persist for some time after formation. STIS

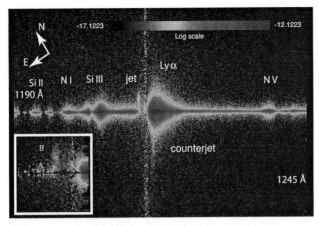

FIGURE 1. Spatially resolved spectroscopy of HD 104237 at Lymanα demonstrating the S-shaped velocity curve characteristic of a jet, togther with the detection of Herbig-Haro knots. The counterjet can be traced no closer than 0.6" from the Herbig Ae star.

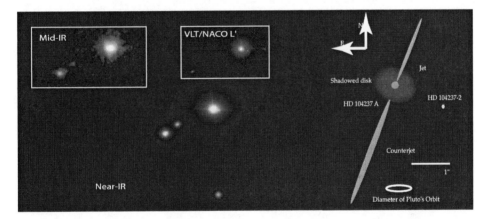

FIGURE 2. The HD 10237 system. left) Near and mid-IR color composite imagery of HD 104237 reveals 2 stars, in addition to the Herbig Ae star which have infrared excesses indicating the presence of circumstellar disks. right) A cartoon of HD 104237 and its nearest companion showing the geometry of the disk and outflow.

coronagraphic imagery show no indication of optically visible nebulosity at $r \geq 0.5$" (65 AU) from HD 104237. However, FUSE spectra for HD 104237 show up to 100 J=0 to J=5 H_2 transitions, corresponding to Log $(N(H_2))=19.4(2)$ cm^{-2}, with a rotational excitation temperature of 300 K [11],[12]. In tandem with Log $(N_{HI})=1.5(5) \times 10^{19}$ cm^{-2} from STIS, the FUSE data show that while the total gas to dust ratio is typical of the diffuse interstellar medium, the ratio of molecular to atomic gas is more appropriate to a dense molecular cloud. The high rotational excitation temperature indicates that the molecular gas is in the immediate vicinity of HD 104237. Similar gas is seen in the line of sight to other low-reddening Herbig Ae stars [13],[14] but is not seen toward older, more centrally cleared debris disk systems, suggesting that the remnants of the natal molecular cloud can be routinely detected over the first 5-7 Myr, but are less commonly seen after 8 Myr [3].

ACKNOWLEDGMENTS

This study made use of HST under programs HST-GTO-9241, HST-GO-5495, and the NASA-CNES-CSA FUSE under GTO programs P163 and P263, and ESO's 3.6m + TIMMI2 under programme ID 71.C-0438, the ESO VLT and NACO under programme ID 71.C-0143, and the ESO 1.52m + FEROS under the ON-ESO agreement for the joint operation of the 1.52m ESO telescope. The study was supported by STIS GTO funding in response to NASA A/O OSSA -4-84 through the HST Project at Goddard Space Flight Center, NASA PO 70789-G, NASA PO S-70502-G, NAS5-32985, NAG5-4808, NAG5-3226, and NAG5-13058, CNPq-Brazil (pr 200356/02-0), NSFC grants 10243004 and 10073021, Deutsche Forschungsgemeinschaft, grant Ste 605/17-2. HST is operated by the Association of Universities for Research in Astronomy, Inc., under NASA Contract NAS5-26555. FUSE is operated for NASA by the Johns Hopkins University under NASA contract NAS5-32985.

REFERENCES

1. van den Ancker, M.E. et al. 1998, A&A 330., 145–154.
2. Feigelson, E.D., Lawson, W.A., & Garmire, G.P. 2003, The ε Chamaeleontis young stellar group and the characterization of sparse stellar clusters, *Astrophys. J* (in press).
3. Grady, C. A. et al. (2003), The Environment of the Optically Brightest Herbig Ae Star: HD 104237, *Astophy.J* (submitted).
4. Brown, A., Tjin A Djie, H.R.E., Blondel, P., Harper, G.M, & Skinner, S.L. 1996, HST GHRS Observations of the Herbig Ae Star HD104237: First UV Observations of a Hot Disk Wind from a Pre-Main Sequence Star, in *Accretion Phenomena and Related Outflows*, editors D.T. Wickramasinghe, L. Ferrario, & G.V. Bicknell, IAU Coll. 163, ASP Conf. Ser. 121, 448–452.
5. Siess, L. Dufour, E., & Forestini, M. 2000, A&A 358, 593–599.
6. Palla, F., & Stahler, S.W. 1993, *Astrophys. J* 418, 414–425.
7. Dominik, C., Dullemond, C.P., Waters, L.B.F.M., Walch, S. 2003, A&A 398, 607–619.
8. Dullemond, C.P. 2003, IAUS 221, E16 (in press).
9. Van Boekel, R. et al. 2003, A&A 400, L21-24.
10. Gullbring, E., Calvet, N., Muzerolle, J., & Hartmann, L. 2000, *Astrophys. J.* 544, 927–932.
11. Herczeg, G.J., Linsky, J.L., Brown, A., Harper, G.M., and Wilkinson, E. 2001, in *The Future of Cool Star Astrophysics*, eds. A. Brown, G. Harper, and T. Ayres, at http://origins.colorado/edu/cs12/
12. Brown, A. Herczeg, G.M., Harper, G.M., Wilkinson, E., Tjin A Djie, H.R.E., and Blondel, P. 2003, GHRS and FUSE Observations of the Herbig Ae Star HD104237, *Astrophys. J*, (in preparation).
13. Roberge, A. et al. 2001, ApJ 551 L97–L100.
14. Lecavelier des Etangs, A. et al. 2003, A&A 407, 935–939.

CO in Disks around Transition Objects

W. R. F. Dent[*], J. S. Greaves[*] and I. M. Coulson[†]

[*]UKATC, Royal Observatory, Blackford Hill, Edinburgh, Scotland EH9 3HJ
[†]JCMT, Joint Astronomy Centre, Hilo, Hawaii 96720, USA

Abstract. Results are described from observations of submm CO emission from isolated young stars with an infrared excess. These include both Vega-excess and isolated Herbig AeBe stars, ie those at the transition between the formation stage and the main sequence. Of those detected, more than 60% show double-peaked profiles, interpreted as emission from rotating gas disks. The spectra of three objects are compared with a basic disk model. HD 141569 has evidence of a double-ring gas structure, with radii of 90 and 250 au - similar to that seen in scattered light. MWC480 has an asymmetric line shape, interpreted as a non-axisymmetric disk temperature distribution. Finally the debris disk HD 9672 shows a broad line from a relatively compact disk or ring.

This molecular gas is found in stars of ages up to 10-20 Myr, and its' removal from the disk signifies the end of the accretion and the start of the collision-dominated phase of disk evolution.

INTRODUCTION AND OBSERVATIONS

Isolated Herbig AeBe stars lie near the end of their formation phase, with typical ages of a few 10^6 yrs. Although they are not normally associated with molecular clouds, their Spectral Energy Distributions (SEDs) show a significant infrared excess coming from a compact circumstellar dusty disk. Vega-excess objects have a similar but generally smaller excess; they lie on the main sequence, but still tend to be relatively young. These two types of objects are thought to mark different evolutionary stages of a stellar/planetary system, an idea which is strengthened by the detection of borderline cases [6]. These systems are often called transition objects.

Most studies of disks around transition objects have concentrated on the dust continuum [10]. Dust is also seen around Vega-excess stars [4], and this is considered to be from secondary material formed in a collisional cascade of larger bodies. However, some transition objects have been detected (and even spatially resolved) in mm-wave spectral lines from molecules such as CO [9]. As well as observationally important because of the extra kinematic information this gives, gas drag in these disks will reduce the relative velocities of the smaller dust grains, affecting any collisional cascade which would otherwise occur. The continuing presence of gas will also have an impact on the growth and composition of larger bodies such as planets and comets.

The results described below were obtained during a more general survey for ^{12}CO emission from transition objects [3]. The J=3-2 line was observed using the 15m James Clerk Maxwell Telescope, and, as the emission regions were smaller than the beam (14"), the resultant spectra have been compared with a basic disk model in order to estimate parameters such as disk size and inclination (see [3] for details).

CP713, *The Search for Other Worlds: Fourteenth Astrophysics Conference*,
edited by S. S. Holt and D. Deming
© 2004 American Institute of Physics 0-7354-0190-X/04/$22.00

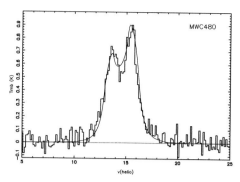

FIGURE 1. Spectrum of MWC 480 in ^{12}CO J=3-2 (histogram), together with non-axisymmetric disk model inclined at 30° to the line of sight (see text for details).

MWC 480

The disk around this isolated 5 Myr-old HAeBe star has been spatially resolved in CO using interferometry [8]. Our new high s:n spectrum (Figure 1) shows emission extending to $\pm 4 kms^{-1}$ from the line centre. By comparing with the model, this implies the gas disk extends from a radius of 250 au down to <20 au. Any gas-depleted inner hole must be smaller than this. Both our data and the interferometric spectrum also show an asymmetric profile, which could be caused by either a density or a temperature asymmetry in the disk. As the ^{12}CO emission is likely to be optically thick, the putative density asymmetry would need to be $\gg 99\%$. However, a temperature difference of only 30% between the two sides of the disk could explain the line shape, and a model for this is shown in Figure 1. Such a temperature difference could be caused by non-symmetric warping of the disk atmosphere, exposing one side to more stellar heating per unit disk surface area. An upgrade of the modelling program will include full radiative transfer and dust heating in order to investigate this effect.

HD 141569

This 5 Myr-old transition object has generated considerable interest because of the extended and complex scattering region seen in coronographic images [2]. Rings or tightly-wound spirals of dust are seen at mean radii of $\sim$200 and 350 au. HD 141569 has a low fractional excess, and CO has been detected in both mm-wave [12] and infrared ro-vibrational lines [1]. It has been suggested that the dust rings could arise from migration of small grains through the gas [11]; the underlying gas distribution is then predicted to be significantly different from the dust.

We obtained a deep CO spectrum (Figure 2), which shows the line to have a relatively broad double-peaked profile. However, a distinctive shoulder is seen, particularly at high relative velocities. This cannot be modelled with a single disk structure, consequently we

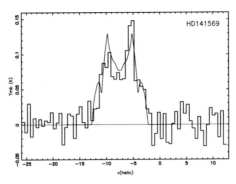

FIGURE 2. Spectrum of HD 141569 (histogram), together with a double-ring model (see text).

adopted a double-ring model. The inclination of the system is well-constrained by the optical images to 52°. Assuming the gas is coplanar, the best fit model has ring radii of 90 and 250 au. The rings also need to be vertically thin (opening angle only 1°) in order to maintain sufficient optical depth to the central star to avoid CO photodissociation. This is consistent with the low fractional infrared excess. The innermost gas ring lies within the coronographic occulting disks, but is similar to the outer limits of the hot disk seen in IR lines [1]. However, the outer CO ring has a radius which places it in the gap between the two dust rings, although our derived radius is very tentative due to limited s:n. Finally there is apparently a significant asymmetry in the low relative velocity peaks, likely indicating a more complex gas geometry. A resolved interferometer image of this disk is required to investigate this structure further.

HD 9672

This 8 Myr-old star (aka. 49 Ceti) is one of only three bone-fide debris disks with $f > 10^{-3}$ (the others being β Pic and HR 4796). CO has been detected from HD 9672 [12], but not the other two [7]. The star lies at only 61pc, and is of some interest as a potential gas-rich debris disk. Although the s:n is low, our new observations confirm the earlier CO detection (Figure 3). Moreover we find the line is broad, with a velocity similar to that of the star (see vertical mark in Figure 3), both of which suggest the emission is from a compact disk.

We also show a model consistent with the data, although the s:n is too low at present to make this interpretation unique. The relative weakness of the line implies a disk of only 16 au radius, although a more inclined, extended ring will also fit the data. Mid-infrared [5] observations indicate a radius of ~50-100 au. Assuming the CO emission region has an outer radius of 75 au, then the model implies a thin ring rather than a disk, with a radial width of 10 au. The ring must also be vertically thin (opening angle 2°), consistent with the low fractional infrared excess. However, a higher s:n spectrum is clearly required to better delineate the line shape.

53

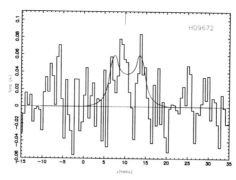

FIGURE 3. Spectrum of HD 9672, together with a compact disk model with inclination 18°.

DISCUSSION: GAS IN TRANSITION DISKS

CO is abundant in young HAeBe disks, but has not been detected in older debris disks (e.g., [7]). Gas drag will significantly affect disk dynamics [11], and the factor describing the degree of gas-dust coupling for grains of size a, known as the stopping parameter, can be written as [11]: $T_{ss} \approx (a/mm) \times (R/100au)^2/(M_{gas}/M_{Jup})$, where R is the disk radius and M_{gas} the gas mass. For $T_{ss} < 1$, the grains are coupled to the gas, have low relative velocities and may not be in a collisional cascade. As ^{12}CO is optically thick, we estimate minimum masses for MWC 480, HD 141569 and HD 9672 of 0.05, 0.008 and $0.001 M_{Jup}$ respectively. Even these conservative limits would mean that grains of $<50, 2$ and 50 μm would be coupled to the gas. If the disks are significantly more massive, then even observationally-bright mm-sized grains may not be collisionally-dominated, and these cannot be regarded as debris disks. Once the gas mass drops below $\sim 10^{-3} M_{Jup}$, this signifies a transition from a coagulating to collision-dominated disk.

REFERENCES

1. Brittain, S.D., Rettig, T.W., Simon, T., Kulesa, C., DiSanti, M.A., Dello Russo, N., 2003, ApJ, 588, 535
2. Clampin, M. et al., 2001, AJ, 126, 385
3. Dent, Greaves & Coulson (MNRAS - submitted)
4. Holland, W.S. et al., 1998, Nature, 392, 788
5. Jayawardhana, R., Fisher, R.S., Telesco, C.M., Pina, R.K., Barrado y Navascúes, D., Hartman, L.W., Fazio, D.C., Lloyd, J., 2001, AJ, 122, 2047
6. Lagrange, A.-M., Backman, D.E. and Artymowicz, P., in Protostars and Planets IV, eds. V. Mannings, A.P. Boss & S.S. Russell, Univ. of Arizona, Tucson, 2000, p. 639
7. Liseau, R., 1999, A&A, 348, 133
8. Mannings, V., Koerner, D.W., Sargent, A.I., Nature, 388, 555
9. Mannings, V. & Sargent, A.I., ApJ, 529, 391
10. Sylvester, R.J., Skinner, C.J., Barlow, M.J., Mannings, V. 1996, MNRAS, 279, 915
11. Takeuchi, T. & Artymowicz, P., 2001, ApJ, 557, 990
12. Zuckerman, B., Forveille, T., Kastner, J.H., 1995, Nature, 373, 494

Early Solar Systems and the Formation of Massive Stars

William K. Rose

Department of Astronomy, University of Maryland, College Park, MD 20742

Abstract. The recent discovery of the 0.8 $M_\odot$ highly iron-deficient star HE0107-5240 indicates that solar systems probably form in the very early universe. We discuss some implications of this discovery. Massive stars are predicted to be more common in the early universe than at the present epoch. We describe previously unpublished calculations of massive protostars.

INTRODUCTION

For some time there has been evidence that type Ia supernovae were rare in the Galaxy when the Fe abundance was below [Fe/H] $\simeq$ -1 [2]. It has been widely accepted that the first generation of stars (i.e. Population III stars) contained only very massive stars because the absence of elements heavier than helium in primordial gas clouds implied that cooling was insufficient for fragmentation to produce low mass stars. Some minimum level of heavy element enrichment is required before low mass stars with solar systems can be formed. Recently [1] have reported the discovery of a star, HE0107-5240, with a mass of 0.8 $M_\odot$ and iron abundance [Fe/H] = -5.3 $\pm$ 0.2. Although HE0107-5240 is the most iron-deficient star known it is relatively rich in carbon and oxygen. Its carbon abundance is [C/H] = -1.3. The most straight forward explanation for the above star interprets it as a second generation star that was formed in the ejecta of a supernova that synthesizes large amounts of lighter elements such as C, N, and O but very small amounts of elements heavier than Mg.

Massive stars with masses of 20 - 130 $M_\odot$ [9] are believed to produce supernovae whose ejecta fall back to form massive black holes after substantial mixing has occurred. Moreover, there is evidence for the existence of very low ^{56}Ni mass type II supernovae [8]. The above supernova namely SN1997D is the prototype for a distinct class of underluminous supernovae with narrow spectral lines. It has also been suggested that a subset of supernovae that form stellar mass black holes are highly energetic and associated with some γ-ray burst sources. These hypernovae produce abundance patterns that are typical of extremely metal deficient stars and are not enhanced in C and N.

Stars more massive than about 130 $M_\odot$ are believed to produce e^- - e^+ pair production supernovae. If the interior of a highly evolved star moves into a region of the ρ-T plane with very high temperature and low density then dynamical instability caused by e^- - e^+ pair production will result. Dynamical instability causes very rapid contraction and temperature increases. The very high temperature sensitivity of the oxygen burning reactions implies that explosive oxygen burning will disperse the entire star into inter-

CP713, *The Search for Other Worlds: Fourteenth Astrophysics Conference*,
edited by S. S. Holt and D. Deming

stellar space [4]. The abundances of the ejecta of pair production supernovae do not explain the observed abundances of the extremely iron deficient star, HE0107-5240. It has, therefore, been suggested that this latter star, which has sufficient metallicity to have a solar system was a second generation star that formed in a gas cloud consisting of primordial gas and gas enriched by ejecta from supernovae whose masses range between 20-130 $M_\odot$. Umeda and Nomoto [9] show that the abundances of HE0107- 5240 are consistent with being produced in the ejecta of a 25 $M_\odot$ progenitor.

Schneider et al. [6] have argued that the Hamburg/ESO prism survey that discovered HE0107- 5240 will discover more stars with [Fe/H] < -5 if the previous lack of detection of very iron deficient halo stars is a consequence of the brighter magnitude limits of previous surveys. The properties of future stars discovered with [Fe/H] < -5 will provide important information about the first generations of star formation. If many very iron deficient stars are discovered then it would follow (as in the present universe) that 1 $M_\odot$ stars were already forming in the second stellar generation. On the other hand, if only a small percentage of very metal poor halo stars discovered in the future have $Z < 10^{-5}$ $Z_\odot$ with the bulk having $Z = 10^{-3}$ - 10^{-4} $Z_\odot$ then the observations would indicate that many low mass stars and their solar systems were able to form only after these latter levels of metallicity had been reached.

RESULTS

Very massive stars were more common and played a more decisive role in the early universe than at the current epoch. Here we briefly describe previously unpublished hydrodynamic calculations of the collapse of 30 $M_\odot$ and 500 $M_\odot$ protostars [5]. The calculations assume spherical symmetry. As discussed above [9] 30 $M_\odot$ stars are predicted to produce supernovae that appear to provide an explanation for the abundance of HE0107-5240 whereas pair-production supernovae, which are predicted to occur in stars more massive than 130 $M_\odot$ produce ejecta that are much more enriched in iron.

When the cores of molecular clouds undergo dynamic collapse dimensional analysis indicates that the mass in fall rate should become approximately equal to the Jeans mass divided by the dynamic timescale (i.e. $\frac{dM}{dt} \sim c_s^3/G$) where c_s is the speed of sound. We note that the above expression for $\frac{dM}{dt}$ is independent of density.

Our 30 $M_\odot$ protostar calculations assumed an initial uniform density and temperature of 10^7 cm^{-3} and 50 K respectively. They confirmed the already well established result that the central regions collapse first and the outer layers follow as pressure support decreases from below. The value of $\gamma = \frac{\rho dP}{P d\rho}\mid_{s=const}$ was initially higher than 4/3 throughout the protostar. However, as collapse proceded became lower than 4/3 throughout the protostar except close to the surface layers. The reduction in γ induced dynamical instability in the protostar and thereby accelerated further collapse. Although the initial density distribution was uniform the resultant density distribution after several thousand years of collapse was approximately $\rho \propto 1/r^2$ (r not small) and therefore similar to an isothermal gas sphere. These results are similar to those reported by Larson and Starrfield [3]. However, their initial conditions were those of an isothermal gas sphere and consequently very different from our initial conditions. In our final models γ remained less than 4/3 except near the center and outer layers of the protostar. Except during the early stages of collapse $\frac{dM}{dt}$ exceeds c_s^3/G given above. The luminosity of the 30 $M_\odot$ protostar was approximately 50 $L_\odot$.

Eta Carina whose mass is $\sim$ 100-200 $M_\odot$ is the most massive known star in the Milky Way. Although stars as massive as 500 $M_\odot$ may not form at the present epoch they are more likely to form in the early universe when the metallicity is much lower than it is today. We describe previously unpublished calculations [5] of the spherical collapse of a 500 $M_\odot$ protostar whose initial temperature and density distributions are uniform. Our calculations cover a time interval of 5000 years. As in the example of the 30 $M_\odot$ protostar described above the central regions of the 500 $M_\odot$ protostar collapsed first and the outer regions followed. The density distribution evolved so as to approximate that of an isothermal gas sphere (i.e. $\rho \propto 1/r^2$). Although γ was initially above 4/3 throughout the protostar it became less than 4/3 except close to the surface. Finally γ became greater than 4/3 at the center, surface and some interior mass elements.

The 500 $M_\odot$ protostar differed qualitatively from the 30 $M_\odot$ protostar in some important respects. Strong fluctuations in temperature and density developed in the core (mass fraction $\sim$ 0.03 - 0.06) of the 500 $M_\odot$ protostar during collapse. Moreover, portions of the core developed temperatures sufficiently high ($\sim 10^7$ - 6 x 10^8 K) for hydrogen and helium burning to occur.

The appearance of shock waves in the core of the 500 $M_\odot$ protostar implies that it is supported by shock waves as well as pressure gradients. Recently T. Abel has reported [7] that in addition to thermal support a calculated primordial cloud is filled with a turbulent cascade of weak shocks. Our hydrodynamic calculations assumed spherical symmetry. It is possible that a cascade of shocks might appear in a lower mass protostar if three dimensional calculations are performed.

Most protostars collapse into pre-main-sequence stars and then evolve quasi-statically onto the main sequence. On the other hand, our 500 $M_\odot$ protostar collapses directly to core temperatures high enough for nuclear reactions to occur. The presence of shock waves in the stellar core indicates that a stable main sequence star is not formed. Even if the protostar contained no initial abundances of C and O, temperatures sufficiently high for the 3α reactions to occur by direct collapse are achieved, and therefore the CNO cycle may become important even in the absence of initial CNO elements. Such high temperatures are realized at densities of $\sim$ 0.1 gm cm^{-3}.

The rapid decrease in H and He abundance between Saturn and Uranus may be a consequence of photoevaporation of the outer solar system before Uranus and Neptune could form. Because massive stars are likely to be very common in the early universe, it is probable that giant planets similar to Jupiter and Saturn are much less frequent.

REFERENCES

1. Christlieb, N. et al. 2002, Nature, 419, 904.
2. Gilmore, G., King, I., and van der Kruit, P. 1990, The Milky Way as a Galaxy (University Science Books, Mill Valley).
3. Larson, R. and Starrfield, S. 1971, A&A, 13, 190.
4. Rose, W. 1973, Astrophysics (Holt, Rinehart and Winston, NY).

5. Rose, W., and Smith, R. 1971, unpublished.
6. Schneider, R. et al. 2003, Nature, 422, 869.
7. Tan, J., and McKee, C. 2003, The Emergence of Cosmic Structures, edited by S. Holt & C. Reynolds (AIP, Melville).
8. Turatto, M. et al. 1998, ApJ, 498, L29.
9. Umeda, H. and Nomoto, K. 2003, Nature, 422, 871.

Evidence for the Turbulent Formation of Stars

N. H. Volgenau[*], L. G. Mundy[*], L. W. Looney[†] and W. J. Welch[**]

[*]Department of Astronomy, University of Maryland, College Park, MD 20742-2421
[†]Department of Astronomy, University of Illinois, Urbana-Champaign, IL 61801
[**]Department of Astronomy, University of California, Berkeley, CA 94720-3411

Abstract. We present high resolution observations of the intensity and velocity fields of the embedded cores NGC 1333 IRAS2, NGC 1333 IRAS4, and L1448N as mapped by the $J = 1 \rightarrow 0$ rotational transitions of $C^{18}O$, $H^{13}CO^+$, and N_2H^+. These lines are optically thin and trace the core kinematics over a range of critical densities. We find that thermal broadening is insufficient to explain the emission line widths. Moreover, the behavior of the line widths as a function of spatial resolution is consistent with the turbulent MHD model clouds created by Ostriker, Stone, and Gammie [1].

INTRODUCTION

Standard star formation theory begins with dense cores that are both quiescent and isolated from energetic events. Despite the theory's many successes, its premises are inappropriate for most star forming regions. First, most stars form in clusters, not alone [2]. Second, protostars generate outflows, which disturb the surrounding cloud material and affect the evolution of neighboring stars [3]. Third, most stars are members of multiple systems [4]. The prevalence of these phenomena must be accounted for in any general star formation theory.

In recent years, several theoretical papers [1, 5, 6, 7] have reinvigorated the idea that turbulent motion is a key component of star formation. If turbulence, which is inferred on large scales from molecular line widths, extends down to scales of thousands or hundreds of AU, then star formation may be a dynamic, rather than quasi-static, process. In a turbulent formation scenario, a dense core forms at the collision interface of turbulent flows, and its evolution is propelled by the local conditions (density, velocity field, magnetic field) at that interface.

OBSERVATIONS

NGC 1333 IRAS2, NGC 1333 IRAS4, and L1448N, three well-known embedded cores in the Perseus Molecular Cloud, were observed with the BIMA millimeter-wave interferometer [8] and the FCRAO 14-meter antenna in the emission from three low-excitation molecular lines. All three lines are expected to be optically thin. The $C^{18}O \; J = 1 \rightarrow 0$ line is an accepted tracer of intermediate cloud densities. The $H^{13}CO^+$ and $N_2H^+ \; J = 1 \rightarrow 0$ lines both have critical densities about two orders of magnitude greater than the $C^{18}O$ line and trace regions of dense gas in the cores [9]. Both $H^{13}CO^+$ and N_2H^+, as ionic

CP713, *The Search for Other Worlds: Fourteenth Astrophysics Conference*,
edited by S. S. Holt and D. Deming

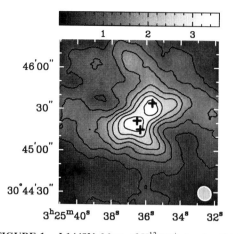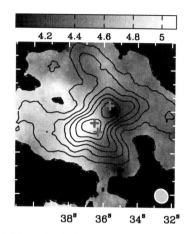

FIGURE 1. L1448N. Maps of $H^{13}CO^+$ $J = 1 \rightarrow 0$ integrated intensity (*left panel*) and velocity (*right panel*) with 10″ resolution. The contours in both panels indicate flux levels of $3, 5, 7, 9, 11, 13, 15 \times \sigma = 0.21$ Jy beam^{-1}·km s^{-1}. The grayscale in the velocity map shows the range of velocities in km s^{-1}. In both panels, the plus symbols indicate the positions of the continuum sources: L1448N(A), L1448N(B), and L1448NW.

species, remain significantly abundant even in cold, dense environments where $C^{18}O$ is expected to freeze out onto dust grains. Chemical models [10] as well as multi-line surveys [11] suggest that N_2H^+ may be among the most robust tracers of regions with high gas-phase depletions.

INTENSITY AND VELOCITY MAPS

The primary advantage of the present study of core velocity fields over previous (single-dish) studies is the addition of the high-resolution BIMA data. Figure 1 displays an example of the maps from our survey: the $H^{13}CO^+$ integrated intensity and velocity maps of L1448N at 10″ resolution. The maps combined BIMA and FCRAO data and were tuned to resolutions of 10″, 5″, and 3″. Because the Perseus Cloud lies at a distance of roughly 300pc [12], the maps show details of the gas distribution in and around the cores on scales from 0.5pc (FCRAO data alone) down to ~ 900AU (BIMA data at 3″ resolution).

Overall, the intensity maps show that the $C^{18}O$ emission is more extended than either the $H^{13}CO^+$ or N_2H^+ emission. The emission peaks are roughly centered on the positions of the continuum sources, except in the highest resolution maps, where they often appear offset by several arc-seconds. The motions of the gas in the cores, as revealed in the velocity maps, defy simple description. In general, the velocity fields lack evidence for simple rotation around an axis of symmetry or simple point symmetry. The fields of IRAS2 and IRAS4 are complex, but similar in that the gas centered on the continuum sources is blueshifted with respect to the surrounding cloud material. This blueshift may be a signature of infall motion, which has been detected in other studies of these cores [13, 14]. In contrast, L1448N shows a velocity difference across the core. In the right panel of Figure 1, the map indicates that the eastern half of the core containing L1448N(AB) is redshifted with respect to the western half containing L1448NW.

VELOCITY FIELD ANALYSIS

We analyze the velocity fields of the cores using the technique suggested by Ostriker, Stone, and Gammie [1], who measure the velocity dispersions in the core at iteratively subdivided spatial resolutions. From their gridded MHD model clouds, they find that the velocity dispersions of 3D grid cells increase steeply as a function of increasing scale length. In contrast, the velocity dispersion of projected (2D) areas show significant scatter at smaller scale lengths. Least-squares linear fits to the correlations yield slopes ranging from 0.07 to 0.12 for projections perpendicular to the mean magnetic field and 0.11 to 0.19 for projections parallel to the mean field. The contrasting trends in the 2D and 3D data suggest that the spread in the velocity dispersion of the projected areas is the result of the superposition of cells of emitting gas, with different mean velocities, along the line of sight.

The largest size scales available in our datasets correspond to the full area of the FCRAO maps ($6' \times 6'$). We subdivide the maps into 2^n intervals, with $n = 0, 1, 2 \ldots n_{max}$, such that $\theta_{map}/2^{n_{max}} \geq \theta_{beam}$. All spectra at a given grid size are independent. The line widths are measured from single Gaussians fitted to the emission profiles in each grid square. Examples of the line width/size relations for the L1448N velocity fields are displayed in Figure 2. Several traits evident in these panels are common to the line width/size relations of all three cores. The number of measurable spectra increases from the largest size scale to smaller scales, until the decreasing S/N ratio limits the spectra

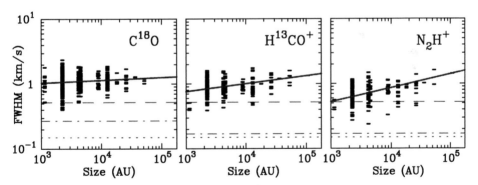

FIGURE 2. **L1448N.** Scale dependence of velocity dispersions from spectra taken through the L1448N datacubes. Solid lines show least-squares linear fits to the data, anchored to the FWHM of the spectrum from the full FCRAO field. Dashed lines show the estimated FWHM for the sound speed of the core envelope: $c_s(T = 14K) = 0.52$km s^{-1}; dash-dot lines show channel widths: $\delta v(C^{18}O) = 0.27$km s^{-1}, $\delta v(H^{13}CO^+, N_2H^+) = 0.17$km s^{-1}; dotted lines show FWHM of purely thermal velocity dispersion: FWHM$(T = 14K) = 0.15$km s^{-1}.

at 3″ resolution. The lower envelope of points shows a steep decline toward smaller sizes, but the upper envelope is roughly flat. The solid lines in each of the panels mark least-squares linear fits to the data, subject to the constraint that the fitted line pass through the spectrum from the full FCRAO field. The slopes of the best fit lines are shallow for the $C^{18}O$ data, steeper for the $H^{13}CO^+$ data, and steepest for the N_2H^+ data. Specifically for L1448N, the slopes are $a(C^{18}O) = 0.042 \pm 0.033$, $a(H^{13}CO^+) = 0.12 \pm 0.04$, and $a(N_2H^+) = 0.21 \pm 0.05$. In Figure 2, the dashed lines show the FWHM of lines broadened solely by sonic turbulence, $c_s(T=14K) = 0.52 \text{km s}^{-1}$, and the dotted lines represent the purely thermal velocity width, $v_t(T=14K) = 0.15 \text{km s}^{-1}$. The thermal line widths are below the lower envelope of measured FWHM, even at the smallest size scale, indicating that the thermal velocities alone are insufficient to produce the observed line widths. In general, our analysis suggests sonic turbulence on smaller scales and supersonic relative motions on larger scales.

The data presented here are a subset from the BIMA survey of embedded cores in the Perseus Cloud [15]. The analysis of the velocity fields in these cores is ongoing. In general, the fields appear too complex to be generated from quiescent clumps of gas undergoing simple rotation. We propose that the observed line width/size relations arise from independent parcels of cloud material seen in projection. This interpretation is consistent with theories of turbulent star formation [1, 5]. The persistence of turbulence on small spatial scales provides a tenable explanation for multiple protostars, which cannot originate from the orderly collapse of a quiescent core.

Acknowledgment. The BIMA array is operated by the Berkeley-Illinois-Maryland Association under grant AST-0028963 from the National Science Foundation.

REFERENCES

1. Ostriker, E. C., Stone, J. M., & Gammie, C. F., 2001, ApJ, 546, 980
2. Clarke, C. J., Bonnell, I. A., & Hillenbrand, L. A. 2000, in Protostars and Planets IV, ed. V. Mannings, A. P. Boss, & S. S. Russell (Tucson: University of Arizona), 151
3. Blake, G. A. 1997, IAU Symposium 178, Molecules in Astrophysics: Probes and Processes, ed. E. F. van Dishoeck, (Dordrecht: Kluwer), 31
4. Mathieu, R. D., Ghez, A. M., Jensen, E. L. N., & Simon, M. 2000, in Protostars and Planets IV, ed. V. Mannings, A. P. Boss, & S. S. Russell (Tucson: University of Arizona), 703
5. Burkert, A., & Bodenheimer, P. 2000, ApJ, 543, 822
6. Klessen, R. S., Heitsch, F., & Mac Low, M. M. 2000, ApJ, 535, 887
7. Klessen, R. S., & Burkert, A. 2001, ApJ, 549, 386
8. Welch, W. J., *et al.* 1996, PASP, 108, 93
9. Turner, B. E., 1995, ApJ, 449, 635
10. Bergin, E. A., & Langer, W. D. 1997, ApJ, 486, 316
11. Benson, P. J., Caselli, P., & Myers, P. C. 1998, ApJ, 506, 743
12. Cernicharo, J., Bachiller, R., & Duvert, G. 1985, A&A, 149, 273
13. Ward-Thompson, D., & Buckley, H. D. 2001, MNRAS, 327, 955
14. DiFrancesco, J., Myers, P. C., Wilner, D. J., Ohashi, N., & Mardones, D. 2001, ApJ, 562, 770
15. Volgenau, N. H., Mundy, L. G., Evans, N. J., II, & SIRTF c2d Legacy Project Team 2003, BAAS, 201, 66.05

Vorticity Generation in Protoplanetary Disks

Josef Koller* and Hui Li*

*Applied Physics Division, X-1, MS P225, Los Alamos National Laboratory, NM 87545

Abstract. Global two-dimensional inviscid protoplanetary disk simulation show the generation of vorticity in the co-orbital region of the planet. Emanating shock waves from the planet lead to non-conservation of vorticity along particle paths. As the system evolves, the potential vorticity (PV or vortensity) profiles eventually develop extrema which renders the disk unstable and vortices emerge. We discuss how these changes occur along particle paths.

INTRODUCTION

The discovery of extrasolar planets induced new interests in the study of planet formation. It is generally accepted that planets form in flattened disks where tidal interactions between planet and the gaseous/dusty disk play a significant role [8,9]. Several authors studied vortices in protoplanetary disks [3,2], but not their formation mechanism. We present here two-dimensional disk simulations showing how a planet and its associated spiral shock waves self-consistently generate vorticity and render the flow eventually unstable.

INITIAL SETUP

We assume a thin protoplanetary disk which can be described by the inviscid two-dimensional Euler equations in cylindrical (r, ϕ) coordinates with vertically integrated quantities. Two gravitating bodies, a central star and a protoplanet, are located at $r = 0$ and $r = 1$, respectively. The disk is modeled in the range $0.4 \leq r \leq 2$. The planet is on a fixed circular orbit at $r = 1$. The self-gravity of the disk is not included.

The mass ratio between the planet and the central star is $\mu = M_p/M_* = 10^{-4}$. Its Hill (Roche) radius is $r_H = (\mu/3)^{1/3} = 0.032$. The disk is assumed to be isothermal with a constant temperature throughout the simulation region (i.e. it is attached to a thermal heat bath). We scaled all parameters by the Keplerian rotation speed v_ϕ at $r = 1$. The sound speed is therefore $c_s/v_\phi = H/r$ where H is the disk scale height and we choose $c_s = 0.05$ for this paper. We select an initial density profile such that the ratio of vorticity to surface density is constant, i.e. $\zeta = (\nabla \times v)_z/\Sigma \approx const$. Since $r_H < H$ the planet is embedded. The gravitational potential of the planet is softened by an approximate three-dimensional treatment [1]. Although there is some accumulation of gas near the planet, we do not allow any disk gas to be accreted onto it. The planet's potential is switched on over 10 orbits to give the disk time for an adiabatic adjustment. We also

CP713, *The Search for Other Worlds: Fourteenth Astrophysics Conference*,
edited by S. S. Holt and D. Deming
© 2004 American Institute of Physics 0-7354-0190-X/04/$22.00

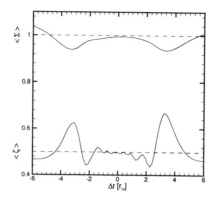

FIGURE 1. Azimuthally averaged surface density $\langle\Sigma\rangle$ and vortensity $\langle\zeta\rangle$ at $t = 50P$ for the sound speed $c_s = 0.005$. The dashed lines represent the initial values for the surface density and the vortensity. In this plot the density is reduced to a flat profile to enhance the features. The simulations where done with an actual density profile of $\Sigma \propto r^{-3/2}$.

made simulations with longer turn-on times but the results presented here do not depend on them.

The simulations are carried out using a code based on FARGO (Fast Advection in Rotating Gaseous Objects) [6, 7]. We allow for inflow and outflow at the inner and outer radial boundary. We studied the influence of several grid resolutions up to $(n_r, n_\phi) = (600, 2400)$. Here we applied a lower resolution of (300, 1200). Again, the results presented here do not significantly depend on the grid size.

SPIRAL SHOCKS AND PARTICLE PATHS

The disk responds to the planet's potential by creating spiral shock waves [4] from dissipated waves launched by the planet. They deposit angular momentum in the disk. Such angular momentum exchange drives disk material away from the planet creating dips in the azimuthally averaged density distribution $\langle\Sigma\rangle$.

Fig. 1 describes how the PV and density profile $\langle\Sigma\rangle$ changes with time. The vortensity shows two very distinct peaks $\Delta r = (r-1)/r_H \approx \pm 3.4 r_H$ and dips at $\Delta r \approx \pm 2.3 r_H$ developing from a constant initial profile. As we described in [5], vortensity dips develop, eventually become unstable and form vortices.

According to the conservation law of vortensity, ζ should be conserved along streamlines in an inviscid disk (see [5]). In this case, however, streamlines of test particles show that for $|\Delta r| > 1.3 r_H$, the material passes through a spiral shock where the dissipation breaks the PV conservation. Fig. 2 shows where the normal velocity component to the shock becomes supersonic. Since the sound speed is $c_s = 0.05$ (indicated by the dashed-dotted line), the shock starts at roughly $|\Delta r| > 1.3 r_H$.

To establish the role of shocks in breaking the PV conservation, we performed several

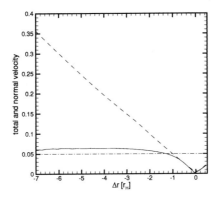

FIGURE 2. Total velocity (dashed) and normal velocity component (solid) in the pre-shock region. Since the sound speed is $c_s = 0.05$, the shock starts at roughly $\Delta r < 1.3 r_H$. Only the inner shock is plotted here. The planet is located at $\Delta r = 0$.

runs with test particles. We placed particles in regions where the vortensity change is the largest, i.e. between $|\Delta r| = 1.5 r_H$ and $4 r_H$. Fig. 3 describes how the vortensity changes which each pass through a shock. Each time a particle crosses the spiral shock, it gains or loses vortensity by redistributing PV in that region. Outside the shocked region the vortensity of the particle stays constant (in agreement with the theory of vortensity conservation). The upper panel in Fig. 3 shows a particle crossing the shock at $r \approx 1.5 r_H$. However, the particle spends most of its orbit at around $r \approx 2.2 r_H$. The deflection is due to the planets gravitational potential. The lower panel describes the evolution of a particle with increasing PV. The particle is also attracted by the planet, however by the smaller amount of $0.45 r_H$ which means the particle crosses the shock around $\Delta r \approx 3 r_H$ but spends the remainder of its orbit at $3.5 r_H$. This change of $r(t)$ near the planet has to be taken into account for explaining why the shock leads to a decrease in PV in some regions and to an increase in others. The results will be published in a forthcoming paper.

CONCLUSION

We carried out high resolution two-dimensional inviscid disk simulation with one embedded planet. We find that the planet excites spiral shocks in the disk which do not conserve potential vorticity. Depending on the region where test particles cross the spiral shock, it can increase or decrease the potential vorticity profile. Eventually inflection points are created which render the flow unstable (see [5]). The growth rate of vorticity is linear however each particle experience a loss or gain of PV only when it crosses the shock. Since most of the time the particle orbits in a smooth region without a shock, it's PV remains constant. This results in a step like function for its vortensity.

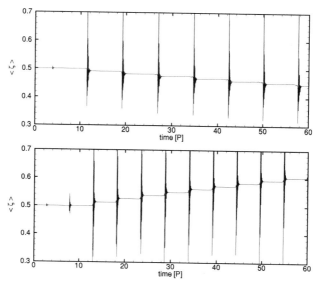

FIGURE 3. Vortensity evolution along a particle path. A particle passing through the shock in a region where vortensity is decreasing (upper panel) or increasing (lower panel). Outside the shocked region the vortensity is constant. Each shock crossing leaves the particle with a residual or loss of vortensity which is accumulated over time.

ACKNOWLEDGMENTS

We wish to thank Neil Balmforth, Darryl Holm, Douglas Lin, and Edison Liang for useful discussions. This research was performed under the auspices of the Department of Energy. It was supported by LDRD (Laboratory Directed Research and Development Program) at Los Alamos and by LANL/IGPP. We are also supported in part by NASA through NAG5-11779 and NAG5-9223.

REFERENCES

1. D'Angelo, G., Kley, W., & Henning, T., *ApJ*, **586**, 540 (2003)
2. Davis, S. S., *ApJ*, **576**, 450 (2002)
3. de la Fuente Marcos, C. & Barge, P., *MNRAS*, **323**, 601 (2001)
4. Goldreich, P. & Tremaine, S., *ApJ*, **233**, 857, (1979)
5. Koller, J., Li, H., & Lin, D. N. C., *ApJL*, **596**, L91 (2003)
6. Li, H., Colgate, S. A., Wendroff, B., & Liska, R., *ApJ*, **551**, 874 (2001)
7. Masset, F., *A&A*, **141**, 165 (2000)
8. Papaloizou, J. & Lin, D. N. C., *ApJ*, **285**, 818 (1984)
9. Ward, W. R., *Icarus*, **126**, 261 (1997)

Planet Embryos in Vortex Wombs

Joseph A. Barranco[*][†] and Philip S. Marcus[**]

Dept. of Astronomy, University of California, Berkeley
†*Current affiliation: NSF Astronomy & Astrophysics Postdoctoral Fellow at the Kavli Institute for Theoretical Physics, University of California, Santa Barbara*
**Dept. of Mechanical Engineering, University of California, Berkeley*

Abstract. One of the enduring puzzles in the formation of planetary systems is how millimeter-sized dust grains agglomerate to become kilometer-sized, self gravitating planetesimals, the "building blocks" of planets. One theory is that the dust grains settle into the mid-plane of the protoplanetary disk (thin, cool disk of gas and dust in orbit around a newly forming protostar) until they reach a critical density that triggers a gravitational instability to clumping. However, turbulence within the disk is likely to stir up the dust grains and prevent them from reaching this critical density. A competing theory is that dust grains grow by pair-wise collisions, forming fractal structures. It is unclear, however, how robust such structures would be to successive collisions. A new and exciting theory is that vortices in a protoplanetary disk may capture dust grains at their centers, "seeding" the formation of planetesimals. We are investigating the dynamics of 3D vortices in protoplanetary disks with a parallel spectral code on the Blue Horizon supercomputer. Some of the lingering questions we address are: What is the structure of 3D vortices in a protoplanetary disk? Are they columns that extend vertically through the disk, through many scale heights of pressure and density? Or are they more "pancake-like" and confined to the mid-plane? Are the vortices stable to small perturbations, such as vertical shear? Are 3D vortices robust and long-lived coherent structures? Do small vortices merge to form larger vortices the way vortices on Jupiter do?

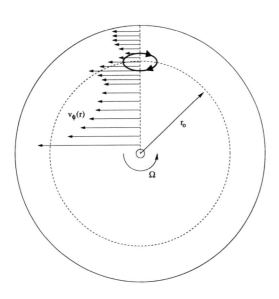

CP713, *The Search for Other Worlds: Fourteenth Astrophysics Conference,*
edited by S. S. Holt and D. Deming
© 2004 American Institute of Physics 0-7354-0190-X/04/$22.00

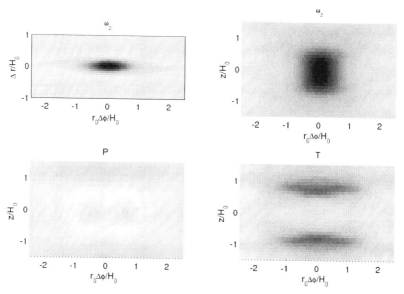

FIGURE 1. Slices through a 3D vortex located at $r = r_0$ in a protoplanetary disk. Vortex turn-around time is roughly 3 years. Clockwise, from upper left: z-component of vorticity in the midplane; z-component of vorticity in ϕ-z plane; temperature in ϕ-z plane; pressure in ϕ-z plane. Anticyclones are regions of high pressure. Note the "cold lids" which are needed to vertically confine the vortex: buoyancy balances vertical pressure gradient.

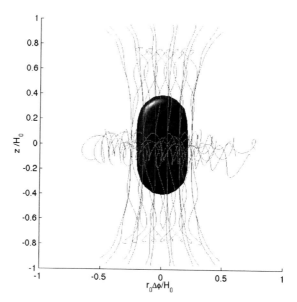

FIGURE 2. Same vortex as in Figure 1. Surface is an isovorticity surface for z-component of vorticity; lines are vortex lines which are everywhere tangent to the vorticity vector.

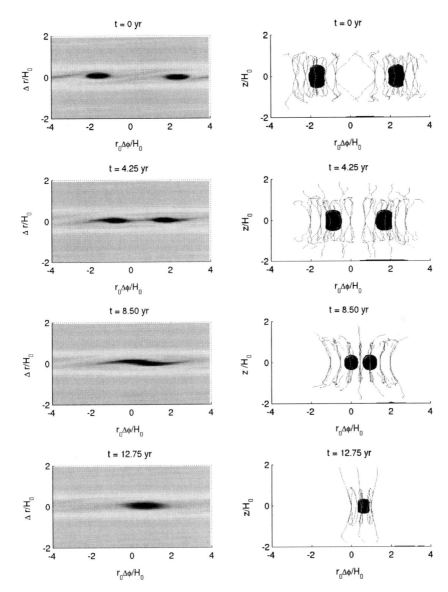

FIGURE 3. Two vortices in the midplane merging to form a new vortex. First column shows z-component of vorticity in the midplane. Second column shows isovorticity surface for z-component of vorticity, and vortex lines.

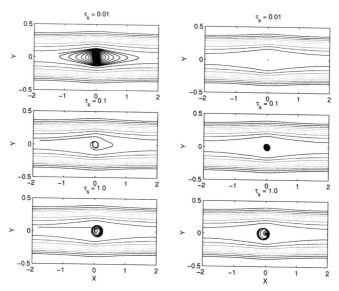

FIGURE 4. Trajectories (projected into the midplane) of individual grains within a vortex in a shear flow. The gray lines of various shades indicate the streamlines of the gas flow around the vortex. The vortex itself is not shown for clarity. Solid black lines are the trajectories of individual grains. $\tau_s \equiv t_s/T_{orb}$ is the stopping time normalized by the orbital period. From top–down, $\tau_s = 0.01, 0.1, 1.0$. First column shows trajectories of grains that were started outside the vortex on Keplerian orbits. Second column shows trajectories of grains started at the center of the vortex.

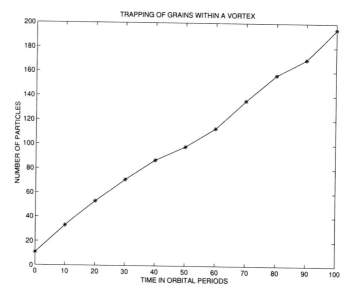

FIGURE 5. Number of particles trapped within vortex as a function of time. The vortex boundary is defined by the boundary of the patch of constant vorticity.

Dusty Disks and Unseen Planets

Timescales for Planetary Formation from Cosmochemistry

John A. Wood

Harvard-Smithsonian Center for Astrophysics, 60 Garden Street, Cambridge MA 02138, USA

Abstract. Meteorites are relicts of the process of planet-building in the solar system. Radiometric dating of selected meteorites reveals the timescales associated with planet formation. 100-km-scale rocky planetesimals formed within 2-3 Myr of initial collapse of interstellar cloud material, during the T Tauri stage of evolution of the sun. By $\sim$10 Myr after collapse, coalescence of small bodies had produced terrestrial planets approaching the size of those in our present planetary system.

INTRODUCTION

"Cosmochemistry" in the title refers to the laboratory study of extraterrestrial samples (so far only solar system samples) by a variety of techniques that are not limited to chemical studies. Particularly important for establishing timescales are isotopic studies and associated radioactive dating of carefully chosen materials; we are fortunate in having access to samples from the full range of time when planets were being formed in the solar system. This paper will review the timescales associated with three stages of formation of terrestrial planets (and gas giants, if they formed around cores of dense material analogous to terrestrial planets):

1. Thermal processing of dust from the interstellar medium in the solar nebula (products are preserved in meteorites).

2. Accretion of processed matter into 100-km-scale planetesimals (samples are also preserved as meteorites).

3. Coalescence of early planetesimals into 1000-km-scale planets.

Radioactive decay transforms a parent nuclide into a daughter nuclide, at a rate (expressed as a half-life or decay constant) that has been measured. The use of radioactive decay for dating depends critically on being able to identify daughter atoms as such. For example, consider the decay of an atom of ^{26}Al, an important nuclide further discussed below, in a hot solar gas. The daughter nuclide is ^{26}Mg. Unfortunately the solar gas already contains abundant ^{26}Mg atoms that were formed by routine stellar nucleosynthesis, and there is no way of identifying the newly-produced ^{26}Mg atom as having ^{26}Al heritage or using this fact to establish a date.

For radioactive dating to work, the radionuclide must have been sited in solid material. A ^{26}Mg atom formed by decay of ^{26}Al in an aluminous mineral will remain trapped in the crystal lattice, even though in its transformed state it is a misfit and would rather be in a magnesian mineral than an aluminous mineral. Its presence in the wrong sort of

CP713, *The Search for Other Worlds: Fourteenth Astrophysics Conference,*
edited by S. S. Holt and D. Deming
2004 American Institute of Physics 0-7354-0190-X/04/$22.00

FIGURE 1. A sawed slab of the Leoville (Kansas) CV3 chondrite (courtesy of Glenn I. Huss, American Meteorite Laboratory). The large irregular light-colored inclusion, as well as several smaller light objects, are CAIs. The numerous small, dark spheroids they are embedded in are chondrules. Figure from [1].

mineral makes it easy to identify as a radioactive daughter and permits its use to date the mineral.

This generalization is true only as long as the mineral remains relatively cool. If it is heated, the misfit ^{26}Mg atom will find it possible to diffuse through the aluminous lattice and eventually find a magnesian mineral where it "fits," but where its identity as an ^{26}Al daughter is unfortunately lost. Thus radioactive dating does not really measure the time since a planetary sample (rock) was formed, but only the time since it last cooled to a *closure temperature* low enough to preclude migration of daughter atoms by lattice diffusion. If a planetary body experiences reheating, its rock is said to have its radioactive clocks "reset" to the time when it cooled.

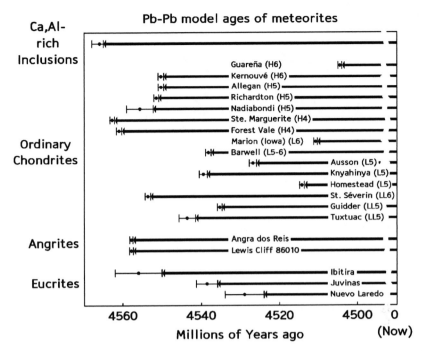

FIGURE 2. Ages (with error bars) of CAIs, bulk chondrites, and planetary igneous rocks (angrites, eucrites) studied by [2]. Names and classes of individual chondrites are given. Figure from [3].

THERMAL PROCESSING IN THE SOLAR NEBULA

Chondritic meteorites consist of fragments of rocky material that formed in the solar nebula. Chondrites (Fig. 1) are aggregations of chondrules (mm-scale) and CAIs (calcium- aluminum-rich inclusions; up to several cm) that were formed by high-temperature processes in the nebula. Chondrites and their constituents can be dated by a variety of radioactive decay schemes. Some of these techniques employ long-lived radionuclides (^{87}Rb, ^{40}K, the isotopes of U; half-lives in Gyr); others involve radionuclides so short-lived (a few Myr) that they have not survived to the present day.

Results of isotopic dating of bulk chondrites and of CAIs in them, by a particularly sensitive technique based on decay of the long-lived U isotopes, are shown in the top part of Fig. 2. (Among the ordinary chondrites letters H, L, and LL identify chemical subgroups, probably derived from discrete parent bodies; numbers denote the degree of thermal metamorphism experienced within parent bodies.) CAIs are seen to be the oldest known objects in the solar system. The bulk chondrites shown appear younger than CAIs in large part because they were reheated and metamorphosed in their parent planetesimals, and their U clocks have been reset. Even the oldest ordinary chondrite shown, Ste. Marguerite, only a few Myr younger than CAIs, shows mineralogical evidence of

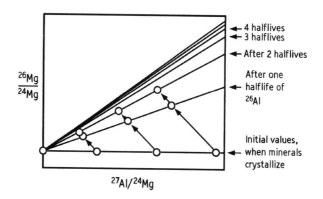

FIGURE 3. Evolution with time of the isotopic compositions of minerals in a primitive meteorite. Arrows show isotopic evolution of individual minerals containing various concentrations of ^{26}Al (proportional to their content of the stable isotope ^{27}Al). The ^{26}Al decays to ^{26}Mg, shown on the ordinate. The stable isotope ^{24}Mg is used to normalize the ratios on both axes.

some thermal metamorphism. Chronology in this earliest time has been clarified by [4], who dated individual chondrules in a chondrite that contains no mineralogical evidence of thermal metamorphism, and found them to be 2.5 ±1.2 Myr younger than CAIs (in a different chondrite).

More can be learned from the study of short-lived radionuclides (e.g., ^{26}Al, ^{41}Ca, ^{60}Fe, ^{53}Mn, and ^{107}Pd) that were present in chondrites when they formed. Discussion here will focus on ^{26}Al, which has been the most extensively studied. Since these nuclides have decayed away entirely, ^{26}Al does not offer a way of dating their host meteorites relative to the present time, but *relative* ages can be determined with great sensitivity. Figure 3 illustrates the use of ^{26}Al as a chronological tool. Points on the horizontal curve at the bottom of the figure represent minerals that had various contents of ^{27}Al (and therefore ^{26}Al) when the meteorite formed. Prior to formation (or resetting) $^{26}Mg/^{24}Mg$ (the ordinate) and $^{26}Al/^{27}Al$ were uniform in the gas, melt, or hot solid that predated the meteorite. The point to the right might represent an aluminous mineral such as anorthite ($CaAl_2Si_2O_8$), where Mg is present only as an impurity; the point at left might represent a magnesian mineral such as forsterite (Mg_2SiO_4), which contains Al only as an impurity. (Formerly minerals like these had to be physically separated for individual mass spectrometric analysis, a laborious task; but in recent years the use of ion microprobes has permitted minerals to be accessed separately within their rock matrix: a focussed beam of high-energy ions sputters atoms from each exposed target mineral, and these are led into a mass spectrometer).

Decay causes a loss of ^{26}Al atoms and a gain of ^{26}Mg atoms, accounting for the sloping arrows in Fig. 3. The more ^{27}Al (hence also ^{26}Al) is present, the greater the effect. The curve joining mineral points rises and rotates about the leftmost point, where there is no ^{26}Al and thus no rise. Finally the curve asymptotically approaches a final position, from the slope of which the initial value of $^{26}Al/^{27}Al$ in the rock can be calculated.

The initial content of ^{26}Al has been determined for a large number of CAIs, and a

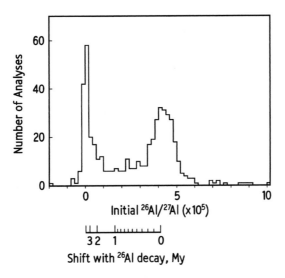

FIGURE 4. Distribution of values of initial $^{26}Al/^{27}Al$ found for CAIs (data as of 1995). Abscissa at the bottom facilitates an interpretation of the dispersion of values about $\sim 5 \times 10^{-5}$ in terms of a range of times in which the CAIs formed. Data from [5], figure from [1].

surprisingly uniform result has been obtained, with $^{26}Al/^{27}Al$ consistently equal to about 5×10^{-5} (Fig. 4). The compactness of the $^{26}Al/^{27}Al \sim 5 \times 10^{-5}$ peak in Fig. 4 points to formation of the CAIs in that peak in a surprisingly short period of time, less than a million years. If allowance is made for experimental error in the data and partial resetting of ^{26}Al clocks in the peak, the period of CAI formation may have been substantially less than 1 Myr. CAIs are a distinctive very-high-temperature set of chondrite components which seem to require distinctive circumstances of formation, and since they are the oldest entities in the chondrites (or anywhere else) I have argued that they were formed during the collapse phase of solar system formation, and identified their short period of formation with the short timescale of free-fall collapse (a few hundred thousand years) [1].

(The $^{26}Al/^{27}Al \sim 0$ peak in Fig. 4 does not correspond to a similarly short time period, of course, since it contains any CAIs that formed more than a few Myr after CAIs in the $\sim 5 \times 10^{-5}$ peak, and also CAIs that have had their ^{26}Al clocks reset by metamorphism.)

ACCRETION INTO 100-KM-SCALE PLANETESIMALS

The chondrules and CAIs in chondrites that have been studied accreted into planetesimals, where they were heated and metamorphosed. The heat came from decay of short-lived ^{26}Al and ^{60}Fe (there is no other plausible heat source). Accretion had to occur within a few half-lives of these radionuclides in order for the planetesimals to be substantially heated by their decay. Figure 5 shows a planetesimal accretion and heating model

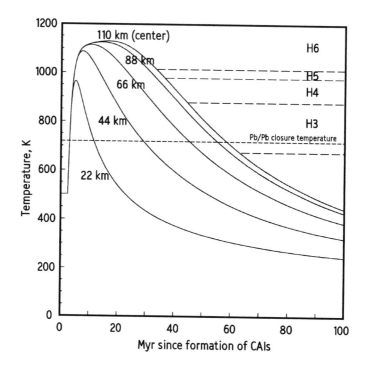

FIGURE 5. Thermal evolution model for an H-chondrite parent body of 110 km radius, calculated by the author. Temperature curves for five depths are shown. Broad dashed lines delineate four temperature ranges in which differing degrees of thermal metamorphism occurred [6]. The closure temperature beneath which Pb daughter isotopes cannot diffuse away from the sites of parent U is also shown. The metamorphic and closure temperatures are approximate, being functions of time as well as temperature.

that is broadly consistent with the peak temperatures reached and cooling histories experienced by H-group chondrites, as estimated from ^{40}Ar-^{39}Ar ages, ^{244}Pu fission-track cooling rates, and metallographic cooling rates [7] as well as Pb/Pb dating. In Fig. 2, the Pb/Pb ages of H5 and H6 chondrites Kernouvé, Allegan, Richardton, and Nadiabondi are (broadly) consistent with residence at a depth of ∼25 km in the planetesimal of Fig. 5. H6 chondrite Guareña was metamorphosed near the center of the planetesimal. H4 chondrites Ste. Marguerite and Forest Dale are harder to square with the thermal model: they cooled only (a nominal) ∼3 Myr after CAIs formed, scarcely longer than the (nominal) ∼2.5 Myr delay before chondrules formed [4] and accretion could proceed! By going to the limits of error bars a more comfortable time interval can be created for the evolution of these two chondrites, but this would entail earlier accretion, more ^{26}Al and ^{60}Fe heating, and a thermal model quite different from that of Fig. 5.

  Planetesimals that accreted significantly earlier than the model in Fig. 5 would have incorporated more live ^{26}Al and ^{60}Fe, enough to heat their interiors to the melting range, and the low-density basaltic lavas created would have risen and erupted onto their surfaces. The basaltic lava that coats the surface of the asteroid 4 Vesta [8] formed in this way. Samples of such lavas are represented in Fig. 2 by the angrite and eucrite meteorite classes. The old ages of Angra dos Reis, Lewis Cliff 86010, and Ibitira are consistent with accretion of their parent bodies earlier than the H-chondrite parent body.

Though not enough is known to fit all the properties of meteorites that have been studied into a completely self-consistent picture of accretion, radioactive heating, and cooling of parent planetesimals, the order of sizes of the latter and the timescale of accretion are clearly defined: radii $\sim$100 km, and accretion 2-3 Myr after CAI formation.

GROWTH OF 1000-KM-SCALE PLANETS

In the third stage of planetary growth listed above, solid materials including 100-km-scale planetesimals accumulated into a set of terrestrial planets similar to those in the present solar system. The process has been modeled numerically, most notably by G. W. Wetherill. In [9] Wetherill treated a planetesimal swarm initially containing 500 bodies rather larger than the 100-km scale (mean radius, 1800 km), and used a Monte Carlo technique to follow the dynamical evolution of the swarm statistically. In a run that produced a set of planets resembling the sun's terrestrial planets, "Earth" aggregated 84% of its final mass in 10.9 Myr and in another 37 Myr it received its final major addition, an impact by a Mercury-size body which might be identified with the hypothetical moon-forming "giant impact."

Assembly of a planet as large as Earth converts copious amounts of mechanical energy into heat, especially in the later stages of accretion; and this permits molten Fe metal, which is immiscible with silicate melt, to sink to its center and form a core. A short-lived radioisotope that is useful for dating core-formation is ^{182}Hf, which decays to ^{182}W with a half-life of 9 Myr. The utility of this decay scheme comes from the fact that Hf and W have different chemical affinities: W, a siderophile element, prefers to be alloyed with Fe metal; Hf, a lithophile element, is soluble in silicate melts and minerals. Thus during core formation most W sinks to the core with molten metal, while Hf remains in the mantle. The crucial thing is that W cannot sink in tiny amounts; it must be collected by droplets of metal of substantial size, and carried along with them when they sink. Once core formation drained the metal from Earth's mantle, this could no longer happen, and any ^{182}W formed thereafter by ^{182}Hf decay must have remained trapped in the mantle. Its abundance today is a measure of when core formation occurred.

Early (1995) research indicated that ^{182}W/^{183}W is almost the same in the silicate portion of the earth as in very old chondritic meteorites, meaning core formation occurred after almost all ^{182}Hf initially present had decayed; $\sim$30 Myr after CAI formation if the core metal separated in single stage of melting, $\geq$50 Myr after CAI formation if core-formation was capped by a moon-forming giant impact [10]. However, more recent work [11,12] has lowered the value of ^{182}W/^{183}W in primitive chondrites, increasing the amount of ^{182}Hf that decayed in the earth's mantle after separation of a core, and pushing back the time when that must have happened to 10-30 Myr after CAIs.

(However, D. J. Stevenson, *pers. comm.*, points out that this measurement of the time of metal/silicate separation does not necessary date the final growth of the earth. Core-formation might have occurred first in a generation of smaller bodies that would later merge to form the earth. If the collisions that merged these bodies melted and thoroughly mixed their substance, then the metal/silicate fractionation "clock would be reset," and the derived time of 10-30 Myr after CAIs would apply to this final merging. However, coalescence is an untidy business, and it is not clear that the objects merging would be so thoroughly mixed that Fe metal droplets would have a chance to scavenge the ^{182}W that had formed in their silicate mantle material. In this case 10-30 Myr only dates initial core formation in the early generation of small bodies; their final coalescence could have occurred at some later time.)

It might seem that measurements of ^{182}W/^{183}W in samples of the moon could clarify the role played in the growth of the earth by a moon-forming giant impact, but on the moon the record is clouded by the fact that the ^{181}Ta$(n, \gamma)^{182}$Ta$(\beta^-)^{182}$W reaction, which is induced in surface materials by the cosmic ray bombardment, produces ^{182}W that is unrelated to ^{182}Hf decay [13].

CONCLUSIONS

Radiometric dating of meteorite samples has established the following timescales associated with planetary formation in the solar system.

1. The oldest objects in the solar system, Ca, Al-rich inclusions, were produced in a short period of time (<1 Myr), perhaps during the collapse phase of protosun formation.

2. Thereafter interstellar dust in the solar nebula was thermally processed on a time scale of a few (2-3) Myr, producing the chondrules abundant in chondritic meteorites.

3. This processed matter accreted immediately into 100-km-scale planetesimals, which were heated by the decay of short-lived radionuclides in them. Basaltic lava erupted on some and cooled a few Myr after CAI formation. Chondritic material in the interiors of other planetesimals that reached metamorphic temperatures but did not melt cooled on a similar timescale.

4. Heirarchical accumulation of the population of planetesimals produced bodies almost as large as Earth and Venus in ~10 Myr. The time when accretion was essentially complete may have been as early as ~30 Myr after CAIs, or it may have taken a few more 10s of Myr.

REFERENCES

1. Wood, J. A. *Geochim. Cosmochim. Acta*, **68**, in press (2004).
2. Allègre, C. J., Manhès, G., and Göpel, C. *Geochim. Cosmochim. Acta*, **59**, 1445–1456 (1995).
3. Wood, J. A. *Space Sci. Rev.*, **92**, 97–112 (2000).
4. Amelin, Y., Krot, A. N., Hutcheon, I. D., and Ulyanov, A. A. *Science*, **297**, 1678–1683 (2002).
5. Macpherson, G. J., Davis, A. M., and Zinner, E. K. *Meteoritics*, **30**, 365–386 (1995).
6. McSween, H. Y. Jr., Sears, D. W. G., and Dodd, R. T. in *Meteorites and the Early Solar System*, edited by J. F. Kerridge and M. S. Matthews, University of Arizona Press, Tucson, 1988, pp. 102–113.
7. Trieloff, M., Jessberger, E. K., Herrwerth, I., Hopp, J., Fiéni, C., Ghélis, M., Bourot-Denise, M., and Pellas, P. *Nature*, **422**, 502–506 (2003).
8. Binzel, R. P. and Xu, S. *Science*, **260**, 186–191 (1993).

9. Wetherill, G. W. in *Origin of the Moon*, edited by W. K. Hartmann, R. J. Phillips, and G. J. Taylor, Lunar &Planetary Institute, Houston, 1986, pp. 519–550.
10. Lee, D.-C. and Halliday, A. N. *Nature*, **378**, 771–774 (1995).
11. Schoenberg, R., Kamber, B. S., Collerson, K. D., and Eugster, O. *Geochim. Cosmochim. Acta*, **66**, 3151–3160 (2002).
12. Yin, Q., Jacobsen, S. B., Yamashita, K., Blichert-Toft, J., Télouk, P., and Albarède, F. *Nature*, **418**, 949–952 (2002).
13. Yin, Q., Jacobsen, S. B., and Wasserburg, G. J. *Lunar Planet. Sci.* **XXXIV**, Abstr. 1510 (2003).

Observations of Dusty Disks

Alycia J. Weinberger

Department of Terrestrial Magnetism, Carnegie Institution of Washington

Abstract. Dusty disks, those that have largely lost their gas as their host stars evolve onto the zero age main sequence, are interesting because of what they tell us about the conditions for planet formation. In the time period of 5-30 Myr, in our own Solar System, the gas giants may still have been forming and terrestrial planets were almost certainly still forming. This proceeding reviews the imaging and spectroscopy of young debris disks in stages analogous to the early Solar System.

INTRODUCTION

It has been nearly twenty years since the first image of a extrasolar disk around the main sequence star β Pictoris [1]. That image dramatically confirmed our theoretical picture of a flattened distribution of dust orbiting a central star. In the intervening years, a new generation of instruments and techniques have been been brought to bear to study in greater detail disks as the birthplaces of planets.

The earliest stages of star and disk formation can be copiously studied in the nearest star formation regions such as Taurus, Scorpius-Ophiuchus, and Scorpius-Centaurus. These studies have revealed that very young stars are indeed surrounded by disks of gas and dust of substantial mass (e.g. [2]). Their geometries have been measured and the effects of their neighborhoods assessed. On the other end of the time sequence, our own solar system planets give themselves up for study in exquisite detail. Careful measurements of asteroidal and cometary material and cratering of rocky bodies has shown the kinds of condensation and collision processes that must have occurred early in our formation history. The intermediate stages of disk evolution are much less well studied.

Where are the disks?

Even now, we are far from having a complete disk census for young main sequence stars. All of the disks known today are inferred from excess infrared emission, characterized by the fraction of the star's luminosity which is re-radiated in the mid-to-far-infrared (L_{IR}/L_*, commonly called τ because it approximates the optical depth of the dust). The IRAS survey provides the most sensitive all-sky catalog of infrared emission, including of order 100 stars with excess (e.g. [3, 4]). The IRAS Faint Source Catalog, complete (SNR$\approx$5) to a flux density limit of $\sim$0.2 Jy at 12 and 25μm, had insufficient sensitivity to probe fully for disks even in the Solar neighborhood. For a fixed number of dust grains, the more luminous the central star, the brighter a disk will be in the infrared. So, our disk

CP713, *The Search for Other Worlds: Fourteenth Astrophysics Conference,*
edited by S. S. Holt and D. Deming
© 2004 American Institute of Physics 0-7354-0190-X/04/$22.00

census is most complete for the luminous A-type stars. In contrast, radial velocity programs survey primarily solar-mass stars. The task of connecting these mature planetary systems with their progenitor disks awaits us. Because of the scarcity of known disks of ages >5 Myr, we know little about the average dissipation rate of disks in later epochs that are very important to the course of planet formation. Hopefully, the Spitzer Space Telescope will change this picture shortly, perhaps even before this proceeding appears in print.

How fast do disks dissipate?

Observations at different wavelengths probe different regimes in the disk based on the temperature that they measure. Most surveys of post T-Tauri star dust have concentrated on hot, $\sim$1000 K dust, observable at L-band. This dust resides at tens of stellar radii only. One set of surveys show that these inner edges disappear at 3–6 Myr, shortly after accretion ends [5]. A survey of the mostly very low mass members of the η Cha cluster, however, finds a high hot disk fraction even at an age of 5–8 Myr [6]. Whether these differences stem from sample selection effects, birth conditions or some other variables remains to be determined. It could be that outer disks, found at lower temperatures and consequently longer wavelengths, last longer, but that does not appear to be the case. However, this is a hypothesis that still needs to be fully tested.

At 10–20 μm, surveys of the 8 Myr old TW Hydrae association show that, while IRAS found three very dusty disks (including TW Hya itself) and one low-mass disk (HR 4796A), few if any lower mass disks reside around the late-type stars [7, 8]. Dust content in TWA is bimodal, with a few stars possessing much dust and most largely devoid of material. In regions analogous to the terrestrial planet region, even at the young age of 5–10 Myr, little material remains or is being generated in collisions around these late-type stars.

A survey of cluster stars with ISO at 60–100 μm showed that from L_{IR}/L_*=0.1 at 5 Myr on, the amount of dust decreases approximately exponentially with time[9]. This corresponds to collisionally dominated dust production from immediately after the cessation of accretion to the current age of the sun (but see [10] for a somewhat dissenting view).

As noted above, most known dusty disks are around A stars. These stars reach the ZAMS by an age of a few million years. The lower mass stars formed along with them, however, will still be pre-main sequence by the onset of the traditional "β Pic-like" stage.

If we look at a planetary formation timeline for our own solar system, we can compare it to observations of nearby young stars to map out the planet-formation stages (Figure 1). Cosmochemists studying meteorites provide very strong constraints on the timing of early planet formation. The earliest dated materials are the Calcium-Aluminum rich Inclusions (CAIs) thought to have formed soon after the first solids condensed in the proto-planetary nebula. Chondrules, the widespread glassy components of most meteorites, date from 1-3 Myr after this. Metamorphism requiring the existence of asteroidal sized bodies seems to take only a few million years to arise. Information on the time of formation of the gas giant planets is non-existent, but must have happened while there

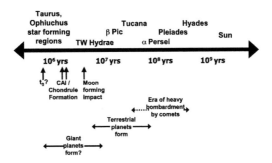

FIGURE 1. A timeline for our planet formation important events in our early Solar System compared to the ages of stars in nearby young associations where we might look for early analogues to the Solar System.

was substantial gas left in the proto-planetary nebula and therefore in less than a few Myr. The terrestrial planets also presumably started to form during this time. The date of the impact on the Earth that likely formed the moon is 30 Myr [11] and this timescale presumably defines the end of the era of terrestrial planet accumulation.

So, to probe the time periods that correspond to planet formation in our own Solar System, we should look at other stars of age 1-30 Myr. Luckily in the last few years, there has been a renaissance in the identification of young stars near the Sun [12, 13]. Over 100 stars closer than 100 pc and younger than 30 Myr are known and are a place to study the planet formation process.

DISK OBSERVATIONS

Detailed studies of disks can be carried out in two wavelength regimes. In the first, we measure radiation scattered off of dust grains in the disks, usually at visible and near-infrared wavelengths. The necessity of high contrast imaging, to detect the faint disk light near the glare of the bright star, has made the Hubble Space Telescope the premier instrument for this work. In the second regime, encompassing mid-infrared through radio wavelengths, we measure thermal radiation emitted by dust grains. The primary instruments are large ground based telescopes necessary to achieve high mid-infrared to radio resolution (e.g. Keck, JCMT, VLA) as well as space borne instruments for surveys and sensitivity (IRAS, ISO, and now Spitzer).

These two regimes provide complementary approaches to measuring grain proper-ties. The shorter the wavelengths, the higher the spatial resolution we can achieve, so detailed imaging is best done in scattered light. However, these same wavelengths pose the greatest observational challenge to separating the disk from stellar diffracted and scattered light, so imaging close to stars in the visible or near-infrared mandates the

TABLE 1. Fundamental properties of three disks and their parent stars

Star	Age (Myr)	Sp. Type	L ($L_\odot$)	L_{IR}/L_* ($\times 10^{-3}$)	D (pc)
HD 141569A	5 ± 3	B9.5Ve	22	8.4	100
HR 4796A	8 ± 2	A0V	20	5	67
β Pic	12^{+8}_{-4}	A5V	7	2.5	19

use of coronagraphs or other high contrast imaging techniques. Scattering efficiency is largely insensitive to dust temperature, so short wavelengths can effectively probe the outer regions of disks while paying only the $1/r^2$ penalty in sensitivity to grains far from the star. Since the stellar output peaks in the visible or near-infrared and dust scatters most efficiently at wavelengths close to the grain size, most scattered light will probe small grains - up to a few microns in size. In contrast, in the thermal emission regime, the star may be negligible in brightness compared to the disk, making it easy to separate the two. Mid-infrared to radio observations will be very temperature sensitive, with each wavelength effectively probing the disk at only certain temperatures and their corresponding radii. Most of the mass in disks is likely in much larger grains, so mass is more effectively probed by long wavelength observations.

Disk Imaging

Since a paper in this proceedings discusses the dynamical evidence for planets in circumstellar disks [14], here I will just briefly review the imaging of young debris disks as a comparison amongst HD 141569A, HR 4796A, and β Pic. Table 1 shows the properties of these three similar A-type stars with well studied circumstellar disks.

For HD 141569A, scattered light imaging at visual and near-infrared wavelengths shows an elongated disk that is clearly evident at radii from 150 to 400 AU [15, 16](Figure 2). At 250 AU an annulus of lower surface brightness,can be seen, though images from HST/STIS show that this "gap" is not complete [17, 18]. HST/ACS images suggest that there may be spiral structure in the disk induced by two nearby companions [19, 20]. Forward scattering by the grains at visual and near-infrared wavelengths and the temperature versus radius of the emitting mid-infrared grains imply that they are fairly small (a few microns) [16, 21]. Mid-infrared images see the disk out to ~150 AU [22] and suggest the presence of an inner hole [21].

For HR 4796A, new STIS images of the disk with 70 mas resolution show that it is entirely located in a very narrow ring with a semi-major axis of 70.8 AU. The dust is confined to a 12 AU ring with steeper inner truncation than outward falloff [23, 24]. The NE side of the ring is brighter than the SW side by ~20% in the visual and a similar level in the near-infrared and mid-infrared [25, 26, 27] (see Figure 3). This non-azimuthally symmetric feature could be due to a non-uniform distribution of grains in the ring, although given the ~300 yr period for grains, one would expect them to be well mixed unless dynamically constrained by a planet.

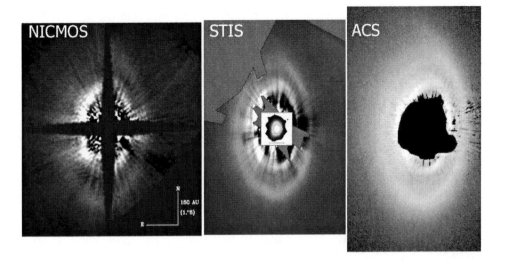

FIGURE 2. Images of the reflected light disk around HD 141569A from three HST instruments: NICMOS (1.1 μm), left[16]; STIS (broad-band visible), center [18], ACS (0.6 μm), right [19]; and the emitted light disk at 18 μm from the Keck Telescope (overlayed, center [21]). Forward scattering by the dust is observed to make the eastern side of the disk brighter. Non radial structure in the form of gaps (interior to 170 AU), arcs and/or spirals is seen in the visible (STIS and ACS) images.

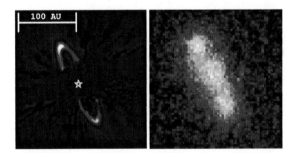

FIGURE 3. Images comparing the scattered and emitted flux from the disk around HR 4796A. Left: STIS image showing visible light reflected from the disk particles. Forward scattering is seen on the eastern side of the disk. The NE ansa is brighter than the SW ansa [24]. Right: Keck Observatory (with the OSCIR instrument) image showing infrared emitted light at 18.2 μm with a similar NE to SW asymmetry [27] (and see [14] for a model).

For the extremely well-studied β Pic, the disk extends from a few to many hundreds of AU and asymmetries in the disk have been observed in scattered and emitted light at all spatial scales (e.g., [28], [29], [30]). In particular, a warp seem at radii 40–120 AU has been explained by the dynamical perturbations induced by a planet within it [29, 31]. Mid-infrared observations taken at the W. M. Keck Observatory now elucidate its structure very close to the star [32, 33].

Images of β Pictoris at 11.7 and 17.9 μm reveal that within a radius of 1″, the disk is tilted 16° west of the previously imaged 1–5″portion, which is itself warped at 3° from the outer disk. Thus the position angle of the disk does not change secularly with increasing radius but rather appears to shift discontinuously and in opposing directions at 1″and ~5″. The disk is also asymmetric at 12 μm, with the SW side brighter than the NE side. Multiple blobs have been seen in the disk at 18 μm that, if real, may signal the presence of multiple planets [33].

Despite the apparent similarity in stellar properties and disk content (as measured by the figure of merit of L_{IR}/L_*), the physical dimensions of the three disks are quite different. For all three, the inferred size of the grains in the disk, <5 μm, implies that radiation pressure blows them away in at least an order of magnitude less time than our derived stellar age. Thus, they must be continuously regenerated, probably through collisions of larger bodies. Radiation pressure and Poynting-Robertson drag are purely radial, and the morphologies of the disks show many non-azimuthally symmetric features (arcs, clumps and warps). It is primarily these features which suggest the dynamical molding of these disks by planets within them.

Composition

The very different bulk compositions of our terrestrial, gas giant, and ice giant planets as well as the Kuiper Belt objects, must arise from some combination of formation processes and segregation of material in the early solar nebula. The history of disk material, i.e. the way dust and gas are processed in circumstellar disks, is therefore key to understanding the resulting planetary systems.

As fascinating as the images described above have been, they provided only limited knowledge of disk composition. Color images in broad-band filters (e.g [34]) combined with modeling of infrared to millimeter SEDs suggest that the disks contain particles of 1 - 1000 μm and can give average albedos. Water ice is the most commonly detected material on surfaces in the outer Solar System and is found everywhere from comets, to the surfaces of satellites, to the small particles in the rings of giant planets [35, 36, 37]. Ices have been proposed to explain two observations made of other disks. First, the visual and near-infrared albedos of large ($>1\mu$m) disk grains are quite high. In the disks around HR 4796A, HD 141569A, and β Pic, the albedos range from 0.2 to 0.5 [38, 23, 16, 39], which is typical for ice-silicate mixtures in our own Solar System. Second, the disk of HD 141569A shows an increase in surface brightness at the ice-condensation radius suggesting that its cold grains are mantled by ice while its warm ones are not [16]. Almost no optical or near-infrared spectroscopy of debris disks has been possible because of the required high dynamic range. Sodium fluorescence has

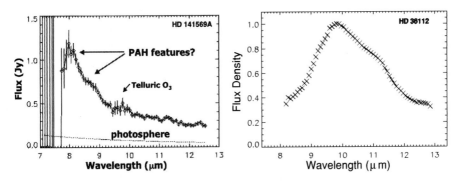

FIGURE 4. Left: Mid-infrared spectrum of HD 141569A taken at the Keck telescope. There is evidence for emission from polycyclic aromatic hydrocarbons at 7.7 and 8.6 μm, however the usually prominent PAH peak at 11.2 μm is not seen. Mid-infrared spectrum of HD 36112, a young Herbig AeBe star with nearly the same age and luminosity as HD 141569, also from Keck. This is a more typical spectrum of a young disk showing mostly amorphous silicates (peaked at $\sim 10\mu$m) with some crystalline silicates (peaked at 11.2μm).

been detected in β Pic [40] and see the results on TW Hya in this proceeding [41].

In contrast, mid-to-far-infrared spectroscopy has been extensively used to probe grain chemistry directly, finding silicates and polycyclic aromatic hydrocarbons similar to those in Solar System comets [e.g. 42]. Waelkens et al. [43] carried out an Infrared Space Observatory (ISO) program on Vega-like and AeBe stars, including about 20 targets, with the Short Wavelength Spectrometer. Some objects show strong silicate and PAH features as well as lines attributed to specific grains such as crystalline forsterite, amorphous olivine and also iron-oxide [e.g. 44]. The crystalline forsterite appears to indicate early heating of the disk similar to that seen in meteorites and comets in our Solar System. The precise locations of the PAH versus silicates are not known for most stars, and indeed spectra usually show one or the other but not both (Figure 4).

In general, of course, spectroscopy has the potential for for finding small amounts of gas, water ice distributions, and the place of crystallization of silicates. However, little of this potential has yet been realized because of the need for high spatial resolution and often very high dynamic range or for edge-on disks.

One disk, however, has been extensively studied, that of β Pictoris (of course). Its proximity, relatively large infrared excess, and nearly edge-on viewing geometry, make β Pic the easiest debris disk to study.

The gas as well as the dust has been shown to be second generation. The line of sight column density of CO in the disk is 6×10^{14} cm^{-2} [45]. Without any assumptions about the spatial distribution of the gas, the lower limit on the CO to H$_2$ ratio is $N(CO)/N(H_2) \geq 6 \times 10^{-4}$ [46] whereas the ratio for translucent interstellar molecular clouds with similar CO column densities is about 1×10^{-6} [47], or 600 times smaller.

Because the disk is not shielded against photodissociaton by interstellar UV photons (which would take a H$_2$ column density of around 10^{22} cm^{-2}[48]), the lifetime of CO in the disk should be about 200 yr, and therefore the CO must be continuously supplied to the disk, possibly by the evaporation of icy comet-like bodies. The evaporation of

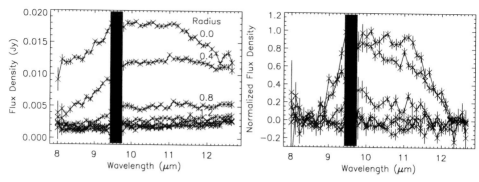

FIGURE 5. Left: Spectra of the β Pic disk after subtraction of the contribution of the photosphere shown as a function of radius (curves shown are, from the top down, for radii of 0.0, 0."4, 0."8, 1."2, 1."6, 2."0, and 2."4). Right: Spectra after subtraction of a continuum fit to the 12.6/8.2 μm color temperature. Silicates, at approximately the same line/continuum ratio, are seen in the first three spectra, but only continuum is seen at larger radii. The vertical black bar blocks the noisy region contaminated by strong Telluric ozone absorption.

icy comet-like bodies naturally explains the large CO/H_2 ratio in the disk: CO is easily trapped in the ices of comets, but H_2 is not (most hydrogen is present as H_2O ice).

Mid-infrared observations taken at the W. M. Keck Observatory now elucidate the dust composition very close to the star [32]. Shown in Figure 5a are 8 – 13 μm spectra of the southwest (brighter) side of the β Pic disk after subtraction of the photospheric component. The spectrum extracted from within 0."2 of the star agrees very well with that observed previously [49] including a flat-topped profile characteristic of a mixture of amorphous and crystalline silicates.

The temperature of the grains at each radius was estimated by the color temperature of the regions on either side of the silicate feature. Temperatures are found in excess of 300 K out to 20 AU, which is very hot even for silicate grains in thermal equilibrium with the star. A Planck function at each color temperature was then normalized to the long wavelength flux density and subtracted from each spectrum. The results are shown in 5. At radii < 1", a clear contribution from silicates is present. For the three spectra with radii <1", the flux to continuum ratios are equal to within the uncertainties. At radii > 1", although flux from the disk is present at a significant level in the spectra, within the uncertainties of the data, it appears to be composed only of continuum. The shape of the silicate peak does not change significantly over the radii where it can be seen at all. This implies that the relative abundances of amorphous and crystalline grains remains approximately constant with radius.

The silicate emission must all come from relatively small (~10 μm) grains. These may be preferentially generated in the inner portion of the disk due to a higher collisional rate of planetesimals and then distributed to larger radii by radiation pressure. At large radii, as they are mixed with a larger, less collisionally dominated grain population, they provide a smaller fraction of the mid-infrared flux and thus a lower line/continuum ratio. Spatial asymmetries in the 12 and 18 μm emission cannot easily be explained by compositional or density differences in the disk; they remain mysterious.

CONCLUSIONS

Most disks appear to enter their dusty debris stage by ages of 5 Myr. This tightly constrains the time available for planet formation, if it is to occur. At very young ages of 5–10 Myr, high spatial resolution disk imaging shows inner disk holes and outer non-axisymmetric structures such as clumps and warps that may be due to the presence of planets. Spatially resolved spectroscopy of disks is just now becoming available and has the potential to illuminate our understanding of the conditions for and evolution of planetary formation as a function of distance from the central stars.

ACKNOWLEDGMENTS

My thanks to Aki Roberge for comments about the gas-to-dust ratio in β Pic, and to Glenn Schneider and Murray Silverstone for providing figures. I am grateful to the NASA Origins of Solar Systems program, NAG5-13145, for supporting my observations of young stars.

REFERENCES

1. Smith, B. A. & Terrile, R. J. 1984, Science, 226, 1421
2. Beckwith, S. V. W. & Sargent, A. I. 1996, Nature, 383, 139
3. Backman, D. E. & Paresce, F. 1993, Protostars and Planets III, 1253
4. Mannings, V. & Barlow, M. J. 1998, ApJ, 497, 330
5. Haisch, K. E., Lada, E. A., & Lada, C. J. 2001, ApJ, 553, L153
6. Lyo, A.-R., Lawson, W. A., Mamajek, E. E., Feigelson, E. D., Sung, E., & Crause, L. A. 2003, MNRAS, 338, 616
7. Jayawardhana, R., Hartmann, L., Fazio, G., Fisher, R. S., Telesco, C. M., & Piña, R. K. 1999, ApJ, 521, L129
8. Weinberger, A. J., Becklin, E. E., Zuckerman, B., & Song, I. 2004, AJ, in press
9. Spangler, C., Sargent, A. I., Silverstone, M. D., Becklin, E. E., & Zuckerman, B. 2001, ApJ, 555, 932
10. Habing, H. J. et al. 2001, A&A, 365, 545
11. Yin, Q., Jacobsen, S. B., Yamashita, K., Blichert-Toft, J., Telouk, P., & Albarede, F. 2002, Nature, 418, 949
12. Jayawardhana, R. & Greene, T., eds, 2001, ASP Conf. Ser. 244: Young Stars Near Earth: Progress and Prospects
13. Zuckerman, B. & Song, I. 2003, Ann. Rev. Astron. & Astrophys., in press
14. Wyatt, M. 2004, in The Search for Other Worlds, S. Holt & D. Deming, eds., AIP (this proceeding)
15. Augereau, J. C., Lagrange, A. M., Mouillet, D., & Ménard, F. 1999, A&A, 350, L51
16. Weinberger, A. J., Becklin, E. E., Schneider, G., Smith, B. A., Lowrance, P. J., Silverstone, M. D., Zuckerman, B., & Terrile, R. J. 1999, ApJ, 525, L53
17. Mouillet, D., Lagrange, A. M., Augereau, J. C., & Ménard, F. 2001, A&A, 372, L61
18. Weinberger, A. J. et al. 2004, in preparation
19. Clampin, M. et al. 2003, AJ, 126, 385
20. Augereau, J. C. & Papaloizou, J.C.B. 2003, A&A, in press (astro-ph/0310732)
21. Marsh, K. A., Silverstone, M. D., Becklin, E. E., Koerner, D. W., Werner, M. W., Weinberger, A. J., & Ressler, M. E. 2002, ApJ, 573, 425
22. Fisher, R. S., Telesco, C. M., Piña, R. K., Knacke, R. F., & Wyatt, M. C. 2000, ApJ, 532, L141
23. Schneider, G. et al. 1999, ApJ, 513, L127
24. Schneider, G. et al. 2004, AJ, in preparation

25. Koerner, D. W., Ressler, M. E., Werner, M. W. & Backman, D. E. 1998, ApJ, 503, L83
26. Jayawardhana, R., Fisher, S., Hartmann, L., Telesco, C., Pina, R. & Fazio, G. 1998, ApJ, 503, L79
27. Telesco, C. M. et al. 2000, ApJ, 530, 329
28. Kalas, P. & Jewitt, D. 1995, AJ, 110, 794
29. Heap, S. R., Lindler, D. J., Lanz, T. M., Cornett, R. H., Hubeny, I., Maran, S. P., & Woodgate, B. 2000, ApJ, 539, 435
30. Lagage, P. O. & Pantin, E. 1994, Nature, 369, 628
31. Augereau, J. C., Nelson, R. P., Lagrange, A. M., Papaloizou, J. C. B., & Mouillet, D. 2001, A&A, 370, 447
32. Weinberger, A. J., Becklin, E. E., & Zuckerman, B. 2003, ApJ, 584, L33
33. Wahhaj, Z., Koerner, D. W., Ressler, M. E., Werner, M. W., Backman, D. E., & Sargent, A. I. 2003, ApJ, 584, L27
34. Weinberger, A. J. et al. 2002, ApJ, 566, 409
35. Irvine, W. M., Schloerb, F. P., Crovisier, J., Fegley, B., & Mumma, M. J. 2000, Protostars and Planets IV, 1159
36. Brown, M. E. & Calvin, W. M. 2000, Science, 287, 107
37. Lebofsky, L. A., Johnson, T. V., & McCord, T. B. 1970, Icarus, 13, 226
38. Jura, M., Malkan, M., White, R., Telesco, C., Pina, R., & Fisher, R. S. 1998, ApJ, 505, 897
39. Pantin, E., Lagage, P. O. & Artymowicz, P. 1997, A&A, 327, 1123
40. Brandeker, A. Liseau, R., Olofsson, G., Fridlund, M. 2003, A&A, in press (astro-ph/0310146)
41. Roberge, A., Weinberger, A. J., & Malumuth, E. 2003, in The Search for Other Worlds, S. Holt & D. Deming eds., AIP (this proceeding)
42. Telesco, C.M. & Knacke, R.F. 1991, ApJ, 372, L29
43. Waelkens, C., Malfait, K., & Waters, L. B.F. M. 1998, Ap&SS, 255, 25
44. Malfait, K., Waelkens, C., Waters, L. B. F. M., Vandenbussche, B., Huygen, E., and de Graauw, M. S., 1988, A&A, 332, L25
45. Roberge, A., Feldman, P. D., Lagrange, A. M., Vidal-Madjar, A., Ferlet, R., Jolly, A., Lemaire, J. L., & Rostas, F. 2000, ApJ, 538, 904
46. Lecavelier des Etangs, A. et al. 2001, Nature, 412, 706
47. Magnani, L., Onello, J. S., Adams, N. G., Hartmann, D., & Thaddeus, P. 1998, ApJ, 504, 290
48. Kamp, I. & Bertoldi, F. 2000, A&A, 353, 276
49. Knacke, R. F., Fajardo-Acosta, S. B., Telesco, C. M., Hackwell, J. A., Lynch, D. K., & Russell, R. W. 1993, ApJ, 418, 440

Modeling the Structures of Dusty Disks: Unseen Planets?

Mark C. Wyatt

UK Astronomy Technology Centre, Royal Observatory, Edinburgh EH9 3HJ, UK

Abstract. One of the legacies of IRAS was to show that some 15% of main sequence stars are surrounded by disks of dust. The dust in these *debris disks* is continually replenished by the destruction of larger planetesimals and so these disks have the potential to tell us a great deal about the outcome of planet formation in their systems. To extract such information the disks need to be resolved. The handful of disks which have had their structure mapped have shown that none of these disks are smooth or axially symmetric. A wide variety of disk structures have been detected, ranging from warped disks of asymmetric brightness to clumpy ring-like disks. Detailed dynamical modeling has identified many of these structures with perturbations from as yet unseen planetary companions. In this paper I review the types of structure that planets can, and indeed have been purported to, impose on debris disks, and look at the evidence we have from modeling of debris disk images that these unseen companions really exist.

CIRCUMSTANTIAL EVIDENCE FOR PLANETS

Before discussing how planets can affect the detailed structure of dust disks, I want to start by considering what information is available from observations of the large scale structure of the disks. Such a discussion is not only necessary to build a coherent picture of what it is that we are observing, but also provides the first circumstantial evidence that planets may have formed in the disks.

Grain Growth

As soon as dust was discovered around main sequence stars it was recognised that this dust cannot be primordial. The reason is that the dust we are seeing is short-lived. Poynting-Robertson (P-R) drag would make the dust spiral onto the star on timescales of around 0.1 Myr. The disks are also so dense that mutual collisions would destroy dust grains on similar timescales. Since main sequence stars have ages in the range 10-10,000 Myr, the dust we are seeing must be continually replenished, presumably by the erosion of an unseen populations of larger planetesimals which have much longer P-R drag and collisional lifetimes.

The nature of this replenishment in the Fomalhaut disk was studied by modeling the spectral energy distribution (SED) of its dust emission [1]. Once the distance of the dust from the star had been constrained by imaging, the SED modeling was able to derive the size distribution of the dust, since smaller grains emit hotter than larger grains at the same distance. This showed that the size distribution is the same as that expected from a

CP713, *The Search for Other Worlds: Fourteenth Astrophysics Conference,*
edited by S. S. Holt and D. Deming
© 2004 American Institute of Physics 0-7354-0190-X/04/$22.00

collisional cascade, in which planetesimals are continually colliding and getting ground down into smaller particles. Consideration of the collisional lifetime of different sized bodies in the disk showed that what we are seeing is just the bottom half of the cascade (7 μm dust to 20 cm pebbles), and that the cascade is fed by planetesimals a few km in diameter. While nothing could be inferred about the number of planetesimals larger than this, because their collisions are too infrequent to be contributing to the dust we are seeing, this does provide circumstantial evidence for planet formation: grain growth has occurred up to at least km sized planetesimals, so perhaps at some location in the disk, presumably closer to the star, grain growth could have proceeded up to planet sized bodies.

Cavities

Further evidence that such planets may have formed closer to the star comes from the lack of warm dust emission in debris disk systems. This implies that the inner regions, some tens of AU in radius, are relatively empty of dust, a finding supported by imaging for the few disks where this has been possible. The argument put forward for this observation supporting the presence of planets is based on the idea that P-R drag would populate these inner regions with dust created in the outer regions of the system. In the absence of inner planets this would result in a disk with a constant surface density extending in toward the star [2]. Introducing planets, however, would cause the dust grains to be scattered or accreted before they reach the inner regions; some would also be captured into its resonances (see next section). However, while this is one explanation for the inner cavities, other interpretations also exist [3, 4] meaning that the presence of a planet may not be a reliable conclusion based on the presence of a cavity.

STRUCTURES CAUSED BY PLANETARY PERTURBATIONS

If there is a planet orbiting in the disk, then its gravitational perturbations will affect the orbital evolution of material in the disk. This would mean that both the planetesimal disk and the dust disk would contain dynamical structure associated with the planet; this structure may be discernible (or even dominant) in observations of the disks. Planetary perturbations can be broken down into one of two kinds: secular and resonant. These two types of perturbation can be identified with specific mathematical terms in the planet's disturbing function [5], and so it is often possible to find an analytical solution to their effect. They also have specific physical meanings and result in two different types of structure. Secular perturbations act on all material in the disk and can be considered as the long term affect of having a planet in the disk. Indeed these perturbations are equivalent to those from the ring of material created by spreading the mass of the planet evenly along its orbit. Since they affect everything in the disk, secular perturbations result in large scale asymmetries such as offsets and warps. Resonant forces act only at specific radial locations in the disk where an object would have an orbital period which is a ratio of two integers, say $p + q$ and p, times the orbital period of the planet. By

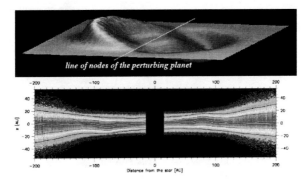

FIGURE 1. Long term effect of introducing a planet on an inclined orbit into a planetesimal disk [10]. Top: orbital planes of planetesimals at different distances from the star. On intermediate timescales planetesimals closer to the orbit of the planet have their orbits aligned with the planet, while those further away retain their primordial orbits. Bottom: model showing how this effect would cause the β Pictoris disk, which is viewed edge-on, to appear warped.

analogy with the dynamical structure of the solar system, resonant orbits can be either over- or under-populated, and a simple discussion of their geometry (see later) shows that such orbits can result in a clumpy disk.

Offsets and Warps: Secular Perturbations

One of the first discoveries of structure within debris disks came from images of β Pictoris. These showed that its disk, which is seen edge-on, is both warped and contains brightness asymmetries, even when asymmetries which can be attributed to observational effects (such as starlight being scattered by dust grains in a forward rather than backward direction) have been taken into account [6, 7]. A brightness asymmetry is also observed in the HR 4796A disk. The emission from this disk comes from a narrowly confined ring, also seen edge-on. As expected, the disk appears in the image as two lobes straddling the star, but one lobe is 5% brighter than the other [8]. Both warps and brightness asymmetries have been modeled as being due to the secular perturbations of unseen orbiting planets [9, 10].

Warps

One of the long term effects of introducing a planet into a planetesimal disk is to make the orbital planes of those planetesimals precess about the planet's orbital plane. After a few precession timescales, the orbital planes of nearby planetesimals are mixed so that the planetesimal disk has its plane of symmetry aligned with the orbital plane of the planet. As the precession is faster for orbits which are closer to the planet, on intermediate timescales the disk can appear warped, with planetesimals close to the planet having their orbits aligned with the planet, while those further away still having

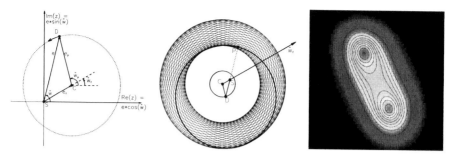

FIGURE 2. The consequence of introducing a planet on an eccentric orbit into a planetesimal disk [9, 8]. Left: the eccentricities, e, and pericenter orientations, $\tilde{\omega}$, of planetesimals precess about a point called the forced eccentricity (defined by e_f and $\tilde{\omega}_f$). Middle: on long timescales the eccentricities and pericenters appear evenly distributed about the circle centered on the forced eccentricity resulting in a disk with a center of symmetry (C) that is offset from the star (S). Right: model of the HR4796A disk showing how when such a disk is viewed edge-on, a brightness asymmetry occurs because one of the disk lobes is closer to the star, and so hotter and brighter, than the other one.

their primordial orbit alignment (Fig. 1 top). This was the basis of a model which showed that the β Pictoris disk warp could be explained by the introduction of a 3 Jupiter mass planet at 10 AU on an orbit inclined by just 3° to the primordial disk [10] (see Fig. 1 bottom). Within the context of this model, it is the product of the planet's mass and its orbital radius squared which are constrained, and the warp is observed because of the relative youth of the β Pictoris system. More recent mid-IR images of β Pictoris indicate that there are at least three planes of symmetry within its disk [11, 12]. Multiple planes of symmetry can also be explained using secular perturbations, since if there is more than one planet in the disk, and these planets orbit on different planes, then the symmetry plane will vary throughout the disk. Warps in multiple (non-coplanar) planet systems arise regardless of the age of the system.

Offsets

The lobe brightness asymmetry seen in the HR4796A disk was explained to be the result of the secular perturbations of a planet on an eccentric orbit in the disk [9]. The reason is that these perturbations affect the eccentricities and pericenter orientations of planetesimals in the disk; they impose a forced eccentricity and forced pericenter orientation on the planetesimals (Fig. 2 left). This means that on long timescales the disk looks like an eccentric ring, with the forced pericenter side being closer to the star than the corresponding forced apocenter side (Fig. 2 middle). While the offset between the two sides was proposed to be very small in the HR4796 disk (1-2 AU), this effect was observed in the mid-IR images, because this caused the lobe on the forced pericenter side to be hotter and so brighter than the other lobe (Fig. 2 right). The interesting thing about this model is that a planet as small as a few Earth masses orbiting with an eccentricity of just 0.02 could be causing this asymmetry. However, the nature of secular perturbations meant that the model could not constrain the mass or orbit of the planet, and it is even

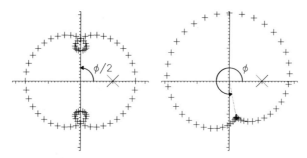

FIGURE 3. Paths of resonant orbits in the frame corotating with the planet [27]: 3:2 (left) and 2:1 (right) resonances. The resonant object is plotted with a plus at equal timesteps along its orbit which has an eccentricity of 0.3. The orientation of the pericenter with respect to the planet, which is shown with a cross, is determined by the resonant argument, ϕ.

possible that the asymmetry is caused by the M star binary companion HR4796B, the orbit of which is as yet unknown. Perhaps the most promising method of testing this model is to look for the offset asymmetry which should exist between the lobes; the orbit of the binary companion must also be constrained to ascertain its effect on the disk structure.

Clumps: Resonant Perturbations

The most common feature of debris disks is that they are clumpy: the Vega [13, 14], ε Eridani [15], and Fomalhaut [16] disks are all clumpy. The only feasible explanations put forward so far for this clumpiness have involved planetary resonances.

To understand why resonances result in clumpy structures one needs to consider two things. First of all, when plotted in the frame corotating with the planet, the pattern traced out by a resonant orbit is such that the object in resonance spends longer at certain longitudes relative to the planet (Fig. 3). This is because the pattern repeats itself after an integer number of orbits, which means that when the object reaches pericenter the planet can be at one of a few longitudes relative to the object. The orientation of those pericenters relative to the planet are defined by the resonant argument, ϕ. The second point is that while in resonance, resonant forces cause the object's resonant argument to librate (i.e., undergo a sinusoidal oscillation). Since the angle about which ϕ librates is the same (or rather varies in a consistent way) for all objects in the same resonance, all such objects are most likely to be found at the same longitudes relative to the planet.

Two different types of model exist to explain the clumpiness of dust disks in terms of planetary resonances, and they differ in the mechanism which puts the dust in the resonances: in one model dust migrates into the resonances by P-R drag, and in the other the parent planetesimals of the dust grains were captured into the resonances when the planet migrated outward.

97

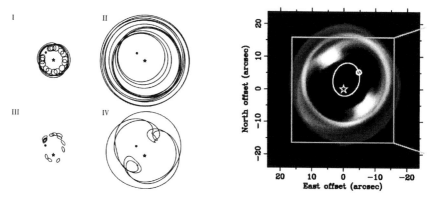

FIGURE 4. Structures resulting from dust migrating into a planet's resonances. Left [22]: four basic structures resulting from (I) low mass planet on a low eccentricity orbit, (II) high mass planet on a low eccentricity orbit, (III) low mass planet on a moderate eccentricity orbit, (IV) high mass planet on a moderate eccentricity orbit. Right [14]: model of Vega's two asymmetric dust clumps resulting from dust trapped into the resonances of a 3 Jupiter mass planet with an eccentricity of 0.6.

Dust Migration into Resonance

After a dust grain is created P-R drag makes a dust grain migrate in toward the star. On its passage inward, if there is a planet on an orbit interior to it, the dust grain will encounter the planet's resonances. Resonant forces can halt the migration causing the particle to be trapped in the resonance. While trapped the particle's eccentricity is pumped up, and eventually it leaves the resonance, then continuing its passage toward the star. The trapping causes a concentration of dust at a distance from the star similar to that of the planet's orbit. For reasons already described the resulting *resonant ring* is clumpy. Such resonant rings exist in the Solar System: one has been detected in the structure of the zodiacal cloud, associated with dust trapped in resonance with the Earth [17], and Neptune is also predicted to have such a ring [2].

This effect was first applied to extrasolar systems to provide an explanation of the radial distribution of dust observed in the β Pictoris disk [18]. Later this effect was used to explain the clumpy structures observed in the Vega and ε Eridani disks [19]. These models showed that it is possible to explain the clumpy structures in terms of dust migrating into the resonances of a 2 Jupiter mass planet orbiting at 40-60 AU from the Vega, and a 0.2 Jupiter mass planet at 55-65 AU from ε Eridani. The models made testable predictions: the planets should be detectable with ground-based observations (none has been detected yet at limits slightly above the masses proposed [20]); and the clumpy structures should orbit the star with the orbital period of the planet, i.e., at $0.6 - 1.6°$/year, motion which should be detectable on timescales of order a decade.

More recently two papers showed how the Vega and ε Eridani dust structures may also be caused by dust migrating into the resonances of planets on eccentric orbits [14, 21]. The proposed masses and orbital radii of the perturbing planets are not significantly different to those proposed previously; the orbits have eccentricities of 0.6 (for Vega's planet, see Fig. 4 right) and 0.3 (for ε Eridani's planet). However, introducing an

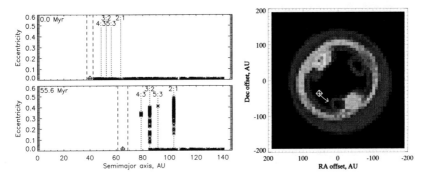

FIGURE 5. Impact of the outward migration of a Neptune mass planet on a planetesimal disk around Vega [27]: orbital distribution of planetesimals at the beginning (top left) and end (bottom left) of the migration; spatial distribution of those planetesimals at the end of the migration (right).

eccentricity into the planet's orbit meant that the dust is trapped into different orbital resonances. This led to the prediction that the structure would also change as it orbits the star, and in the case of Vega's structure, this would orbit at half the orbital speed of the planet. This type of model, and in particular the different types of structure caused by dust migration into the resonances of a high/low mass planet on a circular/eccentric orbit, was summarized in a recent paper [22] (see also, Fig. 4 left).

Planetesimal Trapping by Planet Migration

One problem encountered by the dust migration model is that the disks which have been observed are very dense. This means that their dust grains do not migrate far from their source before they are destroyed in a collision with another dust grain. One paper proposing dust migration into resonance suggested that this may mean that the sources of the dust are themselves already in resonance with the planet [14].

One reason why planetesimals may be in resonance with a planet is evident from the Solar System. Many objects in the Kuiper Belt, including Pluto, are found to be trapped in resonance with Neptune [23], and this configuration is thought to occur because Neptune's orbit actually started much closer to the Sun, at about 23 AU, then migrated outward to its current 30 AU orbit [24]. This migration is thought to have been caused by angular momentum exchange when Neptune scattered the remnant planetesimal disk and to have taken about 50 Myr [25]. As its resonances swept through the primordial Kuiper Belt, many planetesimals became trapped in the resonances and migrated out with Neptune, while at the same time having their eccentricities pumped up. This scenario has been expanded significantly and now explains many of the dynamical features of the current Kuiper Belt [26].

The different kind of dynamical structure you can get from planet migration in extrasolar systems was recently modeled and it was found that with this model it is possible to reproduce the observation of two clumps of asymmetric brightness around Vega [27] (Fig. 5). To do this the model assumed a planet with a mass similar to that of Neptune, and with an orbit that expanded from 40 to 65 AU over a period of about 56 Myr. Within the context of the model, tight constraints were set on these parameters, since only a small range of planet mass and migration rate resulted in the majority of the planetesimals being trapped into the 3:2 and 2:1(u) resonances (Fig. 5 left).

Again the model made testable observational predictions: the orbital motion of the clumps (1.1°/year); the location of the planet (SW of the star if the orbital motion is anticlockwise, as depicted in Fig. 5 right, NE if the motion is clockwise); and the high resolution structure of the disk, which is predicted to include a faint clump on the opposite side of the star from the planet (Fig. 5 right).

Implications for Planet Formation around Vega

The different models for the origin of Vega's clumps (Figs. 4 right and 5 right) have very different implications for how Vega's planetary system formed. All predict the presence of a planet orbiting some 40-65 AU from Vega, but the dust migration models predict this planet to be relatively massive at ~ 3 Jupiter masses, and to have an orbital eccentricity up to 0.6. Such a planet has very different characteristics to any currently known and so would provide an important additional constraint to planet formation models which attempt to explain all the observed endstates of planet formation in a consistent manner ([28]). The planet migration model, on the other hand, predicts a relatively low mass planet, about Neptune mass, on a circular orbit, and with an evolutionary history similar to that proposed for our own Neptune [25, 26]. While the origin of the migration of such a planet remains debatable [25, 28], based on this model it is possible that this system formed and evolved in a similar manner to our own.

Of course, the similarity of the proposed planetary system to our own, or lack thereof, should not be a factor in deciding which model is correct. For that the different observational predictions of the two models need to be borne out. First the orbital motion of the structure needs to be confirmed (a prediction of both models). Then predictions about the higher resolution structure of the disk must be tested. Searches for light from the planet itself will also help determine which model is correct. Finally, it should be pointed out that the physical processes affecting debris disk structure are extremely complicated, with the current models focusing only on specific aspects of those processes. Neither model yet gives a complete description of the disk, and further development of both models, and their implications, is sure to proceed as more observations are obtained.

STRUCTURES NOT CAUSED BY PLANETS

Not all structures in debris disk images are caused by planetary perturbations. To avoid over-interpreting debris disk images it is important that all possible causes of observed structure are explored before a planetary interpretation can be confirmed. Here I consider just four possible sources of structure which fall into one of two categories — either those which have already been purported to have been detected in debris disks (structure from binary companions, or a stellar flyby) and those which are known to have the potential to cause structure (planetesimal collisions, and interaction with the interstellar medium):

- Just as planetary companions can affect the structure of the dust disk, so can the presence of a stellar companion. Many systems with dusty disks are known to be binary systems, but in many cases, particularly in wide binary systems, the orbit is not known. Recently modeling has shown that spiral structure observed in the HD141569 disk can be explained by the secular perturbations of its binary companion [31].
- The passing of a star close to the disk is also a method of causing structure. Clumps which have been detected in the β Pictoris disk have been explained as the consequence of perturbations from a star which passed 700 AU from β Pictoris about 0.1 Myr ago [32]. A variety of structures, including clumps and warps, are possible from stellar flybys [33]. However, the probability of a star passing close enough to a debris disk to cause observable perturbations is thought to be very small given the density of stars in the solar neighbourhood.
- The evolution of the Asteroid Belt is punctuated by sudden brightness increases caused by the destruction of single asteroids. Indeed the products of a few such collisions may make up a significant fraction of the dust in the present day zodiacal cloud [29]. The possibility that the clump observed in the Fomalhaut disk was created in a collision between two large planetesimals was considered [1], but it was shown that this is unlikely to be the case: the very fact that we can see the dust clump means it is very massive, and collisions between planetesimals massive enough to produce the clump should be too infrequent for us to be likely to witness such an event.
- Substantial asymmetries can arise if the erosion of a planetesimal disk is affected by the sandblasting of those planetesimals by interstellar dust grains [30]. However, because of the repulsion from the star of such grains by radiation pressure, this erosion is only thought important at large distances from the star (> 400 AU).

CONCLUSIONS

If there are planets in disks then they will affect the structure of those disks in a variety of ways, introducing clumps, warps and offsets. Modeling the observed structures of debris disks can be used to identify the presence of an unseen perturbing planet and set constraints on its location, mass, orbit and even evolutionary history. However, planets are not the only cause of structure in debris disks: binary companions, stellar flybys, planetesimal collisions and sandblasting by interstellar dust grains all play a role to some extent in shaping the disk. In most cases the contribution of these effects to the observed structures can be ruled out leaving the only feasible explanation as the presence of planets. For this planetary interpretation to be accepted, though, the observational predictions of the models, such as the orbital motion of the clumpy structures, and high resolution structure of the disks, must be confirmed.

REFERENCES

1. Wyatt, M. C., and Dent, W. R. F., *Mon. Not. Roy. Astron. Soc.*, **334**, 589–607 (2002).
2. Liou, J.-C., and Zook, H. A., *Astron. J.*, **118**, 580–590 (1999).
3. Wyatt, M. C., Ph.D. thesis, University of Florida (1999).
4. Takeuchi, T., and Artymowicz, P., *Astrophys. J.*, **557**, 990–1006 (2001).
5. Murray, C. D., and Dermott, S. F., *Solar System Dynamics*, Cambridge University Press, Cambridge, 1999.
6. Kalas, P., and Jewitt, D., *Astron. J.*, **110**, 794–804 (1995).
7. Heap, S. R., Lindler, D. J., Lanz, T. M., Cornett, R. H., Hubeny, I., Maran, S. P., and Woodgate, B., *Astrophys. J.*, **539**, 435–444 (2000).
8. Telesco, C. M., et al., *Astrophys. J.*, **530**, 329–341 (2000).
9. Wyatt, M. C., Dermott, S. F., Telesco, C. M., Fisher, R. S., Grogan, K., Holmes, E. K., and Piña, R. K., *Astrophys. J.*, **527**, 918–944 (1999).
10. Augereau, J. C., Nelson, R. P., Lagrange, A. M., Papaloizou, J. C. B., and Mouillet, D., *Astron. Astrophys.*, **370**, 447–455 (2001).
11. Wahhaj, Z., Koerner, D. W., Ressler, M. E., Werner, M. W., Backman, D. E., and Sargent, A. I., *Astrophys. J.*, **584**, L27–L31 (2003).
12. Weinberger, A. J., Becklin, E. E., and Zuckerman, B., *Astrophys. J.*, **584**, L33–L37 (2003).
13. Holland, W. S., et al., *Nature*, **392**, 788–790 (1998).
14. Wilner, D. J., Holman, M. J., Kuchner, M. J., and Ho, P. T. P., *Astrophys. J.*, **569**, L115–L119 (2002).
15. Greaves, J. S., et al., *Astrophys. J.*, **506**, L133–L137 (1998).
16. Holland, W. S., et al., *Astrophys. J.*, **582**, 1141–1146 (2003).
17. Dermott, S. F., Jayaraman, S., Xu, Y. L., Gustafson, B. A. S., and Liou, J.-C., *Nature*, **369**, 719–723 (1994).
18. Roques, F., Scholl, H., Sicardy, B., and Smith, B. A., *Icarus*, **108**, 37–58 (1994).
19. Ozernoy, L. M., Gorkavyi, N. N., Mather, J. C., and Taidakova, T. A., *Astrophys. J.*, **537**, L147–L151 (2000).
20. Macintosh, B. A., Becklin, E. E., Kaisler, D., Konopacky, Q., and Zuckerman, B., *Astrophys. J.*, **594**, 538–544 (2003).
21. Quillen, A. C., and Thorndike, S., *Astrophys. J.*, **578**, L149–L152 (2002).
22. Kuchner, M. J., and Holman, M. J., *Astrophys. J.*, **588**, 1110–1120 (2003).
23. Jewitt, D. C., *Annu. Rev. Earth Planet. Sci.*, **27**, 287–312 (1999).
24. Malhotra, R., *Astron. J.*, **110**, 420–429 (1995).
25. Hahn, J. M., and Malhotra, R., *Astron. J.*, **117**, 3041–3053 (1999).
26. Levison, H. F., and Morbidelli, A., *Nature*, **426**, 419–421 (2003).
27. Wyatt, M. C., *Astrophys. J.*, **598**, 1321–1340 (2003).
28. Veras, D., and Armitage, P. J., *Mon. Not. Roy. Astron. Soc.*, **347**, 613–624 (2004).
29. Nesvorný, D., Bottke, W. F., Levison, H. F., and Dones, L., *Astrophys. J.*, **591**, 486–497 (2003).
30. Artymowicz, P., and Clampin, M., *Astrophys. J.*, **490**, 863–878 (1997).
31. Augereau, J.-C., and Papaloizou, J. C. B., *Astron. Astrophys.*, **414**, 1153–1164.
32. Kalas, P., Larwood, J., Smith, B. A., Schultz, A., *Astrophys. J.*, **530**, L133–L137 (2000).
33. Larwood, J. D., and Kalas, P. G., *Mon. Not. Roy. Astron. Soc.*, **323**, 402–416 (2001).

Spatially Resolved Spectrum of the TW Hydrae Circumstellar Disk

Aki Roberge*, Alycia J. Weinberger* and Eliot Malumuth†

*Carnegie Institution of Washington, Dept. of Terrestrial Magnetism,
5241 Broad Branch Road, NW, Wahington, DC 20015-1305
†NASA Goddard Space Flight Center, Code 685, Greenbelt, MD 20771

Abstract. We present the first spatially resolved spectrum of scattered light from the TW Hydrae protoplanetary disk. This nearly face-on disk is optically thick, surrounding a classical T-Tauri star in the nearby 8 Myr old TW Hya association. Accretion of disk material onto the central star is still occurring, but there are signs that growth of planetary material has begun.

Our *HST*-STIS spectrum covers the optical bandpass from 5000 Å to 1 μm. After careful subtraction of a PSF star spectrum, spectra can be extracted between 37 AU and about 124 AU from the star. The scattered light spectra have the same color as the star (gray scattering) at all radii, except possibly the very innermost region. This likely indicates that either the scattering dust grains are much larger than 1 μm throughout the bulk of the disk or that the disk remains very optically thick out to 124 AU.

INTRODUCTION

TW Hydrae (spectral type K7 Ve) is a classical T-Tauri star, indicating that gas and dust from a circumstellar (CS) disk are accreting onto the star. This disk has now been imaged at visible, near-IR, and millimeter wavelengths ([1], [2], [3]), and shown to be nearly face-on. A very large fraction of the stellar light is reprocessed by the CS dust to far-IR wavelengths, indicating that the disk is optically thick [4]. These characteristics tend to classify TW Hya as relatively unevolved protoplanetary disk.

However, while $\sim$1 μm grains in the surface layers of the disk can produce the mid-IR spectrum of the disk, very large (mm to cm sized) grains are needed to explain the mm-wavelength spectral energy distribution [2]. The likely presence of crystalline silicates in the disk, which are seen in Solar System meteorites and comets but not in the interstellar medium, also suggests that the growth of planetary material has begun [2].

Previous broad-band photometry indicated that the disk scattering was wavelength-independent and that the grain albedo might be large [2]. This suggests that the disk is composed mostly of icy grains larger than $\sim$1 μm. However, this conclusion is tentative, since it was not based on actual spectra. Also, the photometry provided no information about changes in the scattering with location in the disk. To address these shortcomings, we obtained a spatially resolved *HST*-STIS spectrum of TW Hya, in order to study the size and composition of the dust as a function of location in the protoplanetary disk.

CP713, *The Search for Other Worlds: Fourteenth Astrophysics Conference*,
edited by S. S. Holt and D. Deming
© 2004 American Institute of Physics 0-7354-0190-X/04/$22.00

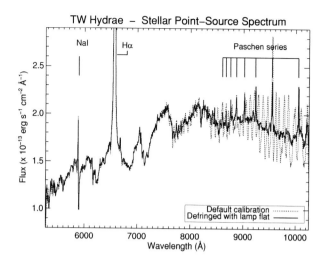

FIGURE 1. *HST*-STIS point-source spectrum of TW Hya. STIS CCD spectra suffer from fringing at long wavelengths (seen in the dotted spectrum). The fringing was removed from all our data using contemporaneous lamp flat images (defringed spectrum shown with the black line).

OBSERVATIONS

We obtained *HST*-STIS G750L long-slit CCD spectra of TW Hya on 2002 July 17. The spectra cover the wavelength range from 5240 Å to 1.027 μm, with a spectral resolution of $\sim$ 500 km s^{-1} (R $\sim$ 500). The spatial resolution is 0.″051 per pixel. One exposure was taken with the 52″ × 0.″2 slit, producing a point-source spectrum of the star (see Figure 1). Additional exposures were taken using the 52″ × 0.″2F2 slit, with the 0.″86 fiducial bar on top of the central star to reduce the amount of scattered stellar light (see left panel, Figure 2). These exposures were repeated for a PSF star (CD-43D2742) of similar spectral type to TW Hya.

The point-source TW Hya spectrum shows H$-\alpha$ and Paschen series emission lines characteristic of accreting T-Tauri stars. These lines are not seen in the spectra of the PSF star. We calculated the ratios of TW Hya to the PSF star at all wavelengths from the point-source spectra. These ratios were multiplied each column of the PSF fiducial image to produce a scaled PSF image with exactly the same color as TW Hya. This scaled PSF was aligned with the TW Hya fiducial image, then subtracted from it, revealing a spatially resolved spectrum of the disk alone (see right panel, Figure 2).

DISK COLOR VS. RADIUS

Spectra of scattered light from the disk can be extracted between 37 AU (0.″66) and 124 AU (2.″19) from the star. Imperfect subtraction of PSF features close to the fiducial bar is the limiting factor on the interior radius. The color and brightness of the scattered

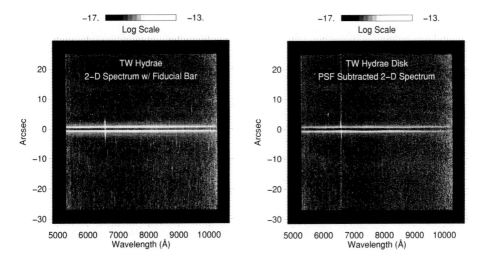

FIGURE 2. (left) Combined and defringed 2-D fiducial spectrum of TW Hya. The dispersion direction is along the x-axis and the spatial direction along the y-axis. The H$-\alpha$ emission line from the star is seen at 6562 Å. (right) Final PSF subtracted 2-D fiducial spectrum of TW Hya. The remaining flux visible is scattered light from the disk. Remaining H$-\alpha$ emission is caused by the fact that the emission line in the point-source spectrum of TW Hya was saturated, while the emission line in the fiducial images was not. This causes the ratio for the scaled PSF at H$-\alpha$ to be underestimated. The large scale factor at H$-\alpha$ also produces a stripe of increased noise in the PSF subtracted image.

light is the same above and below the fiducial bar. Therefore, rows above and below the bar at the same distance from the star were averaged together to produce a folded 2-D disk spectrum. The 1-D spectra extracted at different radii are shown in Figure 3.

In Figure 4, we show the ratio of the disk spectra to the stellar spectrum in Figure 3. Beyond the innermost bin, the disk spectra have the same color as the star, indicating that the scattering is wavelength-independent gray scattering. This confirms earlier results from broad-band photometry [2]. Two explanations for the gray scattering are suggested: 1) the scattering dust grains are larger than 1 μm throughout the bulk of the disk, or 2) the disk is very optically thick all the way out to 124 AU.

In the innermost bin, the flux ratio decreases at long wavelengths. We are not yet completely sure if this is a real feature or is due to imperfect PSF subtraction. However, if it is real, it might indicate blue scattering in the innermost regions of the disk, possibly showing the presence of small grains. Or it might be due to a broad emission feature like the Extended Red Emission between 5000 Å and 9000 Å seen in many interstellar reflection nebulae.

Our planned further work includes : 1) testing of the quality of our PSF subtraction technique to confirm the reality of the change in scattering in the innermost region of the disk and 2) determination of the broad-band radial surface brightness profiles of the disk for comparison to coronagraphic scattered light images.

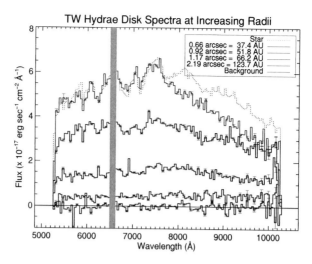

FIGURE 3. Spectra of scattered light at different radii (radius increases from top spectrum to bottom). The much-reduced spectrum of the central star is show at the top with a dotted line. The region around H−α is blocked out (due to the problems scaling the PSF at H−α; see caption to Figure 2).

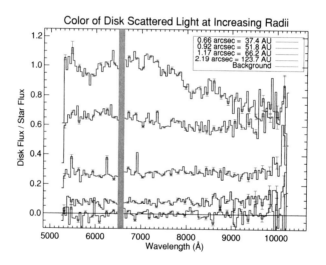

FIGURE 4. Ratio of the disk scattered light to the (much-reduced) stellar flux.

REFERENCES

1. Krist, J. E., Stapelfeldt, K. R., Ménard, F., et al. *ApJ*, **538**, 793–800 (2000).
2. Weinberger, A. J., Becklin, E. E., Schneider, G., et al. *ApJ*, **566**, 409–418 (2002).
3. Wilner, D. J., Ho, P. T. P., Kastner, J. H., et al. *ApJ*, **534**, L101-104 (2000).
4. Adams, F. C., Lada, C. J., & Shu, F. H. *ApJ*, **312**, 788–806 (1987).

HL Tau – The Missing Dust

T. Rettig[*], S. Brittain[*], T. Simon[†], C. Kulesa[**] and J. Haywood[*]

[*]Center for Astrophysics, University of Notre Dame, Notre Dame, IN 46556
[†]Institute for Astronomy, University of Hawaii
[**]University of Arizona, Steward Observatory

Abstract. High-resolution infrared spectra of HL Tau exhibit broad emission lines of ^{12}CO gas phase molecules as well as narrow absorption lines of ^{12}CO, ^{13}CO, and $C^{18}O$. The broad emission lines of vibrationally-excited ^{12}CO are dominated by the hot (T $\sim$1500 K) inner-disk (radius $r <$ 0.2 AU). The narrow absorption lines of CO are found to originate from the circumstellar gas at a temperature of $\sim$100 K. The cooler material indicates a large column of absorbing gas along the line of sight, which indirectly implies a large amount of dust extinction. However, the minimal opacity allowed by our emission line results severely constrains the M-band extinction and suggests that there is much less dust along the line of sight than inferred from the CO absorption data.

INTRODUCTION

HL Tau is an embedded young star located at a distance of $\sim$140 pc [1]. It is likely transitioning from a Class I to a Class II object. It has a $< 1\ M_{solar}$ central star surrounded by an optically thick inner disk that is embedded in a large circumstellar envelope [2].

High-resolution infrared spectra of the narrow absorption and broad emission lines of gas phase CO molecules toward HL Tau can be used to constrain the basic physical properties of the inner accretion disk as well as the surrounding circumstellar envelope. We present high-resolution near-infrared spectra of CO absorption and emission. We show the column density of gas we observe is much larger than one might expect based on the extinction reported for HL Tau. This result can be explained by either grain agglomeration or dust settling to the mid-plane.

CO ABSORPTION LINES

A CO absorption-line study of HL Tau samples conditions in an extended disk over a very long path length. It provides a high angular resolution that is defined by the small physical extent of the background source of illumination (generally a stellar diameter in size) (see Figure 1). The column densities and rotational temperatures for the absorption lines of $^{12}CO(1,0)$, $^{12}CO(2,0)$, $^{13}CO(1,0)$, and $C^{18}O(1,0)$ are presented in Table 1 (see Figure 2). The excitation plot of the fundamental ^{12}CO absorption lines indicates the transitions are optically thick but the isotopes are optically thin which allows us to determine the total column densities. The temperature of the absorbing gas is $\sim$100 K.

CP713, *The Search for Other Worlds: Fourteenth Astrophysics Conference*,
edited by S. S. Holt and D. Deming

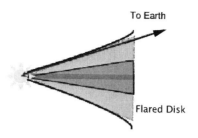

FIGURE 1. Depiction of the line of sight through the flared disk of HL Tau.

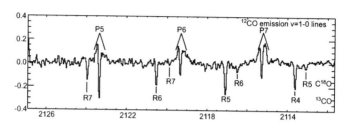

FIGURE 2. Example absortion/emission spectrum.

MISSING DUST?

Our measured ^{12}CO column density of $(7.1 \pm 0.3) \times 10^{18}$ cm^{-2}, coupled with the usual assumption for the CO/H$_2$ abundance ratio of 1.5×10^{-4} (appropriate for cold, dark clouds in which the hydrogen is predominantly molecular; [3]) implies $N(H) = 2 \times N(H_2)$ $= 9.5 \times 10^{22}$ cm^{-2}. For a normal interstellar gas-to-dust ratio $A_V/N_H = 5.4 \times 10^{-22}$ mag cm^{-2} [4] this amount of hydrogen corresponds to an optical extinction of A_V ~ 51 magnitudes. However, previous estimates of the extinction toward HL Tau showed that $A_V \sim 22$-30 magnitudes [5, 6, 7, 2], a factor of more than two lower than our measurement of A_V for the same line of sight. If our observations have over-estimated the extinction, either the dust has settled to the central disk *or* the dust extinction, as a result of grain agglomeration, is much lower at all wavelengths as compared to normal, dense interstellar clouds.

Theoretical models of disks predict that dust preferentially settles to the mid-plane,

TABLE 1. Physical Properties from CO Absorption Lines

Ro-vibrational band	Rotational Temperature (K)	Column Density (cm^{-2})
^{12}CO(2,0)	95 ± 4	$(7.1 \pm 0.3) \times 10^{18}$
^{12}CO(1,0)[a]	84 ± 22	7×10^{18}
^{13}CO(1,0)	72 ± 8	$(7.6 \pm 1.5) \times 10^{16}$
C^{18}O(1,0)	102 ± 7	$(8.8 \pm 1) \times 10^{15}$

leaving behind a dust-depleted atmosphere at higher vertical scale heights [8], but such predictions have not been observationally tested.

ANALYSIS OF THE EMISSION LINES

The excitation plot of the fundamental CO emission lines indicates optically thick transitions. However, the excitation plot of the hotband CO emission lines appear to be optically thin, from which we infer a rotational temperature of 1500 ± 100 K. We calculate the column density of the warm gas in the $v' = 2$ state and use its rotational temperature to extrapolate the column density to all other vibrational energy levels. The effective column density of the CO in all states was determined to be $N(CO) = (1.2 \pm 0.3) \times 10^{13}$ cm^{-2}. It is important to emphasize that we use the CO emission line data (in § below) to constrain the total amount of extinction allowed by the observations. The M-band interstellar extinction is constrained by requiring the corrected column density of CO molecules in the state $v' = 1$ be optically thin, which is required by the linear fit to the $v = 2$-1 excitation plot.

The velocity FWHM of the CO emission lines is 90 km s^{-1}. If we assume an inclination of $40°$ for the disk [9, 10], then the deprojected orbital velocity of the gas is 70 km s^{-1}. Adopting a stellar mass of 0.7 M$_{solar}$ [2] we find that the CO source must be centered at a radial distance of ~ 0.1 AU, so the CO emission is restricted to the inner region of the disk. It is unlikely that this gas extends much beyond 0.2 AU from the 4000 K central star; in the model of Men'shchikov et al. [10], for example, the temperature at $r \sim 0.2$ AU is predicted to be 1300 K and is falling off as $1/r^2$.

LIMITS ON THE M-BAND DUST EXTINCTION

The hot CO emission is expected to originate from a compact, unresolved region in the disk. To determine the physical column density from the CO emission lines, we must correct for both "beam dilution" and dust extinction at the wavelengths of the lines. The angular extent of our beam was 7×10^{-10} sr, which subtends an area of 5500 AU^2 at 140 pc. The total physical column density is related to the effective column density by:

$$N_{physical} = N_{effective} e^{A_M/1.086} (5.5 \times 10^3 / \pi r^2) \tag{1}$$

where πr^2 is the area subtended by the emitting CO, A_M is the extinction in the M-band due to interstellar/circumstellar dust along the line of sight, and the effective density for the state $v' = 1$ is $N_{v=1} = (9.8 \pm 2.1) \times 10^{11}$ cm^{-2}. It is the $v' = 1$ state that must remain optically thin after correcting for extinction and beam dilution.

To correct for beam dilution, we assume the CO emission source is uniformly distributed in the inner disk from $0 < r < 0.2$ AU since, as shown in Section 4 the gas is centered at ~ 0.1 AU. The size of the emitting area of CO we adopt here is likely an overestimation; however, this only makes our upper limit for the M-band extinction more stringent. Thus the physical column density is $N(CO) = (4.3 \pm 0.9) \times 10^{16} e^{A_M}$.

Using a one-sigma deviation from the linear fit around $J = 10$ (the observed $v = 1$-2 absorption line with the greatest opacity at T = 1500 K) requires line center opacities of

$\tau < 0.3$. Performing a curve of growth analysis for a maximum opacity of $\tau = 0.3$ and assuming a line width $= 20$ km s^{-1} shows this opacity is reached at a column density of 2×10^{15} cm^{-2} (v = 1, J = 10). Thus, a maximum column density of CO in state v' = 1 at 1500 K is $N_{v=1} < 1 \times 10^{17}$cm^{-2}, so the extinction in the M-band, A_M, is $< 0.78 \pm 0.23$ mags. Using an extinction law that follows the interstellar reddening curve [4], the extinction expected in the J-band at shorter wavelengths is then $A_J < 8.2 \pm 2.4$. That value is consistent with the extinction measured by Close et al. [2], $A_J = 7.73 \pm 0.42$. The minimal infrared extinction, $A_M < 0.78$, rules out a large population of gray particles along the line of sight, specifically, the 10 magnitudes of gray extinction suggested by Men'shchikov et al. [10].

CONCLUSIONS

The narrow absorption lines from ^{12}CO, ^{13}CO, and C^{18}O are found to originate at large distances from the central star, presumably where the line of sight intersects the flared portion of the disk. The absorbing column of ^{12}CO $(7.1 \pm 0.3) \times 10^{18}$ cm^{-2} indicates a large column of gas along the line of sight.

The CO emission lines are dominated by gas in the hot (T $\sim$1500K) inner-disk < 0.2 AU of HL Tau. The limit on the emission line opacity requires $A_M < 0.78 \pm 0.23$. If we extrapolate to visible wavelengths from the upper limit on A_M (using a normal reddening law), the resulting upper limit on the visible extinction, $A_V < 25\pm7$, is a factor of 2 smaller than the $A_V \sim 51$ mags we inferred from the CO absorption lines. We conclude, therefore, that our analysis of the CO absorption spectrum, which assumes a normal gas/dust ratio, *overestimates* the amount of dust and the amount of dust extinction. From the hot CO gas, the column of absorbing CO gas must instead traverse a line of sight that is depleted of dust and thus reflects a higher than normal gas/dust ratio. Such a finding is consistent with the possibility that we are observing a vertical stratification of gas and dust in which the dust may have settled out of the envelope surrounding HL Tau and collapsed into the mid-plane of its circumstellar disk. This work was supported by NSF-AST-02-05881

REFERENCES

1. Elias, J. H., *The Astrophysical Journal*, **224**, 857–872 (1978).
2. Close, L. M., Roddier, F., Northcott, M. J., Roddier, C., & Graves, J. E., *The Astrophysical Journal*, **478**, 766 (1997).
3. Kulesa, C. A. & Black, J. H., "Abundances of H$_2$, H$_3^+$ & CO in Molecular Clouds and Pre-planetary Disks," in *Chemistry as a Diagnostic of Star Formation*, 2002, p. 60.
4. Mathis, J. S., *Anual Review of Astronomy and Astrophysics*, **28**, 37–70 (1990).
5. Monin, J. L., Pudritz, R. E., Rouan, D., & Lacombe, F., *Astronomy and Astrophysics*, **215**, L1–L4 (1989).
6. Beckwith, S. V. W.& Birk, C. C.,*Astrophysical Journal Letters*, **449**, L59 (1995).
7. Stapelfeldt, K. R. et al., *The Astrophysical Journal*, **449**, 888 (1995).
8. Goldreich, P. & Ward, W. R., *The Astrophysical Journal*, **183**, 1051–1062 (1973).
9. Lay, O. P., Carlstrom, J. E., & Hills, R. E., *The Astrophysical Journal*, **489**, 917 (1997).
10. Men'shchikov, A. B., Henning, T., & Fischer, O., *The Astrophysical Journal*, **519**, 257-278 (1999).

Debris Around Sun-like Stars

J. S. Greaves[†]

UK Astronomy Technology Centre, Royal Observatory, Blackford Hill, Edinburgh EH9 3HJ, UK
Physics & Astronomy, University of St Andrews, St Andrews, Fife KY16 9SS, UK

Abstract. Debris disks constitute the fall-out of collisions between comets, and by imaging these disks around nearby stars we can find out much about the planetary systems they encircle. A new picture is emerging in which cold debris around Sun-like stars is relatively common, with perhaps as many as one-third of systems having far greater dust masses than the Solar System. These systems are largely a distinct set from that of the similar stars with known radial velocity planets. A simple picture is presented in which there is divergent evolution depending on the epoch of planetary core formation versus that of dispersal of the gas disk. This may control the direction of planet migration, producing 'hot Jupiters' or 'cold Neptunes' as extreme outcomes. The Solar System lies somewhere in between the extremes, but the parameter space for finding similar systems is narrowing.

OVERVIEW

It is commonly stated that 15% of main sequence stars possess debris disks while 10% of stars are presently known to have radial velocity planets. These statements need to be somewhat qualified, because the radial velocity spectroscopy technique works only for Sun-like stars, whereas debris disks detected from far-infrared dust emission have a bias towards earlier-type stars. This is because the dust generated by collisions of comets will tend to thermal equilibrium with the star, and hence be warmer in the same orbit around a hotter star. In fact, for Kuiper Belt-like rings at around 45 AU, the 60 micron flux of blackbody dust grains would be 25 times brighter around a 20 $L_\odot$ A dwarf than a 1 $L_\odot$ G dwarf. This observational bias is becoming less with the advent of submillimeter surveys, where emission in the Rayleigh-Jeans tail differs by only a factor of two in brightness. Recently, submillimeter work by [1] has shown that the fraction of stars with debris could be 30%, doubling the suspected population.

Any bias with spectral type needs to be clarified for us to understand the evolution of Solar System analogues. An analysis of IRAS and ISO disk detections by [2] found that 62% of single A stars within 25 pc have debris, compared to only 7% of single G stars. This remains unaccounted for; it is not entirely an age effect because disks were detected for only 18% of the G dwarfs of similar age to the short-lived A stars. However, because of the different disk brightnesses in the far-infrared, the typically cooler G-star disks may not all have been detected. We have very recently begun an unbiased submillimeter survey of G dwarfs between 10 and 15 pc from the Sun, and the preliminary results are pushing the disk fraction up towards the one-third mark suggested by [1]. Given that the radial velocity detections are begining to exceed 10% of surveyed stars, the parameter space for systems like our own, *without* inner giant planets or substantial debris, is narrowing down towards half of all Sun-like stars. This may affect our expectations

CP713, *The Search for Other Worlds: Fourteenth Astrophysics Conference,*
edited by S. S. Holt and D. Deming

of the number of stars where we might find habitable Earths.

THE CLOSEST STARS

The small number of nearby Sun-like stars is not always appreciated. Within 5 pc, if we consider G to mid-K stars there are only 5 such systems other than the Sun, but IRAS and ISO studies reached little further than this in searching for cool debris. Of these systems, α Cen, 61 Cyg and ε Ind all have stellar companions separated by Kuiper Belt-like dimensions, and disks are then not expected because of strong tidal truncation. The two other systems. ε Eri and τ Ceti, both turn out to have much higher debris masses than the Solar System. The former is a young K2 dwarf of about 15% of the Sun's age, surrounded by a dust ring peaking at 60 AU radius containing 0.01 $M_\oplus$ of particles [3]. The population of colliding comets is infered to be very large, and this could be an analagous period to the Earth's epoch of heavy bombardment up to about 0.6 Gyr. However τ Ceti is a G8V star that is *older* than the Sun, in fact 9–10 Gyr old [4] — yet it still possesses around 0.0005 $M_\oplus$ of dust with a cometary population an order of magnitude higher than in the Kuiper Belt (paper in prep.).

The dust masses are expected to decline with time as the colliding bodies grind down towards particles small enough to be blown out of the system by radiation pressure, and some authors have suggested declines as steep as t^{-2}. However these analyses are strongly affected by the abrupt drop in dust mass at the transition from primordial to collisionally-generated dust [1], and [2] find that any systematic decline is not steeper than $t^{-1/2}$ for Sun-like stars up to a few Gyr old. Hence, as the observations are begining to suggest, there could be a large fraction of Sun-like stars at any main sequence age that still possess large dust masses. The Solar System being at the low-dust end of the range is probably fortunate in terms of a smaller parent population of comets that can impact planetary surfaces.

A MODEL FOR DIVERGING EVOLUTION

A major question is *why* there should be divergent evolution around similar stars: why should they not all have hot Jupiters, or all have cold debris disks? The submillimeter survey of [5] found that none of 8 stars with radial velocity planets had cold dust, to a level comparable with that of ε Eri. Conversely, their compilation of 20 stars that are known to have debris, and are also being surveyed in radial velocity experiments, found only one where such a planet has actually been found. (This is in fact ε Eri, the closest star in the sample.) However at least 80% of these dusty systems are infered to have an inner cavity in the disk (from a lack of mid-infrared emission). This is most plausibly swept clear by a planet on a long-period orbit, that has not yet been detected by radial velocity methods.

Although the numbers of systems are small, it is apparent that the presence of inner giants and outer debris disks are not positively correlated; they may be uncorrelated or anti-correlated. A simple hypothesis may explain the latter case, and is sketched

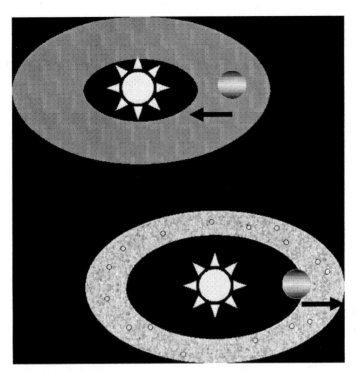

FIGURE 1. Sketch of planetary migration in cases where (top) a core forms early in a gas-rich disk and (bottom) core formation finishes at the time when gas has disappeared.

in Figure 1. The growth of a solid core is often assumed to be necessary to form a giant planet; once it reaches a large enough mass it can then attract the thick planetary atmosphere from the primordial gas disk. However there is evidence that few of these disks survive beyond about 10 Myr. If the core growth time is shorter or longer than this, we may expect different evolution. In our simple model [5], we argue that the core growth time depends on the initial disk mass (or at least, its mass in solids). A massive disk can then produce a large core early on, while the gas disk is still substantial, and this promotes inwards viscous migration. On the other hand, if the core attains a large size at the end of the gas-rich era, there may be time to accumulate only a thin atmosphere (like that of Uranus or Neptune). This planet will then tend to migrate *outwards*, because of angular momentum exchange with planetesimals further out. This simple model could produce either a hot Jupiter or a cold debris disk with a cleared cavity and outer planet, depending only on the initial disk mass and assuming a constant gas dispersal time (the origin of which is still unclear).

CONCLUSIONS

A picture is emerging of divergent evolution around similar Sun-like stars. As many as one-third could have substantially more debris than the Solar System (even at Sun-like ages), while at least a tenth have gas giants much closer in than Jupiter. This has narrowed the parameter space where we expect to see systems like our own. A qualitative model is suggested that could promote either inwards or outwards planetary migration depending on the initial disk mass. It is however based on small numbers of nearby stars, and has one exception, ε Eri, which has both a debris disk and planet at a few AU. New observations, such as those to be made with SIRTF, are important for establishing such a model for diversity of planetary systems.

ACKNOWLEDGMENTS

I thank my collaborators Mark Wyatt, Wayne Holland and Bill Dent for their invaluable input to these studies. The database used for debris disk statistics is publicly available at `http://www.roe.ac.uk/atc/research/ddd/`.

REFERENCES

1. Wyatt, M. C., Dent, W. R. F., and Greaves, J. S., *MNRAS*, **342**, 876–888 (2003).
2. Greaves, J. S., and Wyatt, M. C., *MNRAS*, **345**, 1212–1222 (2003).
3. Greaves, J. S., Holland, W. S., Moriarty-Schieven, G., Jenness, T., Dent, W. R. F., Zuckerman, B., McCarthy, C., Webb, R. A., Butner, H. M., Gear, W. K., and Walker, H. J. *ApJ*, **506**, L133–L137 (1998).
4. Pijpers F. P., Teixeira T. C., Garcia P. J., Cunha M. S., Monteiro M. J. P. F. G., Christensen-Dalsgaard J., 2003, *A&A*, **406**, L15–L18 (2003).
5. Greaves, J. S., Holland, W. S., Jayawardhana, R., Wyatt, M. C., and Dent, W. R. F., *MNRAS*, in press.

Detectability of Debris Disks Around Red Dwarfs at Submillimeter Wavelengths

Jean-François Lestrade

Observatoire de Paris - CNRS, 77 av. Denfert Rochereau, F75014 - Paris - France

Abstract. More than two thirds of the stars in the vicinity of the Sun are red dwarfs (spectral type M) and the frequency of planetary systems around them has not been estimated yet. We show that dust in debris disks around these main sequence stars is cold ($\sim$ 15 K) radiating predominantly at submillimeter wavelengths and has escaped detection as far-IR excess. For the 1809 M dwarfs of the catalog of nearby stars CNS3, we show that the JCMT can presently detect disk with 1 lunar mass or more of dust and that ALMA in the future will be sensitive to 0.001 lunar mass of dust for the closest M dwarfs.

INTRODUCTION

More than two thirds of the stars in the vicinity of the Sun are red dwarfs (spectral type M) and the frequency of planetary systems around them has not been estimated yet. This is due to the precision of the radial velocity measurements which is limited by the high chromospheric activity and faintness of these stars. Only one M star (GL876, type M4) is hosting planets (two) among the 104 stars with planets so far (27/11/03). Another approach to probe the presence of planets around stars is the detection of debris disks by measuring excess fluxes in their SED or by imaging. Debris disks are solar system-sized dust disks with micron-sized grains which are not primordial but by-products of collisions between planetesimals left over from the planet formation process. The Kuiper belt at $\sim$ 50 AU is the debris disk of the Solar System. Relatively warm debris disks have been observed as far-IR excesses (IRAS and ISO) in the SED's of A to K type stars (*e.g.* disk for the A0 star Vega: temp $\sim$ 85 K and size $\sim$85 AU). Images have been produced both of scattered light and of the dust thermal emission from these disks at far-IR and submillimeter wavelengths [1]. A very striking image is the structured disk around the K2 main sequence star ε Eri seen face-on, $\sim$60 AU in size, as observed at 850 μm by JCMT/SCUBA [2]. Among the 530 M stars searched for excess fluxes in IR data, only three have been found [3]. In this contribution, we argue that, M dwarfs being underluminous, their dusty debris disks similar to the Kuiper Belt are cold and radiate predominantly in the submillimeter domain and have escaped detection by IR satellites. The bias towards warm disks in most surveys should be quashed and we study here the detectability of the 1809 M dwarfs of the nearby star catalog CNS3 by the present and future submillimeter facilities. The 3rd edition of the catalog from Gliese and Jarheiss (1991), CNS3 for short, is the most complete catalog for the nearby stars at less than 25 parsecs.

CP713, *The Search for Other Worlds: Fourteenth Astrophysics Conference,*
edited by S. S. Holt and D. Deming

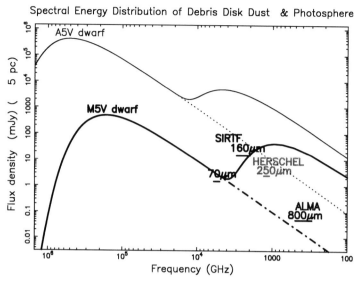

Spectral Energy Distribution of Debris Disk Dust & Photosphere

FIGURE 1. SED of the photospheric emission of an M5 dwarf and of the excess due to a 1 lunar mass dust disk of 14 K. For comparison, we show the SED of an A5 dwarf with a similar but warmer disk of 76 K. Both stars are at 5 pc and have disk radius of 60 AU. SIRTF, HERSCHEL and ALMA 5-σ sensitivities are indicated as horizontal bars over their respective wavelength ranges.

SUBMILLIMETER FLUX DENSITY CALCULATION

The dust particles in a debris disk result from a cascade of collisions between planetesimals and fragments. The resulting dust is made of particles with sizes larger than 10 μm. The equilibrium temperature T_{dust} of such a particle of intermediate size λ_0 and at distance r from a star of luminosity L_* is [4] :

$$T_{dust} = 468 L_*^{1/5} r^{-2/5} \lambda_0^{-1/5} \qquad (1)$$

In our application to the CNS3, the dust temperature T_{dust} is derived in adopting the typical distance $r = 60$ AU and particle size $\lambda_0 = 30\ \mu$m. At submillimeter wavelengths, a debris disk dust is optically thin and so, advantageously, its thermal flux density S_ν is proportional to its dust mass M_d and only slightly dependent on particle sizes. The flux density is :

$$S_\nu = M_d B(\nu, T_{dust}) k_\nu / d^2 \qquad (2)$$

$B(\nu, T_{dust})$ is the Planck function, k_ν is the absorption coefficient per unit mass adopted to be 1.7 cm²/g and d is the star distance in the CNS3 catalog. The photospheric contribution of the star itself at these wavelengths is negligible for the M5 dwarf as seen in Figure 1.

116

TABLE 1. Dust temperature (eq.1) and flux density at 850 μm (eq. 2) for dwarfs M0, M5, M8 at distance $d=5$ pc with $\lambda_0 = 30\mu$m, $r = 60AU$ and $M_d = 1$ lunar mass dust

Spectral type	Luminosity $(L_\odot)$	Dust temp. (K)	S_{350GHz} of dust (mJy)
dM0	$10^{-1.2}$	21	28
dM5	$10^{-2.1}$	14	15
dM8	$10^{-3.1}$	9	6

As it is apparent in Figure 1 for an M5 dwarf, its debris disk would escape detection at IR but be revealed by submillimeter wavelength observations with 1 mJy sensitivity. For comparison, an A5 dwarf is shown and would be detected at IR.

RESULT : DETECTABILITY OF THE 1809 CNS3 DWARFS OF SPECTRAL TYPE M

With the model described above, we have estimated the lowest dust mass M_d detectable at submillimeter wavelengths for the 1809 M dwarfs referenced in the CNS3, in setting the flux density S_v of equation (2) to 5 times the 1-σ noise rms given in Table 2 for the JCMT and ALMA at 850μm and for the Herschel/SPIRE satellite at 250μm. The results are in Figure 2. The collision process in debris disks is expected to produce a clumpy distribution of the dust amenable to high angular resolution observation with ALMA. This clumpiness is already evident in the JCMT image of ε Eri by [2] or in the Plateau de Bure IRAM image of Vega [5].

TABLE 2. Sensitivities of submillimeter facilities

Instrument	type	on-line date	wavelengths (μm)	Integration (min)	Continuum 1-σ noise rms (mJy)
JCMT	15 m	present	850	120	1.5
Herschel/SPIRE	3.5 m space	2007	250, 350, 500	60	0.5
ALMA	64x12 m	2011	850	100	0.010

The 3 histograms of Figure 2 have been plotted with the same scale (10^{-4} to 300 lunar mass) to highlight the improved sensitivity to dust mass from JCMT, presently, to Herschel and ALMA, tomorrow.

CONCLUSION

We have argued that submillimeter observations of M dwarfs might unearth a new population of cold dusty debris disks, signpost of planets around them. We show that for some nearby M dwarfs, circumstellar dust can be probed at the level of 1 lunar mass with the present JCMT and at the level of 0.001 lunar mass for the closest M dwarfs

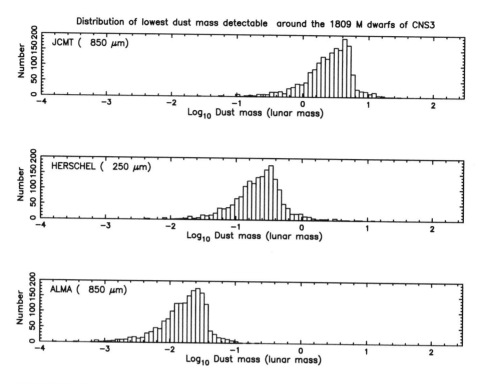

FIGURE 2. Lowest dust masses from debris disks detectable around the 1809 CNS3 M dwarfs by present and future facilities. The flux density S_ν has been set to 5 times the 1-σ rms noise of each telescope to determine this limit.

with the future facility ALMA which is interesting because this is the estimate made for the dust in the Kuiper Belt [6].

Combination of observations from SIRTF/MIPS (60, 170 μm), Herschel (250 μm) and ALMA (850 μm) would provide the SED capable of constraining the geometrical and physical parameters (inner and outer radii, surface density, etc ....) of debris disks around M dwarfs.

REFERENCES

1. Holland, W.S. et al., *Nature*, **392**, 1998, pp. 788–790
2. Greaves, J.S. et al., *Ap.J.*, **506**, 1998, pp. L133–L137
3. Song, I. et al., *A.J.*, **124**, 2002, pp. 514-518
4. Backman, D.E., Paresce, F., *in Protostars and Planets III*, ed. E.H. Levy, J.I. Lumine, Tucson: Univ. Arizona Press, 1993, pp. 1253–1304
5. Wilner, D.J. et al., 2002, *Ap.J.*, **569**, 2002, pp. L115–L119
6. Landgraf, M. et al., 2002, *A.J.*, **123**, 2002, pp. 2857–2861

On the Growth of Dust Particles in a Non-Uniform Solar Nebula

Nader Haghighipour

Department of Terrestrial Magnetism and NASA Astrobiology Institute,
Carnegie Institution of Washington, 5241 Broad Branch Road, Washington, DC 20015

Abstract. A summary of the results of a numerical study of the growth of solid particles in the vicinity of an azimuthally symmetric density enhancement of a protostellar disk are presented. The effects of gas drag and pressure gradients on the rate of growth of dust particles and their settling on the midplane of the nebula are also disucssed.

INTRODUCTION

It has recently been shown that the combined effect of gas drag and pressure gradients causes solid objects to rapidly migrate toward the location of maximum pressure in a gaseous nebula [1, 2]. Such rapid migrations result in the accumulation of objects in the vicinity of local pressure enhancement and may increase the rates of their collisions and coagulations. In a disk with a mixture of gas and small solid particles, the motion of an object is also affected by its interaction with the particulate background. An object may, in such an environment, sweep up smaller particles and grow larger during its gas-drag induced migration. In this paper I present a summary of a study of the growth-rates of dust grains to centimeter-sized objects in a nebula with a local pressure enhancement.

THE MODEL NEBULA

An isothermal and turbulence-free protostellar disk is considered here with a solar-type star at its center. The nebula is assumed to be a mixture of pure molecular hydrogen at hydrostatic equilibrium, and small submicron-sized solid particles. The density of the gas in this nebula is considered to be

$$\rho_g(r,z) = \rho_0 \exp\left\{ \frac{GMm_H}{K_BT} \left[\frac{1}{(r^2+z^2)^{1/2}} - \frac{1}{r} \right] - \beta\left(\frac{r}{r_m}-1\right)^2 \right\}. \tag{1}$$

In equation (1), M is the mass of the central star, G is the gravitational constant, and K_B, T and m_H represent the Boltzmann constant, the gas temperature, and the molecular mass of hydrogen, respectively. The quantities ρ_0, r_m and β in equation (1) are parameters of the system with constant values. The gas density function as given by equation (1) ensures that along the vertical axis, the gravitational attraction of the central star will be balanced by the vertical component of the pressure gradients [2]. It

CP713, *The Search for Other Worlds: Fourteenth Astrophysics Conference,*
edited by S. S. Holt and D. Deming
© 2004 American Institute of Physics 0-7354-0190-X/04/$22.00

also has azimuthally symmetric gas-density-enhanced regions on any plane parallel to the midplane. On the midplane, the density of the gas maximizes at $r = r_m$.

The particles of the background material of a nebula are strongly coupled to the gas and their motions are only affected by the gas drag. For the submicron-sized particles considered here, the rate of the gas drag induced migration is so small that one can assume, at any position in the nebula, the gas and particle distribution functions are proportional. That is, $\rho_{\text{dust}}(r,z) = f\rho_g(r,z)$, where $\rho_{\text{dust}}(r,z)$ is the distribution function of the background material. The solid-gas ratio f is taken to be constant and equal to 0.0034.

EQUATION OF MOTION

For an object larger than the submicron particles of the background material, the above-mentioned gas-particle coupling is less strong. As a result, the object moves faster than the background particles and may collide with them. Such collisions may result in adhesion of the background particles to the moving object and increase its mass.

The rate of change of the momentum of an object due to the sweeping up of the background material is proportional to the rate of the collision of the moving object with those particles. Assuming the sticking coefficient is equal to unity, one can write,

$$\frac{d\mathbf{P}}{dt} = \pi\rho_{\text{dust}}(r,z)\,a^2\,\frac{d\ell}{dt}\,(\mathbf{V} - \mathbf{U}), \tag{2}$$

where a is the radius of the object and $d\ell = |\mathbf{V} - \mathbf{U}|\,dt$. Quantities $\mathbf{V}$ and $\mathbf{U}$ in equation (2) represent the velocities of the object and the particles of the medium. In writing equation (2), it has been assumed that, while sweeping up smaller particles, the density of the object remains unchanged and it stays perfectly spherical. It is important to emphasize that in this equation, the mass of the object, m, and its radius, a, are functions of time, and are related as

$$\frac{dm}{dt} = \pi\rho_{\text{dust}}(r,z)\,a^2\,\frac{d\ell}{dt}. \tag{3}$$

Equation (3) immediately implies

$$\frac{da}{dt} = \frac{1}{4}\left(\frac{\rho_{\text{dust}}}{\rho}\right)\frac{d\ell}{dt}, \tag{4}$$

where ρ is the density of the object.

For small particles such as micron-sized and submicron-sized objects, one can replace $|\mathbf{V} - \mathbf{U}|$ with $\mathbf{V}_{\text{rel}}$, the relative velocity of the object with respect to the gas. The radial, vertical and tangential components of this velocity are, respectively, given by $\dot{r}$, $\dot{z}$, and $r(\dot{\varphi} - \omega_g)$, where the motions of gas molecules along the z-axis have been neglected. Because the gas is at hydrostatic equilibrium, its angular velocity, ω_g, is slightly different from its Keplerian value and is given by

$$\omega_g^2 = \frac{GM}{(r^2 + z^2)^{3/2}} + \frac{1}{r\rho_g(r,z)}\frac{\partial\mathscr{P}_g(r,z)}{\partial r}. \tag{5}$$

In this equation, $\mathcal{P}_g(r,z) = K_B T \rho_g(r,z)/m_H$ is the pressure of the gas.

An object in the model nebula considered here is subject to the gravitational attraction of the central star and the drag force of the gas. The equation of motion of an object in this nebula is, therefore, given by

$$m\ddot{\mathbf{R}} = -\frac{GMm}{(r^2+z^2)^{3/2}}\mathbf{R} - \pi\rho_{\text{dust}}(r,z)a^2\frac{d\ell}{dt}\mathbf{V}_{\text{rel}} - \mathbf{F}_{\text{drag}}, \qquad (6)$$

where $\mathbf{R}(r,z)$ is the position vector of the particle and

$$\mathbf{F}_{\text{drag}} = \frac{2a^2}{3(a+\lambda)}\left(\frac{2\pi K_B T}{m_H}\right)^{1/2}\left[\lambda\rho_g(r,z) + \frac{3m_H}{2\sigma}\right]\mathbf{V}_{\text{rel}} \qquad (7)$$

represents the drag force of the gas. In equation (7), σ is the collisional cross section between two hydrogen molecules and λ is their mean free path.

NUMERICAL RESULTS

The equation of motion of a particle and the growth equation (4) were integrated, numerically, for objects with initial radii ranging from 1 to 100 microns. The mass of the central star was chosen to be one solar mass, $\beta = 1, r_m = 1$ AU, and $\rho_0 = 10^{-9}\text{g cm}^{-3}$. The collisional cross section of hydrogen molecules, σ, was taken to be $2 \times 10^{-15}\text{cm}^{-2}$, and their mean free paths $\lambda\,(\text{cm}) = 4 \times 10^{-9}/\rho_g(r,z)\,(\text{gcm}^{-3})$. At the beginning of integrations, an object was placed at $(r,r/10)$, with a Keplerian circular radial velocity and with no motion along the z-axis. Figure 1 shows the growth of two one micron-sized objects, one migrating radially inward from (2,0.2) AU, and one migrating radially outward from (0.25, 0.025) AU to (1,0) AU, the location of the maximum gas density on the midplane. For the comparison, the radial and vertical motions of these objects without mass-growth have also been plotted. As shown here, by sweeping up the smaller particles of the background, these objects grow to a few centimeters in size and approach the midplane in a time much shorter than the time of similar migration without the mass-growth.

Equations (4) and (6) were also integrated for different values of the object's density and gas temperature. As expected, the rate of growth of the object increased with an increase in the temperature of the gas (Fig. 3). This can be attributed to larger value of the pressure gradients at higher temperatures [1]. Increasing the density of the object while keeping the temperature of the gas constant had, however, an opposite effect. Objects with higher densities tend to grow in size over longer periods of time. The inverse proportionality of the rate of the growth of the object to its density can also be seen from equation (4).

ACKNOWLEDGMENTS

This work is partially supported by the NASA Origins of the Solar System Program under grant NAG5-11569, and also the NASA Astrobiology Institute under Cooperative Agreement NCC2-1056.

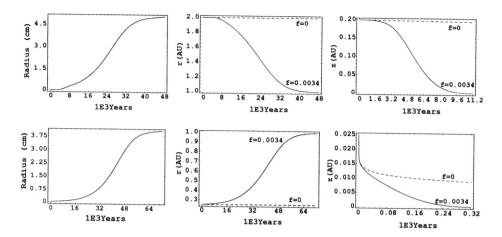

FIGURE 1. Graphs of the radius (left), radial migration (middle), and vertical descent (right) of a 1 micron object with a density of 2 g/cc. The gas temperature is 300 K. The dashed line indicates migration without mass-growth. The middle graphs show an inward radial migration from 2 AU (top), and an outward radial migration from 0.25 AU (bottom) to the location of the maximum gas density on the midplane, (1,0) AU. The initial value of z was taken to be $r/10$.

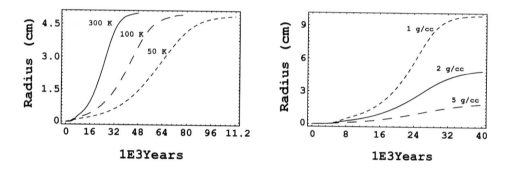

FIGURE 2. The growth of a 1 micron object with a density of 2 g/cc for different values of the gas temperature is shown on the left. The right graph depicts the growth of similar particle with different densities in an isothermal gas with a temperature of 300 K. The particle was initially at (2,0.2) AU.

REFERENCES

1. Haghighipour, H., and Boss, A. P., *Astrophys. J.*, **583**, 996–1003
2. Haghighipour, H., and Boss, A. P., *Astrophys. J.*, **598**, 1301–1311

A Model of the Accretion Disk Around AA Tau

Christophe Pinte* and François Ménard*

*Laboratoire d'Astrophysique de Grenoble, CNRS/UJF UMR 5571, 414 rue de la Piscine, B.P. 53,
F-38041 Grenoble Cedex 9, France

Abstract. The Classical T Tauri Star AA Tau shows quasi cyclic variations of brightness and polarization with a maximum polarization when the system is faintest. Bouvier et al. [1] proposed a model where these variations should be "eclipses" produced by orbiting circumstellar material. The effects of a warp at the inner edge of the accretion disk and of hot spots on the stellar surface on the photometric and polarimetric light curves of AA Tau are studied through multiple scattering Monte-Carlo simulations. Constraints on disk parameters are drawn out, so as to determine whether the warp can be interpreted within the magnetospheric accretion theory. We find that the main features of AA Tau's photopolarimetric variations can be explained by the presence of a warp and that hot spots have a limited influence on them.

INTRODUCTION

Classical T Tauri stars (CTTS) are low-mass pre-main sequence stars surrounded by an accretion disk. Current models suggest that the accretion disk is disrupted at a few stellar radii by the strong stellar magnetic field and that accretion onto the star follows the magnetic field lines, at free-fall velocities. These accretion streams of disk material impact the stellar surface at magnetic poles and produce bright spots.

AA Tau is an especially interesting case to provide some useful insight regarding the inner disk structure. Its particular inclination, almost edge-on, allows to observe quasi-periodic eclipses of the stellar photosphere by circumstellar material. Indeed, AA Tau presents a quite unusual light curve with recurrent deep and wide brightness minima, with a period of about 8.2 days. One the most important characteristics is the lack of color variations as the system brightness varies with a maximum of 0.2 mag for the V-I and B-V indices. In addition to variations on time-scale of years, the eclipses can vary from a period to another and can even almost disappear to reappear the following period, *i.e.* skip an eclipse. The polarization rate increases as the magnitude increases : starting from 0.5% in the bright state, polarization becomes maximum with a level around 2% when the system is faintest. In the following, we present the calculations we have done to fit the light and polarization curves.

CP713, *The Search for Other Worlds: Fourteenth Astrophysics Conference*,
edited by S. S. Holt and D. Deming

MODEL

Origin of photo-polarimetric modulations

Bouvier et al. [1] interpreted the photometric variations as occultations by orbiting circumstellar material, producing a wall which crosses the line of sight. As the wall occults the photosphere, the relative fraction of scattered (and hence polarized) light increases which leads to a stronger polarization at minimum brightness.

Model geometry

The system consists in a star, surrounded by a circumstellar disk and a wall at the inner radius of the disk. The whole system is centrosymmetric relative to the star. The star is modeled by a sphere which radiates uniformly and we consider a flared disk.

We assume the density of the wall to be uniform and we use the same azimuth wall structure as described by Bouvier et al. [1], $h(\phi) \propto \cos(\phi - \phi_0)/2$, this height being radially constant between $r_{in} - \Delta r/2$ and $r_{in} + \Delta r/2$ and zero otherwise.

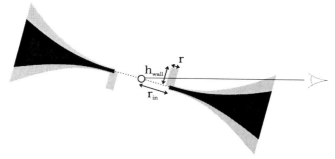

FIGURE 1. Schematic view of the model. The grey parts of the disk represent the volume where the density is of the same order as in the wall and the central black ones show the optically thick zones of the disk.

Results

We focus on the main features of the light curves. The results are presented in Figure 2 : the overall behaviour is reproduced, we find an eclipse with an attenuation of 1.4 mag and the observed width, the polarization increases when the eclipse advances and the color indices V-I and B-V do not show variations larger than 0.2 mag. This is in good agreement with observations and allow us to confirm the interpretation of Bouvier et al. [1].

The polarization vector presents a constant orientation as the system varies : only its length oscillates during the eclipses. This behaviour is represented in Figure 3 by the elongated curve along the Q axis in the Q-U plane. This result confirms the deductions of Ménard et al. [5]: the disk is seen almost edge-on and runs along the east-west direction.

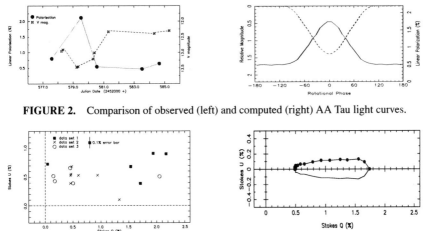

FIGURE 2. Comparison of observed (left) and computed (right) AA Tau light curves.

FIGURE 3. Comparison of observed (left) and computed (right) Stokes parameters. The shift along the U axis can be attributed to interstellar polarization.

Constraints on disk and wall parameters

Wall optical depth. The absence of color variations during the eclipses implies a strong wall opacity. $\tau = 3$ is the smallest optical depth that does not generate reddening at minimum brightness. Polarization rate decreases as the optical depth grows from which we deduce the optical depth should be close to $\tau = 3$.

Maximum size of grains. The grain sizes are distributed according to the power law $n(a) \propto a^{-3.7}$ between a_{min} and a_{max}, which best fits interstellar absorption [4]. The ISM value $a_{max} = 0.9\mu m$ leads to a maximum polarization around 1%. Very small grains ($a_{max} = 0.3\mu m$) increase the polarization but at all rotational phases. The best fits were obtained with a value of $a_{max} = 0.45\mu m$.

System geometry. The light curves can only be reproduced if we see the disk with a grazing incidence. Using the smallest values generally observed for the scale height and mass of other accretion disks we derive a maximum inclination of $i_{max} \approx 76°$. As the inclination decreases, the amplitude of the variations of the U parameter increases, becoming inconsistant with observation at inclination angle lower than $i_{min} \approx 68°$.

NON-STEADY MAGNETOSPHERIC ACCRETION

The observations of Bouvier et al. [2] show that the fadings can sometimes disappear for one rotational period and reappear, almost unmodified on the next cycle. To explain this behaviour as part of the magnetospheric accretion theory, we have to assume that the accretion is non steady and that the wall is not replenished immediately as it empties via magnetospheric funnels. In order for the wall to reappear one period after its destruction, the accretion rate must be sufficiently important. From our new constraints on the wall optical depth, its height and the sizes of grains, we can derive an estimation of the wall mass (gas + dust) of $2 \times 10^{-10} M_\odot$. The necessary accretion rate to generate the wall in one rotational period is then $9 \times 10^{-9} M_\odot.\mathrm{yr}^{-1}$ in agreement with the average accretion rate measured for TT Tauri Stars [3].

CONCLUSION

The AA Tau's light curves are well reproduced by our multiple scattering model. These results allow us to confirm that extinction by circumstellar dust is responsible for quasi-periodic brightness fadings and polarization increases and give us new constraints on the magnetospheric structure in the very first tenths of AU from the central star. AA Tau is a typical CTTS and its quite unusual light curves are the result of its particular inclination. The constraints derived on the structure of the accretion zone should apply to other CTTS and we can expect to detect the presence of a "wall" in other systems, and not only in systems seen close to edge-on.

REFERENCES

1. Bouvier, J. et al. 1999, A&A, 349, 619
2. Bouvier, J. et al. 2003, A&A, 409, 169
3. Hartmann, L., Calvet, N., Gullbring, E., & D'Alessio, P. 1998, ApJ, 495, 385
4. Mathis, J. S. & Whiffen, G. 1989, ApJ, 341, 808
5. Ménard, F., Bouvier, J., Dougados, C., Mel'nikov, S. Y., & Grankin, K. N. 2003, A&A, 409, 163

Radiative Transfer in the Vicinity of a Protoplanet

Hannah Jang-Condell* and Dimitar D. Sasselov*

*Harvard-Smithsonian Center for Astrophysics

Abstract. We calculate the effect of stellar irradiation on the temperature structure of the photosphere of a circumstellar disk in the vicinity of a protoplanet. We restrict our study to small protoplanets that perturb the disk locally but are too small to open a gap. These protoplanets induce a compression of the disk material near it, resulting in a decrement in the density at the disk's surface. Thus, an isodensity contour at the height of the photosphere takes on the shape of a well. When such a well is illuminated by stellar irradiation at grazing incidence, it results in cooling in a shadowed region and heating in an exposed region. Using a method we have developed for calculating three-dimensional radiative transfer on a perturbed surface, we examine the variation of this effect with protoplanetary mass and distance. We conclude that even relatively small protoplanets can induce significant temperature variations in a passive disk, up to $\pm 30\%$ for planets at the gap-opening threshold. Therefore, many of the processes involved in planet formation should not be modeled with a locally isothermal equation of state. Although the temperature perturbation is unlikely to be observable, it may affect accretion of disk material onto the planet or the rate of Type I migration.

1. DISK STRUCTURE

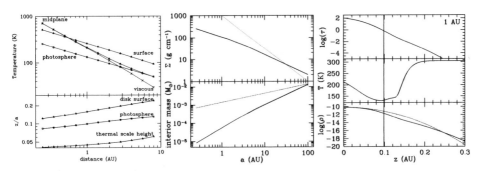

FIGURE 1. The unperturbed disk structure. Left: the temperature at different heights in the disk. Center: the density and mass profiles of the disk, indicated by solid lines. The dotted line shows for comparison a MMSN of 0.01 $M_\odot$. Right: the optical depth, temperature, and density in the z direction at 1 AU.

To calculate the structure of the unperturbed disk, we adopt a method for self-consistently calculating the density and temperature of the disk including heating due to stellar irradiation and viscous heating [1, 2]. The basic equations are as follows:

- Hydrostatic equilibrium: $\nabla P = -\rho \nabla \Phi$
- Temperature: $T(\tau, \mu_0) = [T_\nu^4(\tau) + T_r^4(\tau, \mu_0)]^{1/4}$ where μ_0 is the cosine of the angle of incidence of incoming radiation [1]

CP713, *The Search for Other Worlds: Fourteenth Astrophysics Conference*,
edited by S. S. Holt and D. Deming
© 2004 American Institute of Physics 0-7354-0190-X/04/$22.00

- Optical depth: $\tau(z) = \int_{z_\infty}^z \chi_R \rho(z')dz'$
- Match total integrated surface density with surface density given by central temperature of a viscous disk, $\Sigma = \dot{M}\Omega\mu m_H/(3\pi\alpha k T_c)[1 - (R_\star/a)^{1/2}]$

Given an initial value of μ_0, we solve for $\tau(z), \rho(z), T(z)$ at varying radii and recalculate μ_0. We iterate until μ_0 converges. Figure 1 shows the disk structure given a central star of mass $0.5\,M_\odot$, temperature 4000 K, and accretion rate $10^{-8}\,M_\odot\mathrm{yr}^{-1}$ with $\alpha = 0.01$. The surface is defined to be where the optical depth to incident radiation at stellar frequencies is 2/3, the photosphere is where the optical depth of the disk to its own radiation is 2/3, and the thermal scale height is $h = (c_s/v_\phi)a$ where c_s is the sound speed at the midplane. The dust opacities are taken from D'Alessio et al. [3].

1.1. Perturbed Disk

A protoplanet acts gravitationally on the disk, so that $\Phi = GM_\star/a + Gm_p/r_p$ in the equation of hydrostatic equilibrium. Assuming that the temperature structure is relatively unaffected, $c_s^2\nabla\ln(\rho'/\rho) = -\nabla(Gm_p/r_p)$, where ρ and ρ' are the unperturbed and perturbed density structure, respectively.

1.2. Radiative Transfer

To calculate radiative transfer on a 3-D surface, we sum the radiative flux over the disk surface [4]. The total radiative flux is $F_{irr} = F_r + F_v$ where F_v is the viscous flux, and $F_r = \pi B_{tot} = \Sigma B(\tau, \mu)v\,\delta\Omega$, where $\delta\Omega$ is the solid angle subtended by an area element, and $\cos^{-1} v$ is the angle to the area element. Figure 2 (left) shows the distribution of F_{irr} in the disk's photosphere for an $11\,M_\oplus$ planet at 1 AU. The black circle indicates the Hill radius, and the star is located off to the left.

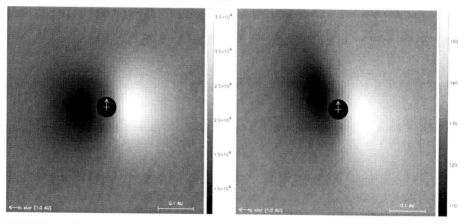

FIGURE 2. Illumination pattern (left) and temperature (right) in the disk photosphere for an $11\,M_\oplus$ planet at 1 AU.

1.3. Differential Rotation

The disk material moves at approximately Keplerian velocities, so that material near the planet does not stay fixed, but moves through the illumination pattern. In order to calculate the temperature structure, we need to take this motion into account along with the heating and cooling rate. We adopt a heating/cooling rate of $C\frac{\partial T}{\partial t} = F_{irr} - \sigma T^4$ where $C = k\Sigma/\mu m_H$ is the specific heat. Figure 2 (right) shows the resulting temperature structure.

2. PARAMETER SPACE

We calculated the effect on the temperature structure of the disk for planets of varying masses and orbital distances for a star of mass $0.5 M_\odot$ and temperature 4000 K accreting at $10^{-8} M_\odot \text{yr}^{-1}$ and $\alpha = 0.01$. We restrict our study to planets which are not massive enough to open a gap in the disk, i.e. planets whose Hill radii (r_H) are less than the thermal scale height of the disk, $h = (c_s/v_\phi)a$. We consider distances of 0.5, 1, 2, and 4 AU, and planet masses such that $\log_2(r_H/h) = 0, -1/3, -2/3, -1, -4/3, -5/3$, and -2. The corresponding planet masses are tabulated in Table 1.

TABLE 1. Planet masses, in $M_\oplus$

r (AU)	r_H/h						
	0.25	0.31	0.40	0.50	0.63	0.79	1.00
0.5	0.596	1.19	2.38	4.77	9.53	19.1	38.1
1	0.689	1.38	2.76	5.51	11.0	22.0	44.1
2	0.838	1.68	3.35	6.70	13.4	26.8	53.6
4	1.13	2.26	4.51	9.03	18.1	36.1	72.2

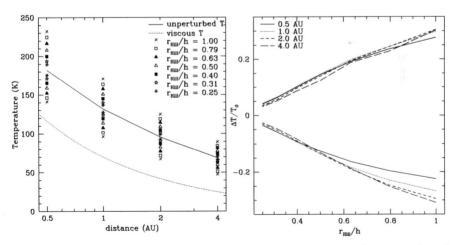

FIGURE 3. Left: Summary of maximum and minimum temperatures at $\tau = 2/3$ near the planet for varying parameters. Right: Fractional temperature variation vs. r_H/h (max and min temperatures).

3. RESULTS

We have found that planets below the gap-opening mass threshold may significantly change the temperature of the disk within a few Hill radii of the planet. Planets can induce temperature variations of several tens of degrees in the photosphere of the disk, a variation which can be as much as $\pm 30\%$, as indicated in Figure 3.

Plotting the fractional temperature variation versus r_H/h, as in Figure 3 (right), the lines lie almost coincident. The divergence of the minimum temperature toward higher r_H/h can be explained by the rise in viscous heating as the distance to the star becomes smaller. Since only stellar irradiation heating is changed by the presence of a planet in our model, viscous heating sets a lower bound to the temperature. The viscous temperature goes as $r^{-3/4}$ while the stellar radiation temperature goes as $r^{-1/2}$, so that close to the star, viscous heating dominates.

4. IMPLICATIONS

While this effect is likely too small to be observable, it can have a number of implications for planet formation. Simulations of planets accreting within disks done to date virtually all employ a locally isothermal equation of state [e.g. 5, 6, 7]. Large temperature pertubations may affect results such as the flow of gas streamlines, the rate of accretion onto the planet, and migration rates.

The perturbed temperature structure may possibly halt or even reverse Type I migration. The balance of inward versus outward resonant torques depends on the temperature gradient and results in net inward migration if the disk temperatures decrease with radius [8]. The effect we have modeled effectively locally reverses the temperature gradient, which may change the net balance of torques to change Type I migration.

Cooling in the shadowed region of the perturbation may change the properties of dust and ice in the disk. The freezing out of volatiles in this region could facilitate planet growth, effectively moving the "snow line" inward. Conversely, the heated region would inhibit the growth of solid particles, resulting in an overall effect of more accretion from the inner disk than the outer disk. We shall explore some of these issues in depth in an upcoming paper [9].

REFERENCES

1. Calvet, N., Patino, A., Magris, G. C., & D'Alessio, P. 1991, ApJ, 380, 617
2. D'Alessio, P., Calvet, N., Hartmann, L., Lizano, S., & Cantó, J. 1999, ApJ, 527, 893
3. D'Alessio, P., Calvet, N., & Hartmann, L. 2001, ApJ, 553, 321
4. Jang-Condell, H. & Sasselov, D. 2003, ApJ, 593, 1116
5. Bate, M. R., Lubow, S. H., Ogilvie, G. I., & Miller, K. A. 2003, MNRAS, 341, 213
6. Kley, W., D'Angelo, G., & Henning, T. 2001, ApJ, 547, 457
7. Bryden, G., Chen, X., Lin, D. N. C., Nelson, R. P., & Papaloizou, J. C. B. 1999, ApJ, 514, 344
8. Ward, W. R. 1997, Icarus, 126, 261
9. Jang-Condell, H. & Sasselov, D., in preparation

Giant Planets

Formation of Giant Planets

David J. Stevenson

Caltech, 150-21, Pasadena CA 91125

Abstract. This is a descriptive and non-mathematical summary of giant planet formation theories. There are two end-member models, core accretion and disk instability. In the core accretion model, several to ten Earth masses of solid (ice and rock) accumulate and this promotes subsequent gas accretion. In the disk instability model, a self- gravitating sphere of gas forms (somewhat analogous to star formation) and a core may arise through condensation and rainout from its low density envelope. The core accretion model may have a time-scale difficulty because the formation of the embryo and infall of gas is not fast relative to the time at which the gaseous disk is removed. The disk instability model suffers from the current theoretical inability to follow the development of these putative instabilities through to the formation of a planet. Observational data on core masses in Jupiter and Saturn do not clearly favor one model. However, the existence and nature of Uranus and Neptune strongly suggest that the formation of appropriate embryos occurs. Moreover, there is considerable flexibility in the elapsed time required to form Jupiter and Saturn by the core accretion model, including times of only a few million years, compatible with disk lifetimes. This suggests that core accretion remains the favored mode in our solar system. It is possible that both mechanisms operate in general and that many extrasolar planetary systems make use of the disk instability mode. Future theory and observations are essential for deciding this issue.

INTRODUCTION

Giant planets are often defined as those bodies that are mostly hydrogen but are distinguished by stars through their inability to undergo fusion. For the purposes of this brief review, I will also include discussion of bodies such as Uranus and Neptune (sometimes called ice giants) because they may provide crucial insight into the formation process. The central question is obvious: How did these bodies form? There are two end-member models that people consider, those where the process has a lot in common with star formation and involves the instability of gas masses, and those where the accumulation of solids precedes the addition of gas. I refer to these as the disk instability and core accretion models, respectively. The disk instability mechanism might be thought of as closely related to the Jeans collapse process for forming low mass sub-stellar companions.

This brief review is descriptive and non-mathematical. Recent more thoroughly documented reviews cover dynamical aspects [1], extrasolar aspects [2] and cosmochemical aspects [3] very well, but do not offer the descriptive comparison of models that I provide here. We are not yet at the point where we can directly observe giant planet formation in other systems. The central question of giant planet formation must therefore be tackled through consideration of the outcome: How is the formation mechanism expressed in the structure and chronology of the observed system? In our solar system, we would seek the answer to this question by looking at the structure of the planets, their dynamical relationships (orbits, etc.), and their satellite systems. But we must also look at the time

CP713, *The Search for Other Worlds: Fourteenth Astrophysics Conference,*
edited by S. S. Holt and D. Deming
© 2004 American Institute of Physics 0-7354-0190-X/04/$22.00

scales of processes, inferred by direct observation of materials (isotopic information) or analog systems (star formation regions). This turns out to be a central but unresolved issue for the success of models. Of course, we should be guided by evidence of other planetary systems, even though the time scale constraints may be less secure in those cases. Is the formation mechanism universal or could it vary from system to system? In this review, I will also discuss what we need to make further progress.

THE STANDARD PICTURE

Although many details are poorly understood, a standard picture has emerged for the star and planet formation process, supported by observation and by theory [4,5,6]. In this picture, here discussed for stars of order a solar mass, collapse of an interstellar cloud of gas and dust proceeds on a short time scale (a few hundred thousand years), leading to a disk that is required by accommodation of the angular momentum budget of the infalling material. A process of angular momentum redistribution within the disk feeds a central protostellar mass. The nature of this process is not well understood but can be thought of as a "viscosity". It causes redistribution of angular momentum (and therefore mass) within the disk on a time scale of one to ten million years, comparable to the total lifetime of the disk. The disk is plausibly of order 100 AU in extent, and typically with a thickness (or scale height) that is about 0.1 of the local radius. The disk mass may be of order 0.1 solar mass, equivalent to a hundred Jupiters. The surface density of the disk declines with radius, but slowly enough that the mass tends to be mostly in the outer parts of the disk. If this disk is dense enough and cold enough then instabilities may occur within it, as it accumulates, especially while the central mass is still low. These are essentially Jeans instabilities but the Keplerian shear as well as the usual competition of pressure and gravity mediate them. If the disk is not very near the stability threshold, then the characteristic time-scale for the development of instabilities can be of order the orbital time, 10 to 1000 yrs.

Independent of this, small (kilometer-sized) solid bodies aggregate quickly, on time scales of a million years or less [7,8]. It is thought that larger bodies may also form quickly, by so-called runaway accretion. This process relies on the small relative velocity of planetesimals and the consequent high gravitational collision cross-section of the largest embryos in the swarm. This cross-section can be as much as a thousand times the geometric cross-section, so that time scales of accumulation can be reduced by a corresponding factor. This lowers the formation time of the largest solid embryos from about a billion years to about a million years or even less at Jupiter orbit, assuming plausible surface densities of solids. The outcome of runaway accretion is to form a set of planetary embryos that have accumulated most of the mass within an orbital radial distance of a few Hill sphere radii. This corresponds to Mars-sized bodies in the terrestrial planet region and Earth mass to ten Earth mass bodies in the giant planet region. In the core accretion scenario of giant planet formation, these embryos are the seeds for accumulating large masses of gas. Massive embryos can also undergo rapid orbital migration in a hundred thousand years or even less by exchanging angular momentum with the much more massive disk. This can cause a rapid inward spiral of the planet and is a natural explanation for the small orbital radii discovered for many extrasolar planets. Apparently, this process was largely avoided in our solar system. At some point, estimated to be at a few million to ten million years after initial cloud collapse, the gaseous disk is removed through the combination of inflow onto the central star and outflow back into the interstellar medium. This event must postdate the formation of Jupiter and Saturn.

GIANT PLANET COMPOSITION

Detailed reviews of giant planet structure include [9,10,11]. The material available to build planets can be divided into three classes: "Gases", "ices" and "rocks". Gas refers to primarily hydrogen and helium, the most abundant elements in the universe. They also happen to be materials that do not condense as solids or liquids under conditions that are encountered during planet formation. The material deep within a planet such as Jupiter can be called a liquid metal but it has a much higher entropy than the value at the conventional critical point of molecular hydrogen where liquid and gas merge. Ice refers to the compounds that form from the next most abundant elements: oxygen, carbon, and nitrogen. These compounds are primarily hydrides (water, methane and ammonia) but also include carbon monoxide, carbon dioxide, and molecular nitrogen among others. The ice label does not mean that these constituents are necessarily found as solid materials, but that is indeed the form in which they would provide the building blocks for planets. However, condensation requires low temperatures, corresponding to the orbit of Jupiter and beyond in our solar system. Ice is roughly two orders of magnitude less abundant than gas, by mass. "Rock" is even less abundant and refers to everything else: primarily magnesium, silicon and iron, and the oxygen that would naturally combine with these elements (far less than the total oxygen because of its much higher abundance). These elementary considerations enable us to understand why it is improbable that we would ever encounter a Jupiter mass planet that is made mostly of iron, say. There would typically be insufficient iron around during the formation of a planetary system to imagine doing this. Our inclination to interpret Jupiter mass extrasolar planets as gas balls (even when there is no information on the size or composition) is not mere solar system prejudice but guided by cosmochemical principles. While it is tempting to argue that planets forming close to stars might tend to have a composition dominated by materials that condense at high temperature, we probably cannot exclude formation of gas balls at small radii. We certainly cannot exclude orbital migration as an important process in the structure of many planetary systems, probably including most of those extrasolar systems thus far discovered, even if it had only a modest role to lay in our own solar system.

There is no doubt that Jupiter and Saturn are mostly hydrogen and helium. No other material of sufficiently low density can explain the global properties of these bodies. However, Jupiter is much closer to cosmic (or primordial solar) composition than Saturn. Roughly speaking, Jupiter and Saturn have similar total amounts of heavy elements (all elements heavier than hydrogen or helium), but Jupiter is over three times the mass of Saturn. We have constraints on the interior structure of these bodies that arise primarily from the observed gravity field (including the gravitational moments caused by planetary rotation) but also from deep atmosphere composition, magnetic field, heat flow and laboratory and theoretical equations of state. Jupiter may have a dense core; Saturn almost certainly has a dense core. It may seem surprising that we can be more precise about Saturn than about Jupiter, since better data exist for Jupiter. However, the putative core of Jupiter is a tiny fraction of the total mass, perhaps about a percent (i.e., three Earth masses), and it accordingly has a very small effect on planetary structure. The common practice of placing a separate core of heavy elements at the centers of these planets is governed by simplicity, rather than by observation. To varying degrees, the "core" could have a fuzzy boundary with the overlying hydrogen-rich envelope. Uncertainties in the hydrogen equation of state continue to be a major source of uncertainty in the interior models.

Both of these planets are enriched in heavy elements throughout (and separate from the presence or absence of a core). In Jupiter, this enrichment is probably about a factor of three relative to cosmic abundance and is readily observed in those volatile components that do not condense out in the observable atmosphere. We cannot be sure that this is the enrichment for water, which condenses out deep and was not observed to be enriched in the presumably dry region that Galileo probe sampled. Water is the most abundant carrier of oxygen and therefore presumably the most abundant heavy material in Jupiter. However, interior models support about ten Earth masses of heavy elements mixed throughout the hydrogen, consistent with this factor of three. It is particularly interesting that this factor of three is even seen in the heavy noble gases, including argon. The threefold enrichment of argon suggests delivery to Jupiter of material that condensed at very low temperatures, probably around 40K, since there is no known way of incorporating argon into solid bodies in large amounts at higher temperature. These planets also supported in situ formation of a satellite system. The Galilean satellite system is particularly impressive and may contain important clues to the last stages of giant planet formation [12]. Ganymede and Callisto are roughly half water ice, and Callisto has most of this ice mixed with rock. It follows that conditions must be appropriate for the condensation of water ice at the location where Ganymede formed, and conditions at Callisto must have allowed formation of that body on a time scale exceeding about 0.1 million years, so that water ice would not melt and lead to a fully differentiated structure.

Uranus and Neptune are far less well understood than Jupiter or Saturn. However, there is no doubt that they are mostly ice and rock, yet also possess two or so Earth masses of gas each. The atmospheres have solar hydrogen to helium ratios (though with large uncertainty because this determination is based on the pressure induced absorption features of hydrogen, a method that has been unreliable for Jupiter and Saturn). The amount of hydrogen extractable from the ices is in principle about 0.2 of the total mass (assuming the hydrogen was delivered as water, methane and ammonia) and this is marginally close to the hydrogen mass required by interior models. Moreover, there is the possibility that methane would decompose into carbon and hydrogen at extreme pressures. However, the atmospheres of Uranus and Neptune are highly enriched in methane (thus limiting massive decomposition of this compound to very deep regions, if any, and there is no experimental or theoretical evidence for extensive decomposition of water or ammonia under the conditions encountered inside these bodies. Consequently, it is not plausible to derive even one Earth mass of predominantly hydrogen gas from the breakdown of hydrogen-bearing ice or rock, even leaving aside the dubious proposition that such decomposed hydrogen would rise to the outer regions of the planet. This gas appears to have come from the solar nebula. Uranus and Neptune must have formed largely in the presence of the solar nebula, a very stringent constraint on the formation of solid bodies.

It is often supposed that the presence or absence of a core in Jupiter (say) can be placed in one-to-one correspondence with the presence or absence of a nucleating body that caused the inflow of gas to form the much more massive envelope. However, there is no neat correspondence between mode of giant planet formation and current presence of a core. One could imagine a core instability model even if there is no core remaining, because the core might become mixed into the overlying envelope by convective processes [11]. One could also imagine making a core in the low-density protoplanet phase by rainout. Making a core by rainout once the material is dense and degenerate is far less likely because the high temperatures and dilution make it thermodynamically implausible. The one exception is helium, which can rain out because of its relatively high abundance and extremely tightly bound electronic states relative to the metallic state of the hydrogen. Even helium is only modestly raining out in Jupiter. It seems likely that whatever model one favors for giant planet formation, it should allow for the formation of a core, since Saturn probably has a core and one must in any event explain Uranus and Neptune. It would be special pleading to attribute a different origin for Jupiter than for the other giant planets. The merits of the two end-member models for giant planet formations should therefore be assessed primarily by their dynamical and time scale predictions, though with attention paid to the nature of the cores that they predict.

CORE INSTABILITY MODEL

The central feature of this model is the existence of a core (ice and rock) mass that does not allow for a hydrostatic equilibrium solar composition envelope that merges smoothly with any plausible solar nebula. The failure of hydrostatic equilibrium can then allow for solutions in which there is a steady inflow of gas. In the simplest picture, there is accordingly a critical mass for the solid embryo above which one may form a giant planet. The giant planets are attributed to solid embryos forming in the outer solar system and reaching this critical mass, but failing do so in the inner solar system. Cameron advanced the idea, but the classic work on this concept was by Mizuno [13]. Popularity for the idea stemmed in large part from the "coincidental" similarity of the critical mass predicted by theory (about ten earth masses) and the range of estimates of actual core masses in Jupiter and Saturn. These core masses were in turn related to the development of planets beyond the "snow line" (the innermost radius in the nebula where water ice condenses), where a higher surface density of solids is likely to exist. Subsequent work has somewhat reduced the attractiveness of the core instability idea for several reasons. First, the actual core masses are in doubt and may not match the requirements for the model. Second, the required critical core mass depends on many parameters and is therefore quite model dependent. We now appreciate that this concept of a critical mass is fuzzy, because its value depends on the dynamical conditions through a dependence on mass inflow rate (i.e., luminosity). Third and perhaps most important, the total elapsed time for the process to run to completion may be many million years and therefore possibly too long relative to the lifetime of the gas disk (cf. [1]).

The most detailed core instability model is Pollack et al. [14]. This model incorporates a particular scenario for the growth of solid embryos and then follows the inflow of gas and solids onto that embryo (using the simplification of spherical symmetry). The model has three stages. Phase I is the formation of the many earth mass embryos by runaway accretion from a planetesimal swarm. Phase II is the longest and corresponds to the slow growth of a gas and solid component, mediated by the cooling time of the growing protoplanet. Phase III begins when the solid and gas masses are comparable and is a rapid inflow of gas, leading quickly to the final mass of the planet. The model does not explicitly consider the mechanism responsible for the truncation of the gas inflow, which could arise by the formation of a tidally truncated gap in the disk. Phase I is modeled in much the same way that many groups now model the formation of the terrestrial planets, apparently successfully. In this game, success for the terrestrial planets can be measured quite well because we have quantitative clocks (isotopic systems) that tell us about that chronology [7, 8]. It is quite likely that the elapsed time for this process is quite accurately estimated at a million years or less for the embryo that led to the formation of Jupiter. In order to get an embryo of order ten Earth masses it is necessary to assume a disk surface density that exceeds by a factor of three or so the "minimum mass solar nebula" (essentially the value needed to explain the total heavy elements in the giant planets). Equivalently, any such model will always make more embryos than are needed to explain the giant planets. One has to assume that the excess embryos are somehow eliminated. At the end of this phase, the embryo has a gas mass of order an Earth mass or so, roughly comparable to the final state of Uranus or Neptune (and thus potentially offering an explanation for bodies of this kind). Phase II takes the longest time and therefore merits the most scrutiny. In this phase, the zone of solid embryo growth is largely exhausted of solid material and there is a gentle arrival of additional solids and gas. Here, one must question both the assumptions that dictate the solid mass inflow as well as the parameters that dictate the inflow of gas. The former depends on the implicit assumptions that there is complete isolation of neighboring zones of embryo growth, probably an artificiality of the model used. The latter depends on the cooling ability of the proto-giant planet and is accordingly sensitive to the opacity of the envelope. The models used assume opacity not much less than that of the interstellar medium. It is possible that the correct opacity is an order of magnitude or more lower, because much of the solid material has already aggregated into large bodies. Of course, there are collisions and condensation that can create additional grain opacity, but there is no reason to expect an interstellar value. The "standard" Pollack model takes 6 or 8 million years for this phase but one could envisage lowering this by a factor of two or three. This might eliminate the time scale problem that mainly motivates the alternative disk instability model. Phase III is rapid and is a positive feedback because the zone of accumulation of gas grows while the protoplanet contracts under gravity. Although this stage may have important consequences for the final properties of the planet, it does not pose a time scale problem. In summary, the core accretion model is perceived to have problems primarily because it might take too long and might require an excessive initial

138

embryo mass. However, these problems are soft in the sense that they are somewhat specific to a particular formulation of the model and not generic to all models in this category. There is a large amount of recent and promising work in this area, most of which is unpublished but has been reported at conferences.

DISK INSTABILITY MODEL

Although there are many papers on disk instabilities, especially by Boss (see references in [1] and [15]), there is not as complete an understanding of how this process would lead to observable giant planets as that offered by the core instability model. The fundamental problem here is that the quite sophisticated 2D or 3D disk evolution models can only follow the development of the system for typically times of tens to hundreds of orbits, far less than the actual time of formation of planets. Although the models may include many relevant processes, including shocks and radiation, they may lack the resolution or input physics needed to include all the processes that might provide for disk evolution. Of course, it could be argued that since disk instabilities are "fast", it does not matter. In this strict and narrow sense, published models may be self-consistent. It does not follow that they are necessarily relevant. The approach to disk instability may be very slow (essentially an accretion time scale) and so the development of an instability may also be slow or even non-existent, if it manifests itself as wave disturbances that redistribute angular momentum and mass. The slow "viscous" processes may actually be competitive with the gravitational instabilities. Even if self-gravitating clumps form, they may not survive over hundreds to thousands of orbits in the Keplerian shear field. This is a very short time-scale compared to others of relevance to the problem of disk evolution and planet formation.

Recently, Boss [15] has extended the disk instability model to a scenario for forming planets such as Uranus and Neptune. He suggests that it might be possible to explain these planets as the remnant of initially Jupiter-mass planets that are stripped of their gaseous envelope by photoevaporation, caused by nearby OB stars. A problem with this model is that it requires a somewhat fortuitous outcome: The observed properties of Uranus and Neptune are very much what you would expect when you form a ten Earth mass solid in the solar nebula. The additional two or so Earth masses of hydrogen is not an accident- it is what the standard theory predicts for that stage [12, 13]. It is unlikely that one could produce an outcome that is fortuitously correct by stripping.

WHAT IS THE CORRECT PICTURE?

If disk instabilities are capable of forming giant planets, it seems likely that they have prevailed in at least some systems. The history of astronomy has taught us that if something can happen, it probably will, at least part of the time. They may even be the most common formation mechanism for giant planets and responsible for the numerous examples of giant planets in the current extrasolar planet catalog. However, the system that we can study most closely is our own and in that singular case, the evidence seems at present to point towards core instability. Despite time scale concerns, there is the likelihood that there exists a reasonable set of parameters and conditions in which the mechanism can work. There is an approximate correspondence between the mass that is required for the embryo and the allowable core masses that planetary models predict. Very importantly, the mere existence of Uranus and Neptune point strongly towards the formation of appropriate embryos, notwithstanding the provocative model of Boss [15]. Uranus and Neptune appear to be excellent candidate embryos for gas giants, and failed to become gas giants only because the nebula was largely removed at the time they approached their final mass. Of course, there is not yet a good model for the formation of Uranus and Neptune! While it is true that the formation of Uranus and Neptune remains poorly understood, we cannot deny the empirical evidence. They exist and cannot be easily explained any way other than by accumulation from smaller bodies. There is sometimes the tendency in science to seek a single explanation for a phenomenon. In the case of giant planets, this would seem to be a false goal because giant planets can come in many forms. We know of two mechanisms that could work. The physical principles are not seriously in doubt for either model, but the precise set of circumstances needed are not well established. Probably, both mechanisms exist, and the phase space of parameters that allow them to work are quite large, perhaps overlapping. The challenge is to decide which mechanism prevails for each system we encounter, and how this is related to parameters such as metallicity, angular momentum budget, proximity of other (possibly high energy) events such as supernovae, mass of central star, and so on.

THE FUTURE

Despite the lack of a close connection between formation mechanism and presence of a core, there is a need to understand better the nature of the giant planet interiors. It is likely that this will test the relative merits of the formation models. For this, we need more experimental and theoretical work on the hydrogen equation of state. It is humbling to admit that despite decades of work, this simplest of systems remains poorly understood. We also need a Jupiter polar orbiter that collects data on gravity and magnetic field close in to the planet. Future missions should also establish the all important water abundance in the deep atmosphere, either through remote sensing techniques (microwave sounding) or through dropping one or more probes into the atmosphere. Eventually, both techniques are needed. The giant planet formation story for our solar system is surely connected to our existence and to the existence of terrestrial planets. Despite the quite successful application of computer simulations to terrestrial planet accumulation, it is essential to fully incorporate the presence of these giant planets into these simulations [16]. It is likely that a fully consistent picture of the origin of Earth depends on the formation mechanism for Jupiter. The explosion of data on extrasolar planets can also be used to test giant planet formation models. If we can establish the location for their formation and orbital migration, then we can probably use this as a diagnostic for the relative merits of the two models. Observations of young systems might provide direct evidence of the formation mechanism, especially for the disk instability model where there is some possibility of directly observing the large non- axisymmetric disturbances that these models predict. It should be stressed however that early observational evidence is unlikely to tell us whether we are observing a giant planet in the making. The challenge of actually observing newly forming giant planets is both achievable and highly exciting. The most important constraint may prove to be Uranus and Neptune, and their analogs in other systems (albeit not yet discovered.) If planets of this kind prove to be common then this will point strongly to a core instability model. The disk instability mechanism may coexist, of course, but we will have at least established empirical evidence that large solid bodies can accumulate on appropriate time scales.

REFERENCES

1. A. P. Boss, Formation of gas and ice giant planets, *Earth Planet Sci Lett.*, **202** 513-523 (2002).
2. P. Bodenheimer and D. N. C. Lin, Implications of extrasolar planets for understanding planet formation, *Ann Rev earth Planet Sci.*, **30** 113-148 (2002).
3. J. I. Lunine, A. Coradini, D. Gautier *et al.*, to appear in *Jupiter*, ed F. Bagenal *et al.*, Cambridge Univ. Press (2004).
4. F. H. Shu, F. C. Adams and S. Lizano, Star formation in molecular clouds - observation and theory, *Ann Rev Astron Astrophys*, **25**, 23-81 (1987).
5. R. B. Larson, The physics of star formation, *Reports on Progress in Physics*, **66**, 1651-1697 (2003).
6. J. J. Lissauer, Planet Formation, *Ann Rev Astron Astrophys*, **31**, 129-172 (1993).
7. G. J. Wasserburg, Isotopic abundances: inferences on solar system and planetary evolution, *Earth Planet. Sci. Lett.*, **86**, 129-173 (1987).
8. A. N. Halliday and D. Porcelli, In search of lost planets - the paleocosmochemistry of the inner solar system, *Earth Planet Sci Lett.*, **192**, 545-559 (2001).
9. T. Guillot, Interiors of Giant Planets Inside and Outside our Solar system, *Science*, **286**, 72-77 (1999).
10. W. B. Hubbard, A. Burrows, & J. I. Lunine, Theory of Giant Planets, *Ann Rev Astron Astrophys*, **40**, 103-136 (2002).
11. T. Guillot, D. Stevenson, W. Hubbard, and D. Saumon, The Interior of Jupiter, to appear in *Jupiter*, ed F. Bagenal *et al.*, Cambridge Univ. Press (2004).
12. D. J. Stevenson, Jupiter and its Moons, *Science*,**294**, 71-72 (2001).
13. H. Mizuno, Formation of the Giant Planets, *Prog. Theor. Phys.*, **64**, 544-557 (1980).
14. J. B. Pollack, O. Hubickyj, & P. Bodenheimer *et al.*, Formation of the Giant planets by Concurrent Accretion of Solids and Gas, *Icarus*, **124**, 62-85 (1996).
15. A. Boss, Rapid Formation of Giant Planets by Disk Instabilty, *Astrophys. J.*, **599**, 577-581 (2003).
16. H. F. Levison and C. Agnor, The role of giant planets in terrestrial planet formation, *Astron J*, **125**, 2692-2713 (2003).

Thoughts on the Theory of Irradiated Giant Planets

Adam Burrows[*], David Sudarsky[*] and Ivan Hubeny[†**]

[*]Department of Astronomy, University of Arizona, Tucson, AZ 85721
[†]NOAO, Tucson, AZ 85726
[**]NASA Goddard Space Flight Center, Greenbelt, MD 20771

Abstract. We have derived physical diagnostics that can inform the direct detection and remote sensing programs of extrasolar giant planets (EGPs) now being planned or proposed. Stellar irradiation of the planet's atmosphere and the effects of water and ammonia clouds are incorporated in a consistent fashion. Whether an EGP is at wide or close-in separations from its parent star, direct detection will soon be possible and will yield centrally important physical and chemical constraints. Our theory of irradiated EGPs is being developed to meet this challenge.

EXTRASOLAR GIANT PLANETS: A NEW SUBJECT IS BORN

To date, more than 110 EGPs (Extrasolar Giant Planets) have been discovered by the radial-velocity technique around stars with spectral types from M4 to F7[1]. These planets have minimum masses ($m_p \sin(i)$, where i is the orbital inclination) between ~ 0.12 M_J and ~ 15 M_J ($M_J \equiv$ one Jupiter mass), orbital semi-major axes from ~ 0.0225 AU to ~ 5.9 AU, and eccentricities from ~ 0 to above 0.7.

Importantly, two EGPs (HD 209458b and OGLE-TR-56b) have been found to transit their primaries [1, 2, 3, 4, 5, 6]. The transit of the F8V/G0V star HD 209458 lasts ~ 3 hours (out of a total period of 3.524738 days) and has an average photometric depth of $\sim 1.6\%$ in the optical. Furthermore, in the first measurement of the composition of an extrasolar planet of any kind, Charbonneau et al.[7] detected sodium (Na-D) in their HD 209458b transit spectrum. This was followed by the detection of atomic hydrogen at Lyman-α and the discovery of a planetary wind[8]. Fortney et al.[9] conclude that the magnitude of the Na-D effect can be explained by the presence of a silicate or iron cloud at a pressure level of ~ 1.0 millibars. Such a cloud would alter the transit depths and effective radii for HD 209458b at other wavelengths as well. Figure 1 depicts their theoretical estimate of the variation with wavelength of the transit radius, with and without such clouds. Measurements at a variety of wavelengths and comparisons with theory such as is represented in Fig. 1 will help constrain the properties of this cloud. With both transit and radial-velocity data, an EGP's mass and radius can be determined, enabling its physical and structural study. However, it is only by direct detection of a planet's light using photometry or spectroscopy that the detailed study of its physical

---

[1] see J. Schneider's Extrasolar Planet Encyclopaedia at `http://www.obspm.fr/encycl/encycl.html` for a reasonable listing, with comments.

CP713, *The Search for Other Worlds: Fourteenth Astrophysics Conference*,
edited by S. S. Holt and D. Deming
© 2004 American Institute of Physics 0-7354-0190-X/04/$22.00

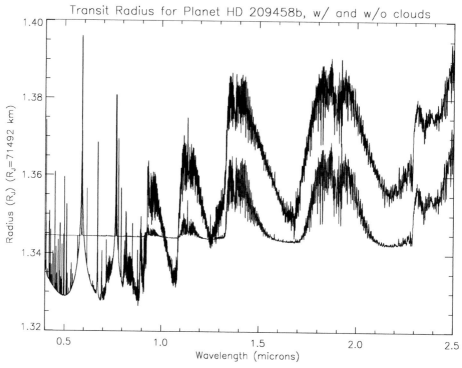

FIGURE 1. Taken from Fortney et al. (2003), this figure shows the theoretical transit radius of HD 209458b as a function of wavelength. The black curve does not include the effects of clouds, while the red (gray) curve does. The curve incorporating the cloud fits the Charbonneau et al. (2002) data at Na-D. See text for details.

attributes can be conducted. By this means, the composition, gravity, radius, and mass of the giant might be derived and the general theory of EGP properties and evolution might be tested[10, 11, 12, 13, 14, 15, 16, 17, 18]. For close-in EGPs, we can anticipate in the next few years wide-band precision photometry from MOST[19], Kepler[20], Corot[21], or MONS[22] that will provide the variations of the summed light of the planet and star due to changes in the planetary phase. In the mid- to far-infrared, the Spitzer Space Telescope (a.k.a. SIRTF, Space InfraRed Telescope Facility[23]) might soon be able to measure the variations in the planet/star flux ratios of close-in EGPs (SBH). Figure 2 shows the planet-to-star flux ratio derived by Sudarsky, Burrows, and Hubeny[17] in the near- and mid-infrared for three close-in EGPs, 51 Peg b, HD209458b, and τ Boo b. At ~8 μm, channel 4 on IRAC might be able to see the variation of the summed planet/star light. At ~24 μm, MIPS might be able to see (with a great deal of informed effort) the corresponding variation in this mid-infrared band. Note that Fig. 2 depicts the phase-averaged ratio and that the planet/star flux ratio at opposition should be 2-3 times higher. For wide-separation EGPs, it is necessary to measure the planet's light from under the

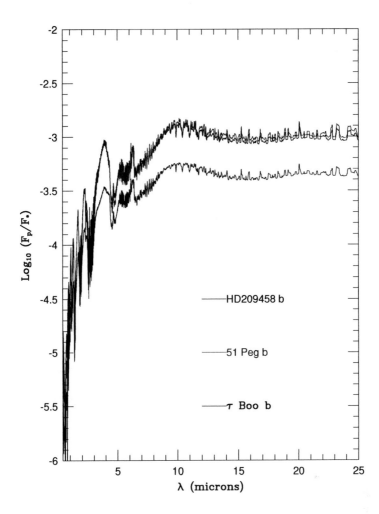

FIGURE 2. Taken from Sudarsky, Burrows, and Hubeny (2003), this figure depicts the phase-averaged planet to star flux ratio as a function of wavelength for the close-in EGPs 51 Peg b, τ Boo b (lowest curve), and HD209458b. At opposition this ratio should be 2-3 times larger and might be detectable in the near future by SIRTF, MONS, MOST, Kepler, and Corot. See text for a brief discussion.

glare of the primary star at very high star-to-planet contrast ratios. To achieve this from the ground, telescopes such as the VLT interferometer[24], the Keck interferometer[25, 26, 27], and the LBT nulling interferometer[28] will be enlisted.

From space, the Terrestrial Planet Finder (TPF[29]) and/or a coronagraphic optical imager such as Eclipse[30, 31] could obtain low-resolution spectra.

To support the above efforts and planning for future programs of direct detection of extraoslar giant planets and to provide the theoretical context for the general analysis of the spectra and photometry of irradiated and isolated EGPs, our group has embarked upon a series of papers of EGP spectra, evolution, chemistry, transits, orbital phase functions, and light curves. The most recent paper in this series (SBH) explored generic features of irradiated EGP spectra as a function of orbital distance, cloud properties, and composition class[13]. In an upcoming paper, Burrows, Sudarsky, and Hubeny[32] calculate the spectral diagnostics of planetary orbital distance, mass and age, using a fully consistent model for the ammonia and/or water clouds that can arise in their atmospheres.

HOW AN IRRADIATED EGP'S SPECTRUM DEPENDS ON ORBITAL DISTANCE

Figure 3 depicts the dependence of the 1-M_J / 5-Gyr EGP's spectrum on distance due to insolation by a G2V star. Orbital distances from 0.2 AU to 15 AU are included.

For the ammonia clouds that form in this orbital distance sequence (at $\gtrsim 6$ AU), the modal particle sizes we find hover near 50-60 μm. The corresponding particle sizes in the water clouds are near 110 μm. These particles are larger than for more massive brown dwarfs. When the chemistry indicates that both cloud types (H_2O and NH_3) are present, we include them both in the atmospheric/spectral calculation. The ammonia cloud is always above the water cloud. Water clouds form around a G2V star exterior to a distance near 1.5 AU, whereas ammonia clouds form around such a star exterior to a distance near 4.5 AU. Note that Jupiter itself is at the distance from the Sun of ~ 5.2 AU. Note also that at 4 AU water clouds form in Jovian-mass objects as early as ~ 50 Myr.

As Fig. 3 shows, in the optical, the flux ratios vary between 10^{-8} and 10^{-10}. In the near infrared, this ratio varies widely from $\sim 10^{-4}$ to 10^{-16}. However, in the mid-infrared beyond 10 μm, the flux ratio varies more narrowly from 10^{-4} to 10^{-7}. Hence, it makes a difference in what wavelength region one conducts a search for direct planetary light.

Gaseous water absorption features (for all orbits) and methane absorption features (for the outer orbits) sculpt the spectra. The reflected component due to Rayleigh scattering and clouds (when present) is most manifest in the optical and the emission component (similar to the spectrum of an isolated low-gravity brown dwarf) takes over at longer wavelengths. For the EGPs interior to ~ 1.0 AU, the fluxes longward of ~ 0.8 μm are primarily due to thermal emission, not reflection. These atmospheres do not contain condensates and are heated efficiently by stellar irradiation. As a result, the Z (~ 1.0 μm), J (~ 1.2 μm), H (~ 1.6 μm), and K (~ 2.2 μm) band fluxes are larger by up to several orders of magnitude than those of the more distant EGPs. Generally, clouds increase a planet's flux in the optical, while decreasing it in the J, K, L' (~ 3.5 μm), and M (~ 5.0 μm) bands. The transition between the reflection and emission components moves to longer wavelengths with increasing distance, and is around 0.8-1.0 μm at 0.2 AU and ~ 3.0 μm at 15 AU.

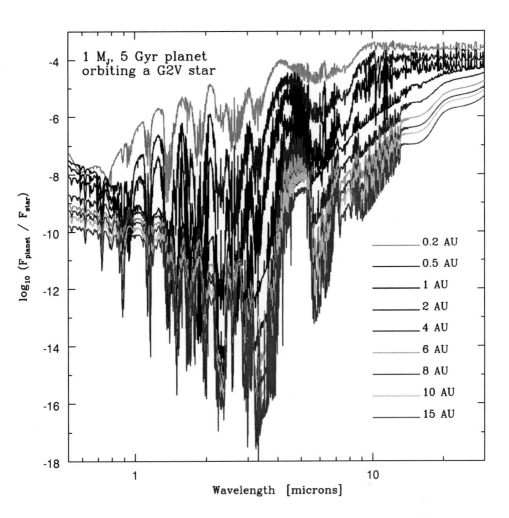

FIGURE 3. Planet to star flux ratios versus wavelength (in microns) from 0.5 μm to 30 μm for a 1-M_J EGP with an age of 5 Gyr orbiting a G2V main sequence star similar to the Sun. This figure portrays ratio spectra as a function of orbital diatance from 0.2 AU to 15 AU. Zero eccentricity is assumed and the planet spectra have been phase-averaged as described in Sudarsky, Burrows, and Hubeny (2003). Note that the planet/star flux ratio is most favorable in the mid-infrared. See text for discussion.

At large distances exterior to ~3.5 AU, the flux longward of ~15 μm manifests undulations due to pressure-induced absorption by H_2. Importantly, there is always a significant bump around the M band at 4-5 μm. The peak of this bump shifts from ~4 μm to ~5 μm with increasing orbital distance. Though muted by the presence of clouds, it is always a prominent feature of irradiated EGPs, as it is in T dwarfs and the Jovian planets of our solar system. Curiously, but not unexpectedly, as Fig. 3 indicates, the planet/star flux ratio is most favorable in the mid-infrared.

Though the major trend is a monotonic decrease in a planet's flux with increasing orbital distance, the 0.2-AU model atmosphere is hot enough that the sodium and potassium resonance absorption lines appear and surpress the flux around Na-D (0.589 μm) and the related K I doublet at 0.77 μm. The result is a lower integrated visible flux that is comparable to that of the otherwise dimmer 0.5-AU model. For greater orbital distances, the atmospheric temperatures are too low for the alkali metals to appear, but the methane features near 0.62 μm, 0.74 μm, 0.81 μm, and 0.89 μm come into their own. Broad water bands around 0.94 μm, 1.15 μm, 1.5 μm and 1.85 μm that help to define the Z, J, and H bands are always in evidence, particularly for the 0.2 and 1.0 AU models that don't contain water clouds. For greater distances, the presence of water clouds slightly mutes the variation with wavelength in the planetary spectra. Hence, smoothed water features and methane bands predominate beyond ~1.5 AU.

HIGH-CONTRAST IMAGING OF WIDE-SEPARATION EGPS

Depending upon orbital distance, age, and mass, spectral features due to methane, water, and the alkali metals are prominent. Furthermore, there is a slight anti-correlation in the effects of clouds in the optical and infrared, with the optical fluxes increasing and the infrared fluxes decreasing with increasing cloud depth. For young and massive irradiated EGPs, there are prominent peaks in the Z, J, H, K, and M bands. Though the optical flux is not a very sensitive function of age, there is a useful age dependence of the fluxes in these bands. Moreover, there is an anti-correlation with increasing mass at a given age and orbital distance between the change in flux in the optical and in the near-IR bands, with the optical fluxes decreasing and the Z, J, H, K, and M band fluxes increasing with increasing mass.

The remote sensing of the atmospheres of EGPs will be challenging, but the detection and characterization of the direct light from a planet outside our solar system will be an important milestone in both astronomy and planetary science. One instrument proposed to meet this challenge is the space-based coronagraphic imager Eclipse[30, 31]. The Eclipse instrument team is predicting a contrast capability of ~10^{-9} for an inner working angle of $0.3''$ or $0.46''$ in the V/R or Z bands, respectively. As Fig. 3 indicates, with such a capability, irradiated EGPs could be detected and analyzed. Many could even be discovered. However, whether such sensitivity is achievable remains to be demonstrated. Be that as it may, further advancement in our understanding of extrasolar planets is contingent upon technical advances that would enable the direct measurement of the dynamically dominant and brighter components of extrasolar planetary systems.

ACKNOWLEDGMENTS

We wish to acknowledge Bill Hubbard, Jonathan Lunine, Jim Liebert, Jonathan Fortney, Christopher Sharp, and Curtis Cooper for conversations and/or previous collaborations that materially improved the work we have summarized here. We also wish to thank NASA for its financial support via grants NAG5-10760 and NAG5-10629 and the Kavli Institute for Theoretical Physics where some of this paper was written. Furthermore, we acknowledge support through the Cooperative Agreement #NNA04CC07A between the University of Arizona/NOAO LAPLACE node and NASA's Astrobiology Institute.

REFERENCES

1. Henry, G., Marcy, G. W., Butler, R. P., & Vogt, S. S. 2000, Ap.J., 529, L41.
2. Charbonneau, D., Brown, T. M., Latham, D. W., & Mayor, M. 2000, Ap.J., 529, L45.
3. Charbonneau, D., Brown, T. M., Noyes, R. W., Gilliland, R. L., and Burrows, A. 2001, Ap.J., 552, 891.
4. Brown, T. M., Charbonneau, D., Gilliland, R.L., Noyes, R.W., and Burrows, A. 2001, Ap.J., 552, 699.
5. Konacki, M., Torres, G., Jha, S., and Sasselov, D. 2003, Nature, 421, 507.
6. Torres, G, Konacki, M., Sasselov, D., and Jha, S. 2003, astro-ph/0310114.
7. Charbonneau, D., Brown, T. M., Noyes, R. W., & Gilliland, R. L. 2002, Ap.J., 568, 377.
8. Vidal-Madjar, A., des Etangs, A., Desert, J.-M., Ballester, G.E., Ferlet, R., Hebrard, G., Mayor, M. 2003, Nature, 422, 143.
9. Fortney, J.J, Sudarsky, D., Hubeny, I., Cooper, C.S., Hubbard, W.B., Burrows, A., & Lunine, J.I. 2003, Ap.J., 589, 615.
10. Burrows, A., Saumon, D., Guillot, T., Hubbard, W.B., & Lunine, J.I. 1995, Nature, 373, 191.
11. Burrows.A., Marley, M.S., Hubbard, W.B., Lunine, J.I., Guillot, T., Saumon, D., Freedman, R., Sudarsky, D., & Sharp, C.M. 1997, Ap.J., 491, 856.
12. Marley, M. S., Gelino, C., Stephens, D., Lunine J. I., & Freedman, R. 1999, Ap.J., 513, L879.
13. Sudarsky, D., Burrows, A., & Pinto, P. 2000, Ap.J., 538, 885.
14. Seager, S. & Sasselov, D.D. 1998, Ap.J., 502, 157.
15. Seager, S. & Sasselov, D.D. 2000, Ap.J., 537, 916.
16. Seager, S., Whitney, B.A, & Sasselov, D.D. 2000, Ap.J., 540, 504.
17. Sudarsky, D., Burrows, A., & Hubeny, I. 2003, Ap.J., 588, 1121 (SBH).
18. Baraffe, I., Chabrier, G., Allard, F., and Hauschildt, P.H. 2003, A. & A., 402, 701.
19. Matthews, J. M., Kuschnig, R., Walker, G. A. H. et al. 2001, in *The Impact of Large-Scale Surveys on Pulsating Star Research*, ed. L. Szabados & D. Kurtz, p. 74.
20. Koch, D., Borucki, W., Webster, L., Dunham, E., Jenkins, J., Marrion, J., & Reitsema, H. 1998, SPIE Conference 3356: *Space Telescopes and Instruments V*, p. 599.
21. Antonello, E. & Ruiz, S. M. 2002, *The Corot Mission*, http://www.astrsp-mrs.fr/projects/corot/corotmission.ps.
22. Christensen-Dalsgaard, J. 2000, http://bigcat.obs.aau.dk/hans/mons/.
23. Werner, M.W. and Fanson, J.L. 1995, Proc. SPIE, 2475, p. 418-427.
24. Paresce, F. 2001, *Scientific Objectives of the VLTI Interferometer*.
25. van Belle, G. & Vasisht, G. 1998, *The Keck Interferometer Science Requirements Document, Revision 2.2, Jet Propulsion Laboratory*.
26. Akeson, R. L. & Swain, M. R. 2000, in *From Giant Planets to Cool Stars*, ed. C. A. Griffith & M. S. Marley, ASP Conference Series, 212, 300.
27. Akeson, R. L., Swain, M. R., & Colavita, M. M 2000, in *Interferometry in Optical Astronomy*, ed. P. J. Lena, Proc. SPIE 4006, 321.
28. Hinz, P. M. 2001, PhD Thesis, The University of Arizona.
29. Levine, B.M. et al. 2003, SPIE, 4852, 221.

30. Trauger, J., Backman, D., Brown, R. A. et al. 2000, AAS Meeting 197, 49.07.
31. Trauger, J., Hull, A. B., & Redding, D. A. 2001, AAS Meeting 199, 86.04.
32. Burrows, A., Sudarsky, D., & Hubeny, I. 2004, submitted to Ap.J..

Astrophysical False Positives Encountered in Wide-Field Transit Searches

David Charbonneau*, Timothy M. Brown[†], Edward W. Dunham**,
David W. Latham[‡], Dagny L. Looper* and Georgi Mandushev**

*California Institute of Technology, MC 105-24, 1200 E. California Blvd., Pasadena, CA 91125
[†]High Altitude Observatory, Ntl Ctr for Atmospheric Research, P.O. Box 3000, Boulder, CO 80307
**Lowell Observatory, 1400 W. Mars Hill Road, Flagstaff, AZ 86001
[‡]Harvard-Smithsonian Ctr for Astrophysics, 60 Garden St., MS-20, Cambridge, MA 02138

Abstract. Wide-field photometric transit surveys for Jupiter-sized planets are inundated by astrophysical false positives, namely systems that contain an eclipsing binary and mimic the desired photometric signature. We discuss several examples of such false alarms. These systems were initially identified as candidates by the PSST instrument at Lowell Observatory. For three of the examples, we present follow-up spectroscopy that demonstrates that these systems consist of (1) an M-dwarf in eclipse in front of a larger star, (2) two main-sequence stars presenting grazing-incidence eclipses, and (3) the blend of an eclipsing binary with the light of a third, brighter star. For an additional candidate, we present multi-color follow-up photometry during a subsequent time of eclipse, which reveals that this candidate consists of a blend of an eclipsing binary and a physically unassociated star. We discuss a couple indicators from publicly-available catalogs that can be used to identify which candidates are likely giant stars, a large source of the contaminants in such surveys.

WIDE-FIELD TRANSIT SURVEYS

More than 20 groups worldwide are now engaged in photometric surveys aimed at detecting Jupiter-sized planets in tight orbits about their parent stars [for a review, see 1, 2]. Several of these projects are small, automated systems with modest apertures (typically 10 cm) and CCD cameras (typically 2k × 2k arrays of ~ 15 μm pixels), which monitor several thousand stars simultaneously in a very wide field-of-view ($6° - 9°$ square). Examples of such instruments are STARE[1] [3, located on Tenerife], PSST [located in northern Arizona], Sleuth[2] [4, located in southern California], Vulcan[3] [5, 6, located in central California] and Vulcan South[4] [to be located at the South Pole], HAT[5] [7, located in southern Arizona], and SuperWASP[6] [8, located on La Palma]. The advantage that all such projects offer over deeper transit surveys is the brightness of the target stars (typically $8 \leq V \leq 13$), which facilitates radial velocity measurements aimed at detecting

---

[1] http://www.hao.ucar.edu/public/research/stare/stare.html
[2] http://www.astro.caltech.edu/~ftod/sleuth.html
[3] http://web99.arc.nasa.gov/~vulcan/
[4] http://web99.arc.nasa.gov/~vulcan/south
[5] http://cfa-www.harvard.edu/~gbakos/HAT/
[6] http://www.superwasp.org/

CP713, The Search for Other Worlds: Fourteenth Astrophysics Conference,
edited by S. S. Holt and D. Deming
© 2004 American Institute of Physics 0-7354-0190-X/04/$22.00

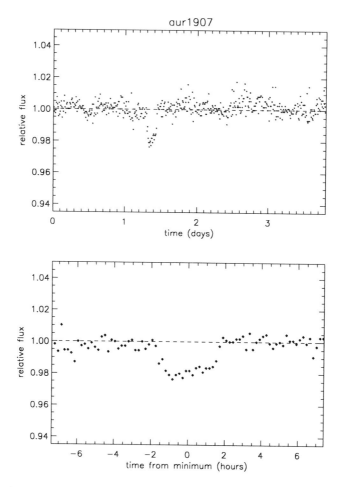

FIGURE 1. (Upper panel) PSST photometric light curve for star 1907 ($V = 11.2$, $B - V = 0.2$) in Auriga, binned and phased to a period of 3.80 days. The light curve is constant outside of eclipse. (Lower panel) The same data near times of transit. The transit is flat-bottomed with a depth of 0.019 mag and a duration of 3.7 hours, consistent with a transit of a Jovian planet across a Sun-like star.

the orbit induced by the planet. Furthermore, it is only for such bright systems that the host of follow-up measurements that are currently being pursued for HD 209458 [9] and other bright extrasolar-planet stars [10] might also be enabled.

The difficulty facing such surveys is not that of obtaining the requisite photometric precision or phase coverage, as several of the aforementioned projects have achieved these requirements. Rather, the current challenge is that of efficiently rejecting astrophysical false positives, i.e. systems containing an eclipsing binary, whose resulting photometric light curves mimic that of a Jupiter passing in front of a Sun-like star.

Brown [11] recently presented detailed estimates of rates of such false alarms for

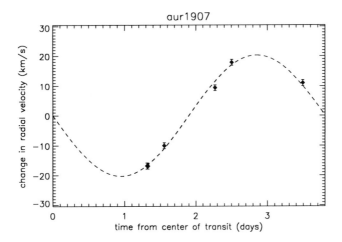

FIGURE 2. Shown are the changes in radial velocity for star 1907 in Auriga (Figure 1) as derived from high-resolution spectra gathered over 3 consecutive nights with the Palomar 60-inch echelle spectrograph. The typical precision is 1 km s^{-1}. The observed variation is consistent with an orbit as predicted by the photometric phase and period. However, the amplitude of the radial velocity orbit is 20 km s^{-1}, indicating a low-mass stellar companion.

these wide-field surveys. The three dominant sources of such signals are (1) grazing incidence eclipses in systems consisting of two main-sequence stars, (2) central transits of low-mass stars in front of large main-sequence stars (typically M-dwarfs passing in front of F-stars), and (3) binary systems undergoing deep eclipses, blended with the light of a third star that falls within the instrumental point-spread-function (the blending star may be either physically associated or simply lying along the line-of-sight). Brown finds that these various forms of astrophysical false positives will likely occur at a combined rate nearly 12 times that of true planetary transits. A direct means to investigate the true nature of each candidate would be high-precision radial-velocity monitoring, but this would impose a hefty burden upon the large telescopes that are required. Thus, all such transit surveys must adopt strategies for efficiently rejecting the majority of such false alarms, to bring the number of candidates to a manageable level.

EXAMPLES OF ASTROPHYSICAL FALSE POSITIVES

For the benefit of researchers pursuing such transit surveys, we present here several examples of astrophysical false positives. These candidate transit signals were identified in a survey of a $6° \times 6°$ field in Auriga by the PSST instrument at Lowell Observatory. A description of the instrument and the results of the survey will be presented elsewhere. Our goal here is to present the follow-up spectroscopy performed with the Palomar 60-inch telescope, which demonstrates conclusively that these candidates are indeed false alarms. Access to telescopes in the range of $1 - 2$ m is widely available. As a result,

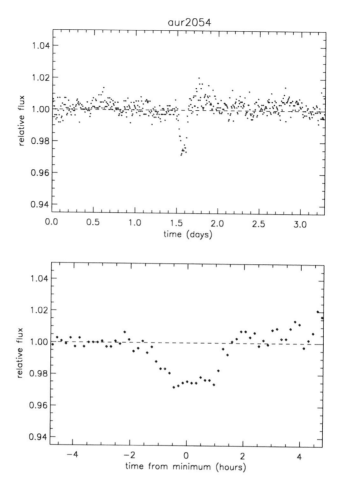

FIGURE 3. (Upper panel) PSST photometric light curve for star 2054 ($V = 11.3$, $B - V = 1.5$) in Auriga, binned and phased to a period of 3.29 days. The light curve is constant outside of eclipse. (Lower panel) The same data near times of transit. The transit is flat-bottomed with a depth of 0.023 mag and a duration of 2.4 hours, consistent with a transit of a Jovian planet across a Sun-like star.

such observations are a reasonable means by which to dispose of such contaminants. Similar work to identify the true nature of candidates arising from the Vulcan survey was presented in [12].

Star 1907: An M-dwarf in eclipse

The PSST photometric light curve (Figure 1) for star 1907 ($V = 11.2$, $B - V = 0.2$) revealed a flat-bottomed transit with a period of 3.80 days, a depth of 0.019 mag and

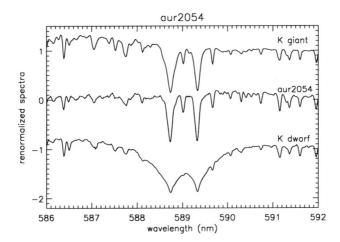

FIGURE 4. The central curve shows a Palomar 60-inch echelle spectrum of star 2054 in Auriga (Figure 3) in the region of the gravity-sensitive sodium D lines. Spectra of two K-stars with approximately the same color are also shown: The lower spectrum is that of the K5-dwarf GJ 380, demonstrating the very broad sodium lines indicative of the high gravity of a dwarf star. The upper spectrum is that of the K5.5-giant HR 5200, where the sodium lines are unbroadened. The semblance of the spectrum of the candidate star to that of HR 5200 indicates that the star is a giant. Since a 3.3-day period would place the object within the physical radius of the giant, this system is likely a blend of K-giant and a fainter eclipsing binary.

a duration of 3.7 hours. The light curve is constant outside of eclipse. However, radial velocities (Figure 2) derived from high-resolution spectra gathered over 3 consecutive nights with the Palomar 60-inch echelle spectrograph (with a typical precision of 1 km s^{-1}) revealed an orbit consistent with the period and phase derived by the photometry, but with an amplitude of 20 km s^{-1}. This large Doppler variation indicates a low-mass stellar companion.

Star 2054: An eclipsing binary blended with a giant star

The photometric time series (Fsigure 3) for star 2054 ($V = 11.3, B - V = 1.5$) shows a flat-bottomed eclipse with a period of 3.29 days, a depth of 0.023 mag, and a duration of 2.4 hours. The light curve is constant outside of eclipse. Figure 4 shows a portion of the spectrum of this object gathered with the Palomar 60-inch echelle in the region of the gravity-sensitive sodium D lines. The unbroadened profile of the sodium features reveals this star to be a K-giant. Since a 3.3 day orbit about a K-giant would place it within the physical radius of the star, we must be seeing a blend of a K-giant and a fainter eclipsing binary (which may or may not be physically associated).

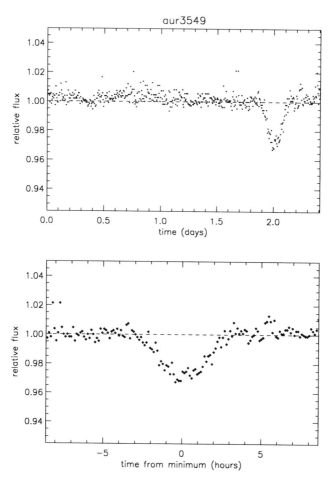

FIGURE 5. (Upper panel) PSST photometric light curve for star 3549 ($V = 11.6$, $B - V = 1.1$) in Auriga, binned and phased to a period of 2.41 days. There is some evidence for a secondary eclipse in the light curve, near a time of 0.4 days. (Lower panel) The same data near times of transit. The transit is somewhat V-shaped, with a depth of 0.028 mag and a duration of 4.3 hours. Although marginally consistent with a planetary transit, these data hint at an eclipsing binary system.

Star 3549: An eclipsing binary blended with an unassociated star

The photometric light curve (Figure 5) for star 3549 ($V = 11.6$, $B - V = 1.1$) shows a mildly V-shaped eclipse with a period of 2.41 days, a depth of 0.028 mag, and a duration of 4.3 hours. Our suspicions were raised by the hint of a faint secondary eclipse in the light curve. Examination of the Digitized Sky Survey image (Figure 6) reveals two sources within the PSF of the PSST instrument. Photometry of the field during the time of a subsequent eclipse (conducted with a 14-inch telescope, which afforded a

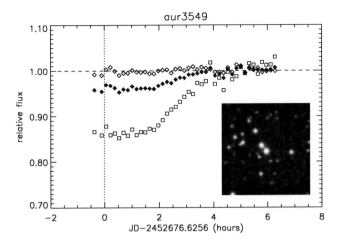

FIGURE 6. The 2 arcmin × 2 arcmin Digitized Sky Survey image (inset) of star 3549 in Auriga (Figure 5) reveals two sources within the PSF of the PSST instrument. Photometry of the field during the time of a subsequent eclipse (the predicted time of center of transit is shown as $t = 0$ in this figure) shows the central brighter source to be constant in time (open diamonds). The fainter star to the NE undergoes a deep (14%) eclipse (open squares). When the light from both stars is summed in a single photometric aperture, the light curve shown with black diamonds results. This photometric time series reproduces that observed by the PSST (Figure 5). Thus, this candidate is the blend of the fainter eclipsing binary to the NE, and the brighter physically-unassociated central star.

significant increase in the angular resolution over the data gathered with the PSST) finds that the central brighter source is constant, whereas the fainter star to the NE undergoes a deep (14%) eclipse (Figure 6). When the light from both stars in summed in a single photometric aperture, the light curve that results reproduces that observed by the PSST. Hence, this system is a blend of an eclipsing binary (the fainter star to the NE), and the physically unassociated brighter star.

Star 4922: A grazing-incidence eclipsing binary

The photometric light curve (Figure 7) for star 4922 ($V = 12.0$, $B - V = 0.4$) reveals a clearly V-shaped eclipse with a period of 1.52 days, a depth of 0.021 mag, and a duration of 2.0 hours. The light curve is constant outside of eclipse, but the shape of the transit already hints at a grazing-incidence eclipsing binary. Palomar 60-inch echelle spectra (Figure 8) gathered on three consecutive nights clearly show two sets of absorption features, and nightly variations in the radial velocity of the components that are consistent with the time scale of the photometric period. The amplitude of the radial velocity orbit is approximately 90 km s^{-1}. Combined with the V-shaped nature of the eclipses, these data lead us to conclude that this is a grazing-incidence main-sequence binary, with a true period that is twice the photometric period.

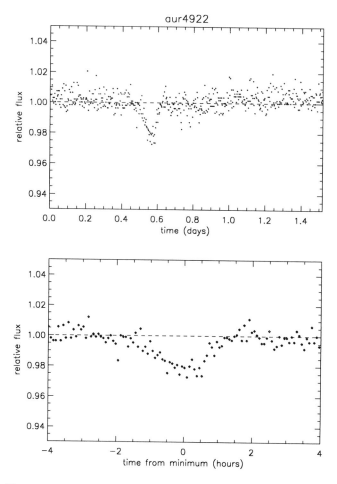

FIGURE 7. (Upper panel) PSST photometric light curve for star 4922 ($V = 12.0$, $B - V = 0.4$) in Auriga, binned and phased to a period of 1.52 days. The light curve is constant outside of eclipse. (Lower panel) The same data near times of transit. The transit is V-shaped, with a depth of 0.021 mag and a duration of 2.0 hours, indicating that the signal is likely due to a grazing-incidence binary system.

EFFICIENT METHODS TO REJECT FALSE ALARMS

The publicly-available USNO-B catalog[7] [13] provides proper motions, which aid in constraining the distance of a star. Typical errors in the proper motions from the USNO-B catalog are ~ 7 mas/yr. Objects with colors consistent with the target spectral types of K, G, or F and a high measured proper motion are likely dwarf stars, and thus good

---

[7] http://www.nofs.navy.mil/projects/pmm/catalogs.html

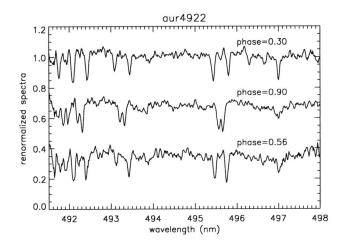

FIGURE 8. Shown are Palomar 60-inch echelle spectra of star 4922 in Auriga (Figure 7) gathered on three consecutive nights. The spectra are clearly double-lined. The variations in radial velocity are consistent with the time scale of the photometric period (the spectra are labeled by the orbital phase predicted by the photometry). The radial velocity orbit is approximately 90 km s^{-1}. Combined with the V-shaped nature of the eclipses, we conclude that this is likely a grazing-incidence binary, with a true period that is twice the photometric period.

candidates. Unfortunately, small proper motions are inconclusive: These stars could either be distant (hence giants), or nearby dwarfs. In addition, colors from the 2MASS catalog[8] aid in separating dwarf stars from giants. Brown [11] illustrates that the $J - K$ colors from the 2MASS catalog provide a useful discriminant: Stars with $J - K \geq 0.5$ are predominantly giants, whereas those with $J - K \leq 0.35$ are almost exclusively dwarfs (Figure 9). The target stars of wide-field transit surveys are those with radii less than 1.3 $R_\odot$. The infrared colors of such stars extend into the intermediate region, so that the $J - K$ color is not a definitive test. Nonetheless, candidates with very red colors ($J - K > 0.7$) are almost certainly giants and can be rejected.

While the spectroscopy described in the previous section provides a reliable means to identify astrophysical false positives, it is labor- and resource-intensive. In order to more efficiently reject such contaminants, L. Kotredes and D. Charbonneau are currently assembling Sherlock, an automated follow-up telescope for wide-field transit searches [14]. Working from a list of transit candidates, Sherlock will calculate future times of eclipse, and conduct multi-color photometry of these targets, with an angular resolution that is significantly improved over the survey instruments. The large majority of false alarms should yield eclipse signals that are color-dependent, whereas planetary transits should be nearly color-independent (except for the minor effects due to limb-darkening). Sherlock will be deployed at Palomar Observatory in early 2004, and will be available to follow up candidates identified by any of the wide-field transit searches.

---

[8] http://www.ipac.caltech.edu/2mass/releases/allsky/

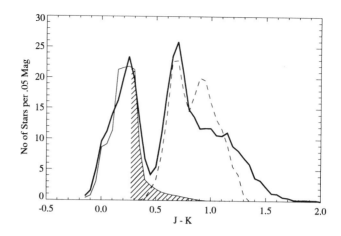

FIGURE 9. Taken from [11], courtesy T. M. Brown. The thick curve shows the histogram of star density vs $J - K$ color from the 2MASS catalog for stars observed by a STARE in a field in Cygnus. The thin solid curve shows the model used by Brown [11] to predict the occurence of main-sequence stars, and the dashed curve shows that for giants. The hatched region shows stars with radii less than $1.3\,R_\odot$. Thus, the $J - K$ color appears to be a useful discriminant against giant stars.

REFERENCES

1. Horne, K. 2003, ASP Conf. Ser. 294: Scientific Frontiers in Research on Extrasolar Planets, eds. D. Deming & S. Seager (San Francisco: ASP), 361, astro-ph/0301249
2. Charbonneau, D. 2003a, Space Science Reviews: ISSI Workshop on Planetary Systems and Planets in Systems, ed. S. Udry, W. Benz, & R. von Steiger (Dordrecht: Kluwer), astro-ph/0302216
3. Brown, T. M. & Charbonneau, D. 2000, ASP Conf. Ser. 219: Disks, Planetesimals, and Planets, eds. F. Garzon and T. J. Mahoney (San Francisco: ASP), 584, astro-ph/0005009
4. O'Donovan, F. T., Charbonneau, D., & Kotredes, L. 2003, AIP Conf. Proc.: The Search for Other Worlds, eds. S. S. Holt & D. Deming, astro-ph/0312289
5. Borucki, W. J., Caldwell, D., Koch, D. G., Webster, L. D., Jenkins, J. M., Ninkov, Z., & Showen, R. 2001, PASP, 113, 439
6. Jenkins, J. M., Caldwell, D. A., & Borucki, W. J. 2002, ApJ, 564, 495
7. Bakos, G. Á., Lázár, J., Papp, I., Sári, P., & Green, E. M. 2002, PASP, 114, 974
8. Street, R. A. et al. 2003, ASP Conf. Ser. 294: Scientific Frontiers in Research on Extrasolar Planets, eds. D. Deming & S. Seager (San Francisco: ASP), 405, astro-ph/0208233
9. Charbonneau, D. 2003b, ASP Conf. Ser. 294: Scientific Frontiers in Research on Extrasolar Planets, eds. D. Deming & S. Seager (San Francisco: ASP), 449, astro-ph/0209517
10. Charbonneau, D. 2003c, ASP Conf. Ser.: IAU Symp. 219: Stars as Suns: Activity, Evolution, and Planets, eds. A. O. Benz and A. K. Dupree (San Francisco: ASP), astro-ph/0312252
11. Brown, T. M. 2003, ApJ, 593, L125
12. Latham, D. W. 2003, ASP Conf. Ser. 294: Scientific Frontiers in Research on Extrasolar Planets, eds. D. Deming & S. Seager (San Francisco: ASP), 409
13. Monet, D. G. 2003, AJ, 125, 984
14. Kotredes, L., Charbonneau, D., Looper, D. L., & O'Donovan, F. T. 2003, AIP Conf. Proc.: The Search for Other Worlds, eds. S. S. Holt & D. Deming, astro-ph/0312432

HST/FGS Photometry of Planetary Transits of HD 209458

A. B. Schultz*, W. Kinzel*, M. Kochte*, I. J. E. Jordan*, F. Hamilton*,
G. Henry†, S. Vogt**, F. Bruhweiler‡, A. Storrs§, H. M. Hart¶, D. Bennum‖,
J. Rassuchine‖, M. Rodrigue‖, D. P. Hamilton††, W. F. Welsh‡‡ and
D. C. Taylor§§

*CSC/STScI, 3700 San Martin Dr., Baltimore, MD 21218
†Tennessee State Univ., Nashville, TN 37203-3401
**UCO/Lick Obs., Santa Cruz, CA 95064
‡IACS/Catholic Univ., Washington, DC 20064
§Towson Univ., Baltimore, MD 21252-0001
¶CSC/Johns Hopkins Univ., Baltimore MD 21218-2695
‖Univ. of Nevada, Reno, NV 89557
††Univ. of Maryland, College Park, MD 20742
‡‡San Diego State Univ., San Diego, CA 92182-1221
§§STScI, 3700 San Martin Dr., Baltimore, MD 21218

Abstract. We present the data, and modeling and analysis results from the photometric monitoring of five planetary transits of HD 209458 using the Fine Guidance Sensors (FGS) onboard the Hubble Space Telescope (HST). We have now included the output from all four FGS photometers in our data reduction and analysis increasing our S/N over our previous results. We have modeled the transits as an opaque spherical planet in a circular orbit about a limb darkened spherical star and simultaneously fit the model to the FGS data and published STIS transit data [1]. The measured light curves show small features, a fraction of the transit depth. Some of these faint bumps and ripples appear to be real. We present an analysis of the FGS transit light curves, showing the results of the model fitting and a search for a possible planetary satellite.

INTRODUCTION

Radial velocity observations of nearby solar-like stars have been used to detect 117 Jupiter-sized planets orbiting 102 stars. Of these Doppler-detected extra-solar giant planets (EGPs), sixteen could be classified as "51 Peg-like" or "roasters." Roaster planets are characterized by small orbital distances ($\sim$0.1 AU), high effective temperatures (900 K $\leq$ T $\leq$ 1500 K), and expanded atmospheres. The Doppler technique used to detect extra-solar planets provides only the minimum mass ($M_p \sin(i)$). The minimum masses for these planets are between $\sim$0.44 and $\sim$1.3 M_{Jup} . Knowledge of the stellar mass combined with modeling the precise photometric transit measurements provides estimates of the basic system parameters such as the orbital inclination and planetary radius. Determining the orbital inclination removes the $\sin(i)$ dependency in the planetary mass estimate. To date, only two planets (OGLE-TR-56b, HD 209458b) have been observed to transit their host stars.

Henry et al. [2] reported the first detection of a transit for HD 209458b. The minimum

CP713, *The Search for Other Worlds: Fourteenth Astrophysics Conference*,
edited by S. S. Holt and D. Deming
© 2004 American Institute of Physics 0-7354-0190-X/04/$22.00

mass for HD 209458b as determined by the Doppler technique is $M_p\sin(i) \sim 0.62\ M_{Jup}$ [3, 4]. The transit fixes the orbital inclination at $i = 86.6^o \pm 0.14$ [1], which leaves HD 209458b with a true mass of $\sim 0.63\ M_{Jup}$: quite comfortably a planet-sized body.

OBSERVATIONS AND DATA REDUCTION

HD 209458b is the only transiting extra-solar planet that is observable with the FGS. Five transits were observed: June 11, 2001, September 11, 2001, November 10, 2001, January 16, 2002, and September 30, 2002. FGS1r with its F550W filter ($\lambda = 5500$ Å, $\delta\lambda = 5100 - 5875$ Å) was used to obtain the data. An FGS contains four standard photomultiplier tubes (PMTs) used as high-speed photometers operating in a counting mode at 40 Hz (0.025 sec) [5]. The expectation for each observation was to begin before the predicted time of ingress (and egress), and to capture the ingress and egress turnover points. Total integration times would depend on the available on-target time per HST orbit minus the time for FGS setup and acquisition of guide stars. Each PMT yielded $\sim 6,500$ counts per 0.025 sec sample per PMT (S/N ~ 80).

This was the first use of the FGS to observe a bright source (HD 209458, V=7.64) for more than a few minutes. The uncalibrated data display a time dependency in the FGS response. In addition, HST orbit phase dependent lower level variations were captured in the data, probably caused by HST breathing. The FGS dead time correction was applied to the data, drop outs were removed, and the data were placed into one second bins. To remove the time variable telescope/PMT response, a 5th order Chebyshev polynomial was fit to the out-of-transit data for each transit and each PMT. The resulting curve was divided into each visit's data. For each time point, the 4 channel data were combined using a weighted average, where the weight is the square of the RMS from the out-of-transit data Chebyshev fit. Finally, the data were placed into 80 second bins prior to model fitting to match the STIS sampling time. The normalized data for the 5 transits are presented in Figure 1.

The June 2001 observation consisted of three contiguous HST orbits, to verify the observing strategy and to capture the mid-transit point for fitting purposes. The September 2001, November 2001, January 2002, and September 2002 observations consisted of two non-contiguous HST orbits, each with one non-HD 209458 orbit between the two orbits on HD 209458. The September 2001 and January 2002 observations caught the ingress and egress turn over points.

LIGHTCURVE MODELING

We have simultaneously fit the FGS and STIS transit data [1], via χ^2 minimization, to a model consisting of an opaque spherical planet in circular orbit, transiting a limb-darkened star. There are seven free parameters: time of transit center T_0, stellar radius R_*, planetary radius R_p, orbital inclination (i), period (P), and the stellar limb darkening parameters u_1 and u_2. The star was assumed to have a mass of $1.1 \pm 0.1\ M_\odot$ [6] and

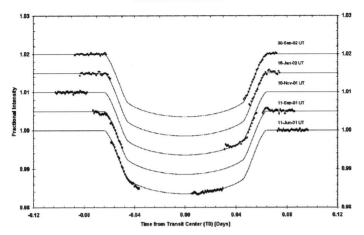

FIGURE 1. Five planetary transits of HD 209458. The in-transit data were normalized to the out-of-transit data. An arbitary offset was added to individual transits for display purposes.

quadratic limb darkening of the form:

$$I(\mu)/I(1) = 1 - u_1(1 - \mu) - u_2(1 - \mu)^2$$

where μ is the cosine of the angle between the line of sight and the stellar surface normal. The parameters were fit to the time of transit center T_0 on November 10, 2001.

The uncertainty in each model parameter value was estimated using the technique given in Press et al. [7]. Using the parameter values obtained from the fit to the FGS and STIS data, 100 simulated data sets were generated. Each simulated datum was varied from the predicted normalized flux value by the addition of random noise whose magnitude was based upon the RMS of the fit to the data. Each simulated data set was then fit using the same model and free parameters as before. In addition, for each fit, the assumed mass of the star was varied based upon its uncertainty ($\pm 0.1 M_\odot$ [6]). The standard deviation of each parameter obtained from the fits to the 100 simulated data sets is used as the uncertainty for that parameter. The results of this fit are listed in Table 1. The limb darkening coefficients for the STIS and FGS data are different due to the different bandpasses used.

ANALYSIS AND CONCLUSIONS

The fit results are consistent with previous observations [1, 3, 4, 8, 9] The inclusion of the data from all four FGS PMTs allowed for a more consistent data set which is better fit by the model as compared to previously reported results [10]. In addition, they decreased the parameter uncertainties in some cases by a factor of 10.

The STIS data for each transit was obtained using 5 contiguous HST orbits. The continuous pointing allowed HST to stabilize. All but the first transit observed with

TABLE 1. Fit to the HST FGS & STIS Observations of HD 209458.

	Data Fit	
	value	uncertainty
T_0	2452223.895819	0.000031
R_* ($R_\odot$)	1.154	0.036
R_p (R_J)	1.367	0.043
inclination i	86.º525	0.º054
Period (days)	3.52474408	0.00000029
u_1(FGS)	0.549	0.054
u_2(FGS)	−0.043	0.081
u_1(STIS)	0.347	0.040
u_2(STIS)	0.228	0.066
χ^2	2.7	

the FGS only used 2 non-contiguous HST orbits. For each of these transits, HST was pointed at targets unrelated to this program during the mid-transit times. The changing HST attitude did not allow the temperatures of the telescope to stabilize and made it difficult to calibrate the PMT gain changes. Thus, some of the individual features in the FGS data at or below about 0.1% may be calibration artifacts.

To date, none of the analyses have indicated the presence of a moon. A moon of 2-Earth radii in size with a 1.5 day period would cause a ∼0.1% dip in the light curve, which would have been easily detected in the data. Tides would cause a moon of this size to decay into the planet in a time scale of order 10^5 years. One of the remaining tasks is to determine the upper limits for the radius and mass of an undetected hypothetical moon.

This work was supported by NASA through HST General Observer grant GO-09171.01-A from the Space Telescope Science Institute (STScI).

REFERENCES

1. Brown, T. M., et al. 2001, *ApJ*, **552**, 699.
2. Henry, G., Marcy, G., Butler, R. P., and Vogt, S. S. 1999, IAU Circ. 7307.
3. Henry, G., Marcy, G., Butler, R. P., & Vogt, S. S. 2000, *ApJ*, **529**, L41.
4. Charbonneau, D., Brown, T.M., Latham, D.W., and Mayor, M. 2000, *ApJ*, **529**, L45.
5. Nelan, E., and Makidon, R., *HST Data Handbook for FGS*, **4.0**, (2002), URL http://www.stsci.edu/instruments/fgs.
6. Mazeh, T., et al. 2000, *ApJ*, **532**, L55.
7. Press, W., et al. 2002, *Numerical Recipes in C++*, **Section 15.6**, Cambridge University Press.
8. Castellano, T., et al. 2000, *ApJ*, **532**, L51.
9. Robichon, N. and Arenou, F. 2000, *A&A*, **355**, 295.
10. Schultz, A.B. et al. 2003, ASP Conf. Ser. **294**, p. 479.

OGLE-TR-56

Guillermo Torres*, Maciej Konacki†, Dimitar D. Sasselov* and
Saurabh Jha**

*Harvard-Smithsonian Center for Astrophysics
†Caltech, Department of Geological and Planetary Sciences
**University of California–Berkeley, Department of Astronomy

Abstract. In early 2003 our team announced the discovery of the second extrasolar transiting planet, around the faint star OGLE-TR-56 ($V = 16.6$), based on the detection of small changes in the radial velocity of the primary. The star was originally identified as a candidate by the OGLE team from the shallow and periodic dips in its brightness. We present here new precise radial velocity measurements that confirm the variation measured earlier, supporting the conclusion that the companion is indeed a planet. Additional photometric observations are also available, which combined with the spectroscopy yield improved parameters (mass and radius) for the planet.

INTRODUCTION

One of the most successful searches for transiting extrasolar planets to date has been the OGLE survey, which has uncovered a total of 137 candidate transiting planets in several fields toward the bulge of the Galaxy and in the constellation of Carina [1, 2, 3, 4]. All of these stars show small dips in brightness at the level of a few percent, suggesting the presence of a low-mass companion in orbit. The stars are typically much fainter than the usual targets of transit surveys, ranging in brightness from about $V = 14$ to $V = 19$. The fields toward the Galactic center are also quite crowded.

In 2002 we began a systematic spectroscopic follow-up campaign in the bulge sample to weed out "false positives". Examples of false positives include a stellar (as opposed to a substellar) companion orbiting a large star (B-A main sequence star, or a giant), grazing eclipses in a stellar binary, or contamination by the light of a fainter eclipsing binary along the same line of sight (referred to as a "blend"), so that the otherwise deep eclipses of that binary are diluted by the other star and look very shallow (transit-like). As a result of that campaign, we were able to rule out up to 98% of the original candidates as false positives, and we then focussed on the remaining handful of good candidates for higher-precision work.

In early 2003 our team announced the discovery of the planet around the star OGLE-TR-56 ($V = 16.6$), with a period of only 1.21 days [5]. So far this is the only known transiting extrasolar planet aside from the well-known case of HD 209458b [6, 7].

We report here additional observations that confirm the original radial velocity variation detected by [5], and improve the estimate of the mass of the planet. We also report an updated light curve solution based on new OGLE photometry, yielding an improved planetary radius.

CP713, *The Search for Other Worlds: Fourteenth Astrophysics Conference*,
edited by S. S. Holt and D. Deming
© 2004 American Institute of Physics 0-7354-0190-X/04/$22.00

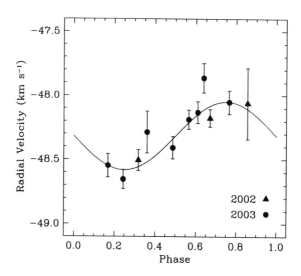

FIGURE 1. Radial velocity observations and velocity curve for OGLE-TR-56, as a function of orbital phase (fixed from the photometry).

SPECTROSCOPIC OBSERVATIONS AND ORBITAL SOLUTION

High-resolution spectra were obtained with the Keck I telescope using the HIRES instrument (High-Resolution Echelle Spectrometer) on two runs during August 1–3 and August 11–12 of 2003. This is the same instrumentation used by [5] for the original observations that led to the planet discovery. The resolving power achieved is $\lambda/\Delta\lambda \approx 65,000$. The wavelength reference was established using exposures of a Thorium-Argon lamp before and after each stellar exposure.

Radial velocities were obtained by cross-correlation against a synthetic template for OGLE-TR-56 with parameters matching those of the star. The uncertainties in the radial velocities are limited by systematics at the level of about 100 m s^{-1}, as described in more detail by [8]. Internal errors are smaller, ranging from 70 to 90 m s^{-1} in most cases.

Because the object transits the parent star every 1.21 days, the ephemeris is well known from the light curve analysis (see below). We have adopted this ephemeris for the spectroscopic solution, and so the only parameters that need to be adjusted are a radial velocity offset and the velocity semi-amplitude (K), since the orbit is assumed to be circular.

Our radial velocity curve is shown in Figure 1, where the circles represent the 8 new measurements (August 2003) and the triangles are based on the original 3 spectra by [5], re-reduced for uniformity.

The new measurements confirm the variation detected originally, although with a larger amplitude than the first estimate, which was based on only 3 observations (with two free parameters). The significance of the determination is now much greater, as

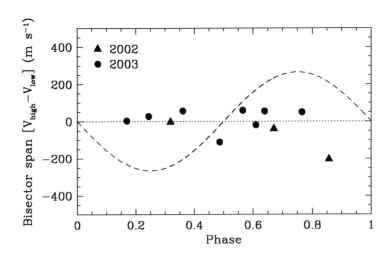

FIGURE 2. Bisector span used as a proxy for line asymmetry for each of our spectra of OGLE-TR-56, as a function of orbital phase. Over-plotted for reference is the velocity curve from Fig. 1, which shows that there is no correlation of the asymmetries with phase.

can be seen visually in the figure, and the errors are better characterized because of the increased number of observations. The measured K amplitude is 265 ± 38 m s^{-1}, and the RMS residual from the fit is 114 m s^{-1}. The minimum mass for the planet is $M_p \sin i = 1.33 \pm 0.21 \times 10^{-3}$ M$_\odot \times (M_* + M_p)^{2/3}$, where M_* is the mass of the parent star and i is the inclination angle of the orbit. The absolute mass of the planet is derived below.

TESTING FOR BLENDS: THE SPECTRAL LINE BISECTORS

Spurious radial velocity measurements can result if the spectral lines of the main star are blended with those of another fainter object, which can introduce slight asymmetries in the line profiles. Changes in the asymmetry correlating with orbital phase are a sign that the velocity variations are not real, and thus that there is no planet.

We have tested this for OGLE-TR-56 by computing the spectral line bisectors [e.g., 9] for each of our spectra directly from our cross-correlation functions, which are representative of the average spectral line profile. The difference in velocity between the bisectors at two different levels in the correlation function ('high' and 'low') is defined as the "bisector span", and is used as a measure of the asymmetry of the mean spectral line.

Figure 2 shows the bisector span as a function of orbital phase. For reference we display also the orbital solution from Figure 1. There is no significant correlation with phase, either positive or negative. This means that the velocity variations measured above are real, and thus confirms that we have a planet.

TABLE 1. Orbital and physical parameters for OGLE-TR-56b.

Parameter	Value
Orbital period (days)	1.2119189 ± 0.0000059
Transit epoch (HJD−2,400,000)	52075.1046 ± 0.0017
Center-of-mass velocity (km s^{-1})	−48.317 ± 0.045
Eccentricity (fixed)	0
Velocity semi-amplitude (m s^{-1})	265 ± 38
Inclination angle (deg)	81.0 ± 2.2
Stellar mass ($M_\odot$) (adopted)	1.04 ± 0.05
Stellar radius ($R_\odot$) (adopted)	1.10 ± 0.10
Limb darkening coefficient (I band)....	0.56 ± 0.06
Planet mass (M_{Jup})	**1.45 ± 0.23**
Planet radius (R_{Jup})	**1.23 ± 0.16**
Planet density (g cm^{-3})	**1.0 ± 0.3**
Semi-major axis (AU)	0.0225 ± 0.0004

RESULTS AND CONCLUSIONS

Additional photometric transits have been detected by the OGLE team in 2003 (for a total of 13), and refinements in the analysis have removed small systematic errors in the original photometry. We have carried out a new light curve solution resulting in an improvement in the photometric parameters.

The results for OGLE-TR-56 are given in Table 1. The planet is twice as massive as HD 209458b, but has approximately the same size and orbits at only half the distance to its parent star. This makes it particularly interesting from the theoretical point of view.

Several other good transit candidates from the lists released by the OGLE team are currently being followed up spectroscopically with Keck. With the measurement and analysis techniques we have developed, we expect to be able to discriminate with a high degree of confidence between true planets and false positives.

REFERENCES

1. Udalski, A., Paczyński, B., Żebruń, K., Szymański, M., Kubiak, M., Soszyński, I., Szewczyk, O., Wyrzykowski, Ł., & Pietrzyński, G. 2002a, Acta Astron., 52, 1
2. Udalski, A., Żebruń, K., Szymański, M., Kubiak, M., Soszyński, I., Szewczyk, O., Wyrzykowski, Ł., & Pietrzyński, G. 2002b, Acta Astron., 52, 115
3. Udalski, A., Szewczyk, O., Żebruń, K., Pietrzyński, G., Szymański, M., Kubiak, M., Soszyński, I., & Wyrzykowski, Ł. 2002c, Acta Astron., 52, 317
4. Udalski, A., Pietrzyński, G., Szymański, M., Kubiak, M., Żebruń, K., Soszyński, I., Szewczyk, O., & Wyrzykowski, Ł. 2003, Acta Astron., 53, 133
5. Konacki, M., Torres, G., Jha, S., & Sasselov, D. D. 2003a, Nature, 421, 507
6. Henry, G. W., Marcy, G. W., Butler, R. P., & Vogt, S. S. 2000, ApJ, 529, L4
7. Charbonneau, D., Brown, T. M., Latham, D. W., & Mayor, M. 2000, ApJ, 529, L45
8. Konacki, M., Torres, G., Sasselov, D. D., & Jha, S. 2003b, ApJ, 597,1076
9. Gray, D. F. 1992, The Observation and Analysis of Stellar Photospheres, 2nd Ed. (Cambridge: Cambridge Univ. Press), 417

First Results From Sleuth:
The Palomar Planet Finder

Francis T. O'Donovan*, David Charbonneau* and Lewis Kotredes*

*California Institute of Technology, M/C 105-24, Pasadena, CA 91125

Abstract. We discuss preliminary results from our first search campaign for transiting planets performed using Sleuth, an automated 10 cm telescope with a 6 degree square field of view. We monitored a field in Hercules for 40 clear nights between UT 2003 May 10 and July 01, and obtained an rms precision (per 15-min average) over the entire data set of better than 1% on the brightest 2026 stars, and better than 1.5% on the brightest 3865 stars. We identified no strong candidates in the Hercules field. We conducted a blind test of our ability to recover transiting systems by injecting signals into our data and measuring the recovery rate as a function of transit depth and orbital period. About 85% of transit signals with a depth of 0.02 mag were recovered. However, only 50% of transit signals with a depth of 0.01 mag were recovered. We expect that the number of stars for which we can search for transiting planets will increase substantially for our current field in Andromeda, due to the lower Galactic latitude of the field.

ACQUISITION AND ANALYSIS OF SLEUTH OBSERVATIONS

Sleuth[1], located at Palomar Observatory in Southern California, is the third transit-search telescope in our network which comprises STARE[2] (PI: T. Brown, located in Tenerife), and PSST (PI: E. Dunham, located in northern Arizona). Sleuth is an f/2.8 lens with a 10 cm aperture that images a $6° \times 6°$ field of view onto a 2048×2048 back-illuminated CCD camera. Sleuth conducts nightly observations with an SDSS r' filter, but also gathers color images in g', i' & z' during new moon. Sleuth automatically adjusts the focus for changes in temperature and filter. A separate f/6.3 lens feeds the guide camera. The automated observations, including operation of the clamshell enclosure, are controlled by a workstation running Linux. In the event of threatening weather, the on-site night assistant for the 200" telescope can close the system remotely, and an observatory weather station provides additional protection. At dawn, the night's data are automatically compressed and sent by ftp to our workstation at Caltech.

Between UT 2003 May 10 and July 01, we monitored approximately 10,000 stars $(9 < R < 16)$ in a field in Hercules. Figure 1 shows the calculated rms error in our photometry. We applied the STARE photometry code (written by T. Brown [HAO/NCAR] and with adaptations by G. Mandushev [Lowell Obs.]) to calibrate and perform weighted-aperture photometry upon the images. We subsequently combined the time series for each star for all nights. These light curves were then processed through a

---

[1] http://www.astro.caltech.edu/~ftod/sleuth.html
[2] http://www.hao.ucar.edu/public/research/stare/stare.html

CP713, *The Search for Other Worlds: Fourteenth Astrophysics Conference,*
edited by S. S. Holt and D. Deming
© 2004 American Institute of Physics 0-7354-0190-X/04/$22.00

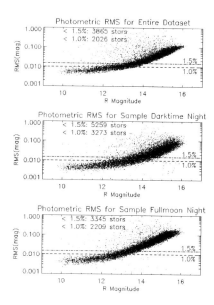

FIGURE 1. The calculated rms as a function of R magnitude for the 10,000 brightest stars in our Hercules field for three subsets of data: (upper panel) all observations; (center panel) observations from the night of June 29th UT during new moon; (lower panel) observations from the night of June 14th UT during full moon. The numbers of stars with rms below 1.0% and 1.5% for each dataset are shown.

cleaning pipeline to remove discrepant data and correlated variations in magnitude. Finally, the data were averaged in 15 minute bins to produce final light curves.

We then used the box-fitting algorithm of Kovács et al. [1] to search these light curves for transits with periods ranging from 1.5 to 7.5 days. This program assigned a Signal Detection Efficiency (SDE – see [1]) to the star, based on the significance of the transit detection. Figure 2 shows a histogram of the SDEs of the thousand brightest stars in the Hercules field, and compares it with the SDEs of a thousand simulated light curves with a Gaussian noise distribution of the same mean and variance as the original timeseries.

We also tested the transit search code by inserting simulated transits into our photometric data and attempting to recover these signals – an example of such an injected transit is shown in Figure 3. The recovery rate of these transits is about 85% for transits with a depth of 0.02 mag, and about 50% for 0.01 mag transits.

THE NEXT STEPS

Throughout September and October 2003, we gathered photometric data on a field in Andromeda. The lower Galactic latitude of this field ($b = -16°$ versus $b = +40°$ for the Hercules field) will lead to a significantly larger number of stars that we can search

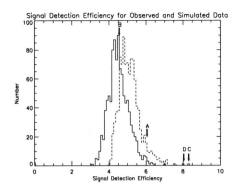

FIGURE 2. The solid line shows the histogram of derived values for the Signal Detection Efficiency (SDE) for the best-fit transit signal for each of the 1000 brightest stars. The histogram of derived SDE values for our simulated data set is plotted as the dashed line. The phased time series for two stars with large SDE values (indicated as C and D on the plot) are shown in Figure 4. We also inserted transits into the data and attempted to recover these: phased light curves of two such recovered systems are shown in Figure 3, and the resulting SDE values are indicated on the plot for the actual data (A) and the simulated data (B).

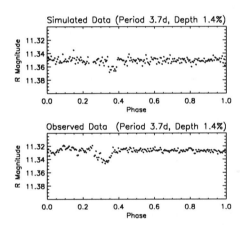

FIGURE 3. As a test of our ability to recover transit signals, we injected transits of varying period and depth into the data, as well as into simulated data. The recovered 1.4%-deep transits for a 3.7 day period for two such stars are shown for the simulated data (upper panel) and actual data (lower panel). The SDE values for these recoveries are shown in Figure 2.

for transiting planets. The Andromeda data also benefit from a firmware upgrade to our guider, which removed the large night-to-night pixel offsets that the Hercules data were subject to. Whereas the typical offsets in Hercules were 20 pixels, in our current field these offsets rarely exceed 5 pixels, and guiding within a single night is often good to 3 pixels.

We will perfom more detailed tests of our transit detection algorithm to calibrate and

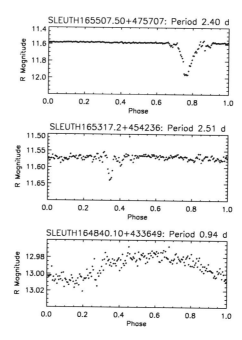

FIGURE 4. The phased light curve for: (C – upper panel) a grazing-incidence eclipsing binary system, and (D – center panel) a low-amplitude eclipsing system. The SDE values for these systems are shown in Figure 2. The phased light curve of a detected low-amplitude variable stars is also shown (lower panel).

improve the detection probability. We are concerned that the cleaning pipeline itself may be introducing photometric correlations between the target stars, and thus reducing our detection capability. We intend to introduce simulated transits into the data before cleaning, and verify that this processing does not remove these events.

Example Sleuth light curves of eclipsing binaries are shown in Figure 4. Figure 4 also shows the light curve of a variable star. We intend to compile a catalog of such binaries and variables. A dominant concern for any transit survey is the rejection of false positives, i.e. systems containing an eclipsing binary that mimic the photometric light curve of a transiting planet. (The eclipse depths in Figure 4 are too great to be mistaken for planetary transits.) To minimize resource-intensive spectroscopic follow-up work, we are planning to conduct our own high angular-resolution, multi-color photometry of transit candidates. To this end, we are currently building Sherlock (Kotredes et al. [2]), an automated follow-up telescope for wide-field transit searches.

REFERENCES

1. Kovács, G., Zucker, S., & Mazeh, T. A&A, **391**, 369-377 (2002).
2. Kotredes, L., Charbonneau, D., O'Donovan, F. T., & Looper, D. (2004), these proceedings.

Sherlock: An Automated Follow-Up Telescope for Wide-Field Transit Searches

Lewis Kotredes, David Charbonneau, Dagny L. Looper
and Francis T. O'Donovan

California Institute of Technology, M/C 105-24, 1200 E California Blvd, Pasadena, CA 91125;
ltk,dc,ftod@astro.caltech.edu, dagny@caltech.edu

Abstract. The most significant challenge currently facing photometric surveys for transiting gas-giant planets is that of confusion with eclipsing binary systems that mimic the photometric signature. A simple way to reject most forms of these false positives is high-precision, rapid-cadence monitoring of the suspected transit at higher angular resolution and in several filters. We are currently building a system that will perform higher-angular-resolution, multi-color follow-up observations of candidate systems identified by Sleuth (our wide-field transit survey instrument at Palomar), and its two twin system instruments in Tenerife and northern Arizona.

INTRODUCTION

Wide-field photometric surveys for transits of short-period gas-giant planets consist of several months of single-band observations of typically 5000 targets in a six-degree-square field of view. A number of such surveys, including the network consisting of Sleuth [1] (Palomar, PI: D. Charbonneau, see [1]), STARE [2] (Tenerife, PI: T. Brown) and PSST (Lowell, PI: E. Dunham) have produced several candidates with light curves very similar to that of HD 209458 [2]. The most significant challenge facing such surveys is not the difficulty of obtaining the requisite precision and phase coverage, but rather the ability to rule out the large number of false positives that are typically encountered. These false positives can be removed through high-resolution spectroscopy, as was done in the case of the OGLE-III transit candidates [3, 4]. However, for relatively bright stars, there is an easier way to reject such candidates.

Brown [5] discusses various transit-like signals typically identified by searches similar to Sleuth and classifies them into a number of specific types. There are three primary sources of these false positives. The first of these are binary stars undergoing a grazing eclipse (MSU). The second and third types involve blends between a binary star undergoing a deep eclipse and a third star, either a foreground object (MSDF) or a third member of the system (MSDT). Using a typical transit campaign for the STARE instrument, Brown calculates that for every 10000 stars observed, 0.39 planets will be detected with three transits. However, 2.27 false positives of type MSU will also appear, 1.26 false

---

[1] http://www.astro.caltech.edu/~ftod/sleuth.html
[2] http://www.hao.ucar.edu/public/research/stare/stare.html

CP713, *The Search for Other Worlds: Fourteenth Astrophysics Conference,*
edited by S. S. Holt and D. Deming
© 2004 American Institute of Physics 0-7354-0190-X/04/$22.00

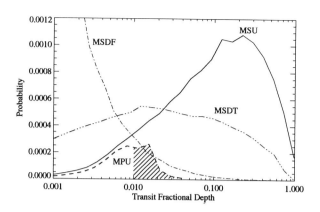

FIGURE 1. Taken from [5], the marginalized probabilities of detection of a transit per unit log transit depth. The shaded area is represents planets detectable in current surveys. Figure labels are explained in the text.

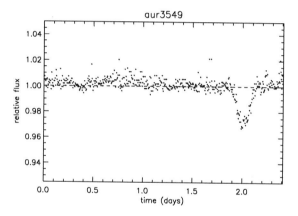

FIGURE 2. Phased R-band light curve of a star in Auriga as observed with the PSST by E. Dunham & G. Mandushev (Lowell Observatory). Based on this light curve alone, this appears to be a reasonable candidate, as the period, duration, and depth are consistent with the passage of an inflated gas-giant across a Sun-like star. However, follow-up photometry shows this object to be a blend containing an eclipsing binary (see Figure 3).

positives of type MSDF will be present, and 0.98 candidates of type MSDT will be observed. Combining these numbers, Brown finds that a transit survey will observe more than 10 false positives for every true planet. This is demonstrated by Figure 1, which shows the probability of finding different classes of object at a given transit depth. In the regime of 1-2 % eclipse depths that we study, these three sources of false positives are at least as plentiful as the planets we seek.

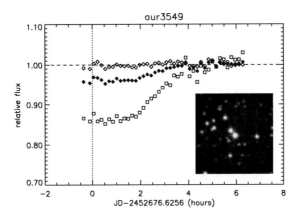

FIGURE 3. The inset Digitized Sky Survey image shows this target to consist of a central bright source and an adjacent fainter star to the NE. Both of these stars are contained within the PSF of the PSST. We carried out follow-up photometry of this field with a 14-inch telescope located on the roof of the Caltech Astronomy building, during a night when an eclipse was predicted to occur. Observations began during mid-eclipse. The predicted time of eclipse occurs at t=0 of this figure. The open diamonds show the light curve for the central star, which is evidently uneclipsed. The open squares show the light curve for the star to the NE, which undergoes a deep (14 %) eclipse. When the light from both stars is summed in a single aperture, the apparent eclipse depth is reduced to $\sim$ 3 %, reproducing the PSST light curve (see Figure 2).

SHERLOCK: REMOVING THE ECLIPSING BINARIES

We are currently assembling a telescope named Sherlock that will be dedicated to examining transit candidates from Sleuth, as well as a number of other transit surveys which monitor stars brighter than V=13. Highly automated and observing in custom RGB filters with a better angular resolution (1.7 arcsec/pixel) than Sleuth (10 arcsec/pixel), Sherlock will be able to reject most of the contaminants from these transit surveys. In so doing, Sherlock will greatly reduce the rate of false positives, bringing the number of transit candidates to a manageable level. Figures 2 & 3 show an example of how blends of eclipsing binaries can be separated from transiting planets. PSST identified a star in Auriga undergoing 2.5 % deep eclipses with a period of 2.4 days (Figure 2). Examination of the Digitized Sky Survey images (Figure 3) revealed two sources within the PSF of the PSST instrument. Photometry of the field during a subsequent eclipse with a 14-inch telescope on the rooftop of the Caltech Astronomy building demonstrated that the fainter star was undergoing deep (14 %) eclipses. When the light from both stars was summed within a single photometric aperture, the transit shape and depth as observed by the PSST was reproduced (Figure 3).

Measurement of the color dependence of the transit depth should remove both grazing incidence binaries (due to the effect of limb-darkening on the eclipse depth), and blends of eclipsing binaries (due the change in relative brightness of the blended and occulted stars as a function of color). In contrast to this, planetary transits should be nearly constant with color. Furthermore, the increased angular resolution will separate the light from physically unassociated blended stars, and subsequent photometry will reveal

Meade LX200GPS 10" f/6.3 Schmidt-Cassegrain Telescope
Apogee 1024 x 1024 pixel back-illuminated CCD camera
Filter wheel containing custom RGB and clear filters
SBIG STV Autoguider
Automated operation controlled by Linux workstation
Cloud cover monitored by Snoop, the Palomar All-Sky Camera

which object is undergoing eclipses. Sherlock will not be able to reject all sources of false detection. Central eclipses by very dim stellar objects (notably M dwarfs) will not show detectable variation with color. In addition, blends wherein the occulted and blending star have the same color will not be distinguished. Even with these exceptions, however, Sherlock is expected to reduce greatly the ratio of false positives from these surveys. Multi-epoch spectra of viable candidates will be gathered with high-resolution spectrographs on 1-2 m class telescopes, which should rule out the presence of stellar or brown dwarf companions. Surviving candidates will be monitored with Keck HIRES to determine the radial velocity orbit induced by the planetary companion.

SHERLOCK SPECIFICATIONS

Table 1 lists the specifications for Sherlock. The system will be located in the same clamshell enclosure as Sleuth, our primary transit search instrument. Weather decisions are made by the on-site 200-inch telescope night assistant, with additional protection provided by a weather station capable of closing the clamshell roof. Sherlock will be completely automated, calculating future times of eclipse for all active candidates, and observing the highest priority object in eclipse each night. The automated nature of the system is an advantage over comparatively labor- and resource-intensive multi-epoch spectroscopic follow-up. In addition, this dome hosts the new all-sky camera Snoop [3], which provides weather monitoring for the observatory. Given that observing time on this system should be plentiful, we invite teams conducting wide-field transit surveys to contact us regarding follow-up of their candidates.

REFERENCES

1. O'Donovan, F. T., Charbonneau, D., and Kotredes, L., 2003, AIP Conf Proc: The Search for Other Worlds, eds. S. S. Holt & D. Deming, (Springer-Verlag), astro-ph/0312289
2. Charbonneau, D., Brown, T. M., Latham, D. W., and Mayor, M., 2000, ApJ, 529, 245
3. Konacki, M., Torres, G., Jha, S., and Sasselov, D. D., 2003, Nature, 421, 507
4. Konacki, M., Torres, G., Sasselov, D. D., and Jha, S., 2003, ApJ, 597, 1076
5. Brown, T., 2003, ApJ, 593, L125

---

[3] http://snoop.palomar.caltech.edu

EXPLORE/OC: A Search for Planetary Transits in the Field of the Southern Open Cluster NGC 6208

Brian L. Lee*, Kaspar von Braun†, Gabriela Mallén-Ornelas**,
H. K. C. Yee*, Sara Seager† and Michael D. Gladders‡

*Dept. of Astronomy & Astrophysics, Univ. of Toronto, 60 St. George Street, Toronto,
Ontario M5S 3H8 Canada
†Department of Terrestrial Magnetism, 5241 Broad Branch Road NW, Washington, DC 20015 USA
**Harvard-Smithsonian Center for Astrophysics, 60 Garden Street, Cambridge, MA 02138 USA
‡Carnegie Observatories, 813 Santa Barbara Street, Pasadena, California 91101 USA

Abstract. The EXPLORE Project expanded in 2003 to include a campaign to monitor rich southern open clusters for transits of extrasolar planets (EXPLORE/OC). In May and June 2003, we acquired precise, high-cadence photometry of the second open cluster in our campaign, NGC 6208. Here, we present preliminary results from our *I*-band survey of over 60000 stars in the field of NGC 6208, around 5000 of which were monitored with photometric precision better than 1%.

INTRODUCTION

The EXPLORE collaboration has previously monitored Galactic plane fields for evidence of transiting close-in extrasolar giant planets [1]. Using the experience built up in the deep Galactic plane searches, we are now conducting a complementary study of open clusters.

While the potential number of monitorable stars in the Galactic plane fields is generally higher than in open clusters, an open cluster offers a large sample of stars sharing a common age, metallicity, distance, and reddening. By trading off the number of stars that can be monitored, we thus gain control over the properties of our stellar sample (although, since open clusters are concentrated in the Galactic plane, many of the stars will be from the field and not part of the cluster sample; also, we may see differential reddening across the field). In surveying a specific cluster, we accumulate statistics for stars sharing a common set of relatively easily measured properties, and we can target a specific range of spectral types by adjusting the exposure time. Our campaign will be useful in order to constrain the occurrence of planets as a function of age and metallicity, and will also allow us to compare the frequency of planets in a cluster environment to that in the Galactic field.

NGC 6208 (distance=1.10kpc, E(B-V)=0.35, [Fe/H]=0.00 [2]; age=1.00 Gyr [3]) was chosen out of a list of potential targets because of its observability, its high star count, and a distance and reddening such that with five-minute exposures, cluster G and K dwarfs would typically be monitored with better than 1% photometry in *I*. We conduct our survey in the *I*-band (rather than *V* or *R*) to minimize the effects of limb-darkening on the

CP713, *The Search for Other Worlds: Fourteenth Astrophysics Conference*,
edited by S. S. Holt and D. Deming
© 2004 American Institute of Physics 0-7354-0190-X/04/$22.00

slope of a transit's ingress and egress, allowing for better discrimination of good transit candidates from other low-amplitude variations (such as blends and grazing binaries) [4].

Here, we present some of our observations, data reduction techniques, and preliminary results from our survey of NGC 6208. Preliminary results of the study of the first target in the EXPLORE/OC campaign, NGC 2660, are presented in a companion paper [5] in this volume.

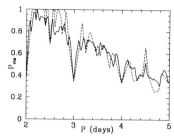

FIGURE 1. Probability P_{vis} of detecting an existing transiting planet, assuming the signal is large enough to detect, as a function of orbital period P in days, and averaged over all phases. A detection requires two transits to occur during the nights of an observing run. The solid line shows the total probability for our actual NGC 6208 run; the probability is similar to that achieved for a hypothetical perfect run of only 18 consecutive 10.8 hour nights, shown by the dashed line. Transits with integer periods are statistically difficult to detect because for approximately half the phases, the transits will always occur during the day. The mean P_{vis} for 2-5 day periods for our run was 62%.

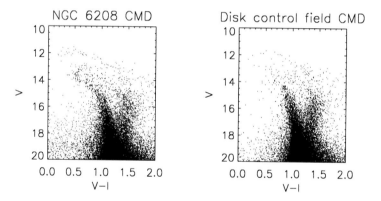

FIGURE 2. Colour-magnitude diagrams of NGC 6208 ($l = 333.7°$, $b = -5.8°$) and an off-cluster comparison field ($l = 334.7°$, $b = -5.8°$). The bright end of the cluster main sequence appears clearly on top of the Galactic field population in the NGC 6208 colour-magnitude diagram.

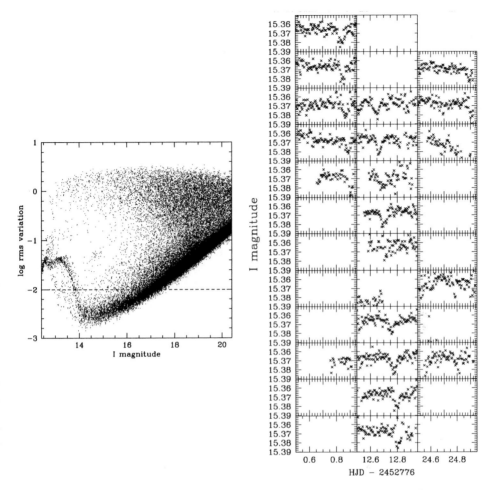

FIGURE 3. Left panel: Precision of one night of *I*-band photometric monitoring of the field of NGC 6208. Approximately 5000 stars show RMS variation less than 1% (indicated by the dashed line) on this night. **Right panel:** Light curve of an $I = 15.4$ star in the field, with a periodic transit-like photometric signal at the $\sim$2% level. The light curve from the first night (HJD=2452776) of the run is in the bottom left; light curves from successive nights are shown progressing upwards to the top of the page (then wrapping to the bottom of the next column). No monitoring was done on nights 1, 2, 4-7, 23-26, 31, and 32.

OBSERVATIONS

From May 16 to June 19, 2003, we observed NGC 6208 ($l = 333.7°$, $b = -5.8°$) using the Swope 1-m telescope at Las Campanas Observatory, Chile. We obtained 1730 *I*-band monitoring frames over $\sim$ 22 clear nights out of a run one-third interrupted by poor winter weather (the effect of this loss on transit detection is illustrated in Figure 1). When monitoring NGC 6208, we took successive 300 sec. I-band exposures of a $15' \times 23'$ field centred on the cluster, imaged onto 2048×3350 CCD pixels, with 130 seconds readout. On a photometric night, we also obtained a handful of BVR frames for photometric calibration and construction of colour-magnitude diagrams, and $BVRI$ frames of a nearby off-cluster comparison field centred at ($l = 334.7°$, $b = -5.8°$). We have collected low-resolution ($R \approx 2000$) spectra of selected variable and selected typical stars from our field (~ 50 in all) for MK classification; these data were taken on July 27 and 28, 2003, using the B&C spectrograph on the Magellan Baade 6.5-m telescope at Las Campanas, using the 1200 lines/mm grating blazed at 4000Å, providing a spectral coverage of 3607Å-5239Å and a dispersion of 0.8Å/pixel.

ANALYSIS AND RESULTS

The $BVRI$ data were used to construct colour magnitude diagrams of the NGC 6208 field and the nearby Galactic field (see Figure 2). The cluster main sequence is clearly distinguishable from the Galactic field population for $I < 13.5$. There, the star count for the cluster field is 21% above the count for the comparison field. The percentage of Galactic field stars per magnitude bin grows towards fainter magnitudes.

The monitoring data were analyzed using the existing EXPLORE photometric pipeline [1]. This fast pipeline allowed us to make light curves of a given night's data with one day's turnover, and hence to begin looking for transit-like features while still at the telescope. Figure 3 shows the root mean square variation in the I magnitude over one night's monitoring, and demonstrates that we achieve 1% precision from $I \approx 17$ up to saturation of the detector near $I \approx 14.2$.

We examined, by eye, all of the best light curves (unphased photometric precision better than $\sim 1.5\%$, ~ 15000 light curves). To flag a light curve as a candidate transit, we required the detection of at least two low-amplitude eclipse-like features. Our window function for detecting two transits is shown in Figure 1.

In our data, we found one light curve featuring a transit-like signal (see Figure 3). Preliminary spectroscopic data indicate this star's spectral type is late G; however, because the ingresses and egresses of the eclipses are relatively shallow, it is unlikely to be a planet [4]. While this star may not harbour a transiting planet, it is one example of our ability to rapidly detect variability at the 1% level in our open cluster surveys.

REFERENCES

1. Mallén-Ornelas, G., Seager, S., Yee, H. K. C., Minniti, D., Gladders, Michael D., Mallén-Fullerton, G. M., Brown, T. M., *ApJ*, **582**, 1123 (2003).
2. Twarog, B. A., Ashman, K. M., and Anthony-Twarog, B. J., *AJ*, **114**, 2556 (1997).
3. Piatti, A. E., Clariá, J. J., and Abadi, M. G., *AJ*, **110**, 2813 (1995).
4. Seager, S., and Mallén-Ornelas, G., *ApJ*, **585**, 1038 (2003).
5. von Braun, K., Lee, B. L., Mallén-Ornelas, G., Seager, S., Yee, H. K. C., and Gladders, M. D., this volume (2004)

EXPLORE/OC: A Search for Planetary Transits in the Field of the Southern Open Cluster NGC 2660

Kaspar von Braun[*], Brian L. Lee[†], Gabriela Mallén-Ornelas[**],
Howard K. C. Yee[†], Sara Seager[*] and Michael D. Gladders[‡]

[*]*Department of Terrestrial Magnetism, Carnegie Institution of Washington,*
5241 Broad Branch Road, Washington, DC 20015
[†]*Dept of Astronomy and Astrophysics, University of Toronto,*
60 St. George St., Toronto, Canada M5S 3H8
[**]*Harvard-Smithsonian Center for Astrophysics,*
60 Garden Street, MS-15, Cambridge, MA 02138
[‡]*Carnegie Observatories,*
813 Santa Barbara St., Pasadena, CA 91101

Abstract. We present preliminary photometric results of a monitoring study of the open cluster NGC 2660 as part of the EXPLORE/OC project to find planetary transits in Galactic open clusters. Analyzing a total of 21000 stars (3000 stars with photometry to 1% or better) yielded three light curves with low-amplitude signals like those typically expected for transiting hot Jupiters. Although their eclipses are most likely caused by non-planetary companions, our methods and photometric precision illustrate the potential to detect planetary transits around stars in nearby open clusters.

INTRODUCTION

As part of the EXPLORE[1] Project [1], we have recently begun a survey of southern open clusters (OCs) with the aim of detecting planetary transits around cluster member stars (EXPLORE/OC[2]). Probing cluster populations provides a complement to our ongoing deep monitoring studies of rich Galactic fields [1].

Open cluster monitoring provides the following advantages and incentives:

- In general, metallicity, age, distance, and foreground reddening are either known or may be determined for cluster members (more easily than for random field stars). Thus, planets detected around cluster stars will readily represent data points for any statistic correlating planet frequency with age or metallicity of the parent star.

- The planet-formation processes, and hence planet frequencies, may differ between the open cluster and Galactic field populations. This study allows the EXPLORE Project to compare these two different environments.

---

[1] http://www.ciw.edu/seager/EXPLORE/explore.htm
[2] http://www.ciw.edu/seager/EXPLORE/open_clusters_survey.htm

CP713, *The Search for Other Worlds: Fourteenth Astrophysics Conference,*
edited by S. S. Holt and D. Deming
© 2004 American Institute of Physics 0-7354-0190-X/04/$22.00

- Specific masses and radii for cluster stars may be targeted in the search by the choice of cluster and by adjusting exposure times for the target. In general, smaller stars offer better chances to detect the low-amplitude transit signal.

The difficulties and challenges involving open cluster surveys are:

- The number of monitored stars is typically lower than in rich Galactic fields, reducing the statistical chance of detecting planets.
- Determining cluster membership of stars in the open cluster fields without spectroscopic data is difficult due to the contamination by Galactic field stars. Since the clusters are typically concentrated toward the Galactic disk, this contamination may be significant.
- Significant differential reddening across the cluster field and along the line of sight can make isochrone fitting (and subsequent determination of physical parameters such as age, distance, and metallicity) difficult.

Note that both of the latter two difficulties may at least in part be circumvented by obtaining spectra. In this work, we illustrate some of the points mentioned above, describe our observing and data-reduction strategies, and show some of our preliminary results of the southern open cluster NGC 2660. Preliminary results of the study of our second target, NGC 6208, are presented in a companion paper [2] in this volume.

DATA

Our observations of NGC 2660 ($l = 265.9°; b = -3.01°$) were obtained with the Carnegie Institution's Swope 1m Telescope at the Las Campanas Observatory in Chile during the nights of 2003 Feb. 10–28 which were partially hampered by poor weather conditions. Our field of view was approximately 23' × 15' in size. Only I-band data with a cadence of around 7 minutes were obtained for monitoring. In that filter, a transit will be most easily distinguishable from contaminants such as grazing binaries because color-dependent limb-darkening results in steeper ingress and egress as well as a flatter eclipse bottom in I than in other bands [3]. We show in Fig. 1 our observing window function which indicates the likelihood of our detecting *existing* transits and measuring their periods to within an aliasing factor (we require at least two transits). A necessary condition for detecting existing transits is, of course, high-precision photometry [1]. In Fig. 2, we show our photometric precision as a function of magnitude.

RESULTS

Around 7000 unphased photometric light curves of the stars with the best photometric precision were visually inspected to detect transit-like signals. We show in Fig. 3 three examples of light curves with the low-amplitude signature typically expected for transiting hot Jupiters. Although the eclipses visible in the light curves are most likely not due to planets, we are currently analyzing spectral data on them to determine spectral types of the parent stars as well as their variations in radial velocities. Regardless of whether or not we are able to find a planetary transit in NGC 2660, we have illustrated the potential of our methods to detect (unphased) amplitude variations of less than 1%, the most fundamental requirement to detect planetary transits.

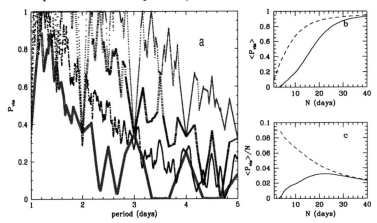

FIGURE 1. Probability P_{vis} of detecting existing transiting planets with different orbital periods. P_{vis} is calculated with the requirement that two transits must be observed. **Panel a:** P_{vis} of detecting 2 transits of an existing transiting planet with a period between 2 and 5 days after 21 (top curve), 14 (second curve from the top) and 7 (bottom curve) consecutive, uninterrupted nights of observing (10.8 hours per night). The difficulty of detecting some phase angles is shown by the dips in the curves (e.g., orbital periods of an integer number of days may always feature their transits during the day and are therefore statistically harder to detect). All phases are considered for each period. The second curve from the bottom shows the real P_{vis} for our monitoring study of NGC 2660 (19 nights of 7-8 hours per night, with interruptions due to weather and telescope scheduling; see Fig. 3). **Panel b:** The mean P_{vis} as a function of number of consecutive nights in an observing run (10.8 hour nights). The solid line is for the requirement to detect two transits and the dashed line for one transit. This figure indicates how much the likelihood of finding existing transits grows with an increasing number of nights of observing. **Panel c:** The efficiency of $<P_{vis}>$ per night. For the two-transit requirement (solid line) and 10.8 hour long nights, an observing run of 21 nights is most efficient. For the single transit requirement, the efficiency decreases monotonically with the number of nights since additional nights have progressively lower probabilities of detecting "new" transits.

ACKNOWLEDGMENTS

We would like to express our gratitude to the Las Campanas Staff for their unparalleled helpfulness and dedication to optimizing every little aspect of our observing runs.

REFERENCES

1. Mallén-Ornelas, G., Seager, S., Yee, H. K. C., Minniti, D., Gladders, Michael D., Mallén-Fullerton, G. M., & Brown, T. M. 2003, AJ, 582, 1123
2. Lee, B., von Braun, K., Mallén-Ornelas, G., Yee, H.K.C., Seager, S., & Gladders, M. 2004, this volume
3. Seager, S., & Mallén-Ornelas, G. 2003, ApJ, 585, 1038

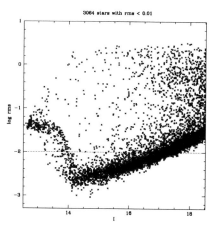

FIGURE 2. Photometric precision of night 1 of our monitoring run of NGC 2660. Of the roughly 21000 monitored stars, slightly more than 3000 have photometry of precision 1% or better. This rms is measured as the scatter around the mean magnitude of the star under investigation. The 1%-photometry stars cover a magnitude range of slightly more than 2.5 mags. By adjusting the exposure time, one can therefore target OC member stars of a range of certain spectral types to maximize the likelihood of detecting a transit (taking into account distance to the cluster and foreground reddening).

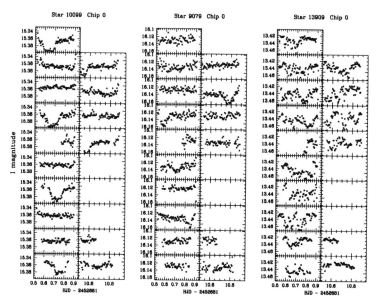

FIGURE 3. Examples of real-time (unphased) light curves from our monitoring run of NGC 2660. Every panel represents data taken during one night, starting on the bottom left with night 1. Night 2's data are directly above it, night 3 above that and so on. We did not obtain any data during nights 13-15. All three panels display the low-amplitude signal that we are looking for in the search for planetary transits even though they are most likely caused by a larger-sized companion (left panel) or grazing binaries (middle and right panels) [3].

KELT: The Kilodegree Extremely Little Telescope

Joshua Pepper*, Andrew Gould* and D. L. DePoy*

*Department of Astronomy, 4055 McPherson Lab, 140 West 18th Ave., Columbus, OH 43210-1173

Abstract. Transits of bright stars offer a unique opportunity to study detailed properties of extrasolar planets that cannot be determined through radial-velocity observations. We propose a technique to find such systems using all-sky small-aperture transit surveys. The optimal telescope design for finding transits of bright stars is a 5 cm "telescope" with a $4k \times 4k$ camera. We are currently building such a system and expect to detect ~ 10 bright star transits after one year of operation.

INTRODUCTION

Transit surveys are widely believed to provide the best means to discover large number of extrasolar planets. At the moment, all ongoing transit surveys are carried out in relatively narrow pencil beams. They make up for their small angular area with relatively deep exposures. These surveys are potentially capable of establishing the frequency of planets in various environments, but they are unlikely to find the kinds of transits of bright stars that would be most useful for intensive follow-up analysis. Although some of the surveys of field stars are considered "wide field", their total survey areas are small compared to 4π steradians.

OPTIMAL TELESCOPE DESIGN

It can be demonstrated that the best way to find HD209458b - type transits is with an all-sky photometric survey. According to Pepper, Gould & DePoy [1] (hereafter PGD),

$$\frac{dN_t}{dM_V} = 5\frac{\Omega}{4\pi}F(M_V)\frac{f(M_V,r,a)}{0.75\%}\left(\frac{a}{a_0}\right)^{-5/2}\left(\frac{r}{r_0}\right)^6\left(\frac{\gamma}{\gamma_0}\right)^{3/2}\left(\frac{\Delta\chi^2_{min}}{36}\right)^{-3/2} \quad (1)$$

where γ is the number of photons collected from a fiducial $V = 10$ mag star during the entire experiment, $\Delta\chi^2_{min}$ is the minimum difference in χ^2 between a transit lightcurve and a flat lightcurve required for candidate detection, and $F(M_V)$ is the function:

$$F(M_V) = \left[\frac{n(M_V)}{n_0}\right]\left[\frac{L(M_V)}{L_0}\right]^{3/2}\left[\frac{R(M_V)}{R_0}\right]^{-7/2} \quad (2)$$

In equation (1) we have adopted $\gamma_0 = 1.0 \times 10^7$, $a_0 = 10R_\odot$, $r_0 = 0.10R_\odot$, and in equation (2) we have made our evaluation at $M_V = 5$ mag (i.e. $R_0 = 0.97R_\odot$, $L_0 =$

CP713, *The Search for Other Worlds: Fourteenth Astrophysics Conference*,
edited by S. S. Holt and D. Deming
© 2004 American Institute of Physics 0-7354-0190-X/04/$22.00

$0.86 L_{\odot}$, and $n_0 = 0.0025\,\mathrm{pc}^{-3}$). Note that $\gamma_0 = 1.0 \times 10^7$ corresponds to approximately 500 20-second exposures with a 5 cm telescope and a broadened (V+R) type filter for one $V = 10$ mag fiducial star. Since N_t depends on the characteristics of the survey primarily through the total photon counts, γ, and only logarithmically (through $\Delta\chi^2_{\min}$) on the sampling strategy and the size of the explored parameter space (see PGD), telescope design must focus on maximizing γ, which is given by

$$\gamma = \frac{KEL^2 T}{\Omega F^2} \tag{3}$$

where E is the fraction of the time actually spent exposing, L is the linear size of the detector, T is the duration of the experiment, F is the focal ratio of the optics, and K is a constant that depends on the telescope, filter, and detector throughput. (For these calculations, we will assume $K = 40\,\mathrm{e^- cm^{-2} s^{-1}}$, which is appropriate for a broad V+R filter and the fiducial $V = 10$ mag star.) Interestingly, almost regardless of other characteristics of the system, the camera should be made as fast as possible. We will adopt $F = 1.8$, beyond which it becomes substantially more difficult to design the optics. A more remarkable feature of equation (3) is that all explicit dependence on the size of the primary optic has vanished: a 1 cm telescope and an 8 m telescope would appear equally good.

PGD shows that two additional considerations (read-out time, scintillation noise) drive one toward smaller apertures, while two others (sky noise and focal-plane distortion) prohibit going beyond a certain minimal size. Combined, these four effects imply a sweet spot,

$$D \approx (5\mathrm{cm})\, 10^{0.2(V_{\max} - 10)} \tag{4}$$

where $V_{\max}$ is the targeted magnitude limit of the survey.

THE KILO-SQUARE-DEGREE EXTREMELY LITTLE TELESCOPE (KELT)

Motivated by the logic of the "KELT equation" (3), we have begun building the Kilo-degree Extremely Little Telescope (KELT). As suggested by equation (4), KELT has a D = 4.2 cm, f/1.9 lens (a Mamiya medium-format photographic lens) mounted on a 4096×4096 CCD with 9 μm pixels (an Apogee Instruments AP16e camera using a Kodak KAF-16801E detector). The images are Nyquist sampled over the entire field, as required to simultaneously minimize the problems of sky noise and intra-pixel variations. Field tests show that the focal-plane distortions are manageable even in the corners of the $26° \times 26°$ field. Test images were taken with this initial KELT system in June 2003 in Ohio. The images are roughly as expected for the optical performance of the Mamiya medium-format camera lens used. Figure 1 shows representative light curves for several of the stars. In general, we achieve 1-4% photometric accuracy for stars with $V = 6 - 10$ mag. These estimates are based on simple aperture photometry against a $V = 19$ mag arcsec^{-2} sky, and we expect a factor of 2-4 improvement using DoPhot data reductions of photometry at Kitt Peak.

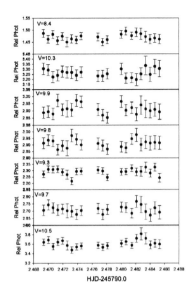

FIGURE 1. Relative light curves of 7 stars located near the center of the field. These data were taken at a site in Ohio in June 2003. The approximate visual magnitudes are given in each panel. In general, we achieve 1-4% photometric accuracy for stars with $V = 6 - 10$ mag. These estimates are based on simple aperture photometry against a $V = 19$ mag arcsec^{-2} sky. We expect a factor of at least 2-4 improvement using DoPhot data reductions of photometry at Kitt Peak.

Data Acquisition

We plan to install KELT at a host site in New Mexico, where the telescope will be in an enclosure that will be opened every night. The telescope will execute a standard cycle of observations in fixed terrestrial (alt-az) coordinates, covering the areas of the sky within about 45° of the zenith. Each 30 s exposure will be tracked, and pointing to the next field will be executed during the 30 s read-out time. The region within 45° of zenith could be covered in 10 separate pointings, but the two most southerly ones will be duplicated and one of the two most northerly deleted in order to equalize sky coverage over the long term. That is, each cycle will require 11 minutes, after which it will be repeated on the same section of the terrestrial sky (which has now moved 3.75° to the east in celestial coordinates). In this way, the observations will (over the course of entire year) obtain roughly uniform coverage of about 2π of the sky, roughly the northern hemisphere.

We anticipate $E = 50\% \times 20\% = 10\%$, where the first factor accounts for read-out time and the second for time lost to daylight, weather, and instrument problems. We will employ a broad (V + R) filter for which we expect $K = K_0 = 40\,\mathrm{e}^-\mathrm{cm}^{-2}\mathrm{s}^{-1}$. Hence, according to equation (3), during one year of observations, $\gamma = 7 \times 10^7$ photons will be collected from each $V = 10$ mag star. That is $\gamma = 7 \times \gamma_0$. Recall from equation (1) that γ_0 photons were required to probe $V_{max} = 10$ mag stars for the transit of Jupiter-sized planets.

However, equation (1) is based on source-photon statistics alone. According to PGD, the photon requirements should be increased by a factor $1^2 + (4.2/5)^{2/3} + 0.75(4.2/5)^{-2}(1.9/1.8)^{-2} = 2.8$. Hence, in 1 year, there is a "margin of safety" (to allow for unanticipated and/or unmodeled problems) of a factor 2.5 in photon counts, corresponding to a factor 1.6 in S/N. For three years of operation, the margin of safety is about a factor 2.7 in S/N. We believe that this is adequate to ensure success.

Data Analysis

Photometry of each image will be carried out using a modified DoPhot package that is already well tested on microlensing data. The positions and magnitudes of all target stars brighter than 10 mag (and indeed several mag fainter) are already known from Tycho-2. Where necessary, this can be supplemented with USNO-A data to fainter mags. Thus, DoPhot can operate in its more efficient fixed-position mode. The main modification required will be to take account of clouds. In traditional "small-field" (< 30 arcmin) monitoring, accurate relative photometry is not seriously affected by clouds because the clouds dim all stars by approximately the same amount. This will certainly not be the case for our $\sim 25°$ field. We will attempt to mitigate this problem by measuring star brightnesses only relative to nearby stars. In the final analysis, however, we expect to lose more time to clouds than is lost in smaller-field monitoring.

Identification of transit candidates should be fairly straightforward. At $V = 10$ mag, there will be 1% errors (allowing for both scintillation and sky noise). Of course, this is in itself not good enough to plausibly identify the 1% transits due to Jupiter-sized planets. However, a 2-hr transit should yield 11 such measurements and hence a 3σ signal. Hence, it is not necessary to test all possible folds of the lightcurve to search the very large (period/phase) parameter space: one can focus first on the restricted space consistent with all subsets of the 2σ individual transit detections.

In the Northern sky, there are approximately 153,000 stars with $V \leq 10$ mag. Gould & Morgan [2] showed that about 104,000 of these can be identified as being significantly evolved or early type (and so too large to be useful for planetary transits) using a Tycho-2 reduced proper motion (RPM) diagram. Rejection of evolved stars will not only speed up the data processing, it will also remove one of the major sources of false candidates, namely K giants blended with eclipsing binaries (whether forming an associated triple or not). Vetting of the remaining candidates by follow up photometry during predicted transits using larger telescopes and by RV measurements will be much easier than for transit candidates from other surveys, simply because the KELT candidates will be extremely bright.

REFERENCES

1. Pepper, J., Gould, A., and DePoy, D. L., *Acta Astronomica*, **53**, 213 (2003)
2. Gould, A., and Morgan, C. W., *The Astrophysical Journal*, **585**, 1056 (2003).

Transit Spectroscopy of the Extrasolar Planet HD 209458b: The Search for Water

Patricio Rojo[*][†], Joseph Harrington[**], Dara Zeehandelaar[**],
John Dermody[**], Drake Deming[‡], David Steyert[‡], L. Jeremy Richardson[§]
and Günter Wiedemann[¶]

[*]*Cornell University, Ithaca, NY 14850-6801*
[†]*e-mail:pato@astro.cornell.edu*
[**]*Center for Radiophysics and Space Research (CRSR), Cornell University, Ithaca, NY 14850-6801*
[‡]*Planetary Systems Branch, Code 693, NASA's GSFC, Greenbelt, MD 20771*
[§]*NASA's GSFC, Code 693, Greenbelt, MD 20771 and LASP/U, Colorado, 1234 Innovation Drive, Boulder, CO 80303*
[¶]*Universitäts Sternwarte Jena, Schillergässchen 2, 07745 Jena, Germany*

Abstract. We are developing a technique to measure the atmospheric composition of extrasolar planets through transit spectroscopy. Current observational capabilities have not yet reached enough sensitivity to detect the Earth-like planets that life as we know it requires to evolve. We anticipate, however, that this technique will detect constituents of the atmospheres of Earth-like planets once future space-based observatories become sensitive enough.

We are currently using our methods on the extrasolar close-in giant planet HD 209458b. Bulk and orbital parameters are well constrained for this planet and there are measurements of atmospheric sodium [1], hydrogen [2], carbon and oxygen [3]. However, nothing is known about the abundances of molecules relevant to life.

We are studying the modulation of the stellar spectrum as the planet transits in front of the star. Different wavelengths become extinct at different levels in the exoplanet, causing the occulting area, and therefore the modulation, to be wavelength-dependent. This dependency allows us to identify atmospheric constituents. Signal-to-noise estimates show that data we have obtained from the Very Large Telescope (VLT), the Infrared Telescope Facility (IRTF), Palomar, and Keck are sensitive enough to measure or place useful limits on the atmospheric abundances of water and maybe even of carbon monoxide and methane.

In order to detect the very weak expected modulation (~ 4 parts in 10000), we are building a detailed radiative-transfer model to cross correlate with the data. This model will accept atmospheric temperature, density, composition, and cloud distributions profiles for the planet. We will then be able to measure the abundance of molecules relevant to life in the atmosphere of an extrasolar planet.

TRANSIT MODULATION

HD 209458b is, as of today, the only transiting system bright enough for spectroscopic analysis with current instruments. Hence, its atmosphere is currently under intense scrutiny: [1] detected sodium, [2] detected and extended hydrogen exosphere, and just recently, [3] also detected carbon and oxygen in the exosphere. A search for carbon monoxide was unsuccesful in its first attempt [4], as was a search for a secondary eclipse signature [5, 6].

CP713, *The Search for Other Worlds: Fourteenth Astrophysics Conference,*
edited by S. S. Holt and D. Deming
© 2004 American Institute of Physics 0-7354-0190-X/04/$22.00

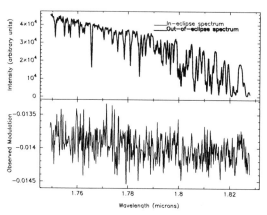

FIGURE 1. Simulated modulation in the data from VLT observations

As the planet transits, it blocks a section of the star equal to the area of the planet. This area depends on the wavelength we use to observe the transit. A light ray passing through the planet's atmosphere becomes dimmer at wavelengths located in the cores of absorption features than at wavelengths where the atmospheric molecules do not absorb much. Hence, the planetary radius at which, say, 90% of the light is absorbed varies with wavelength. Thus, the opaque area of the planet varies with wavelength according to the spectral features of the atmospheric molecules. Because the spectral signature is unique, we can identify the molecules responsible for the modulation. Furthermore, the strength of this modulation will allow measurements of molecular abundances.

The upper panel in Figure 1 shows an observed out-of-transit spectrum and a simulated in-transit spectrum with the expected noise level, while the lower panel shows one minus the ratio of those two spectra.

The expected noise level is high as can be seen when comparing the lower panel of Figure 1 to the noiseless modulation model in Figure 2. Wavelengths in the continuum and wavelegths in the core of some absorption features have a contrast of up to 0.04% between them. This translates to a required S/N of 7500 for a 3σ detection of the effect.

If we cross correlate the data with the model, the individual pixels in each feature and the multiple features all contribute towards a single cross-correlation peak, increasing the S/N proportionally to the square root of the number of pixels to be combined. There are about 20 excited features of water vapor in the wavelength range of our observations and about 3 pixels per feature, giving us an equivalent S/N above our minimum needs. Figure 3 shows the cross-correlation result from the data and model in Figs 1 and 2.

We are building a detailed radiative transfer model, which will include temperature, pressure, molecular abundances, and cloud profiles for the planet. Profiles will be tested against radiative equilibrium models from various sources [7, 8, 9, 10]. Opacity information for water will be obtained from [11], while one of us (Steyert) is conducting precise laboratory measurements for methane.

The model will include spherical layered symmetry for the planet and effects like ray bending (due to varying atmospheric index of refraction) and stellar limb darkening.

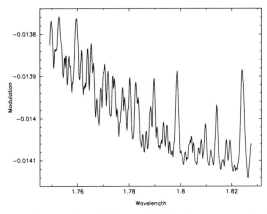

FIGURE 2.   Transit model (Seager 2003, private communication).

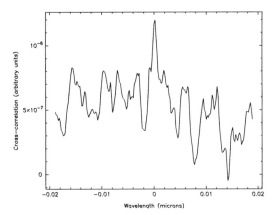

FIGURE 3.   Cross-correlation result from simulated data

OBSERVATIONS

We are focusing on the detection of water, carbon monoxide, and methane, all of which are relevant to organic life. In order to succeed in this task we require data quality at the limits of current capabilities. Also, identification of water features in the infrared (IR) requires wavelength resolution of $\lambda/\Delta\lambda \sim 1500$ or more.

Table 1 summarizes the observations undertaken and instruments used. A total of 12 transit nights are available to us now. However, instrument problems and bad weather will limit the validity of some of those nights. Although the Hubble Space Telescope (HST) has an optimal location, it lacks the required spectral resolution in the infrared.

It is of prime importance to discover and cancel any systematic effects that could ruin the result. Shifting the position of the star along the spectrograph slit (nodding) allows us to measure and subtract sky emission. Also, because telluric water vapor usually varies on timescales of minutes in an unpredictable way, we limit exposures to 2 minutes.

TABLE 1. Observations and Instruments details

System	Area (m^2)	$\frac{\lambda}{\Delta\lambda}$	CD?*	Array length	Nights Trans	Nights Non-	Qty (GB)
Palomar/HNA	20	850	N	256	2	2	0.56
Keck/NIRSPEC	79	25000	Y	1024	4	0	10.28
VLT/ISAAC	50	4000	N	1024	3	1	7.88
IRTF/SpeX	7	1500	Y	1024	3	1	5.06
Total					12	4	23.77

* Is the spectrograph cross dispersed?

Optimal extraction of the spectra from the reduced frames is a critical step. Adapting the algorithm of [12], we obtain the most accurate spectra from the observations. We combine spectra only from the same telescope whenever the setup was kept the same (in order to avoid systematic errors and rescaling). We anticipate S/N of at least 200 per pixel per spectrum, or 2000 per pixel per transit after combination.

CONCLUSIONS

Detection of life-relevant molecules on extrasolar planets such as H_2O, CO and CH_4 constituents is possible through transit spectrospcopy. This will be of prime interest once the sensitivity of future space-based observatories reaches levels capable of measuring life-bearing planets.

Even a negative result for this study will be meaningful; it will either place strong constraints on models or it will help to justify the need for space-based instruments.

REFERENCES

1. Charbonneau, D., T. Brown, R. Noyes, R. Gilliland 2002, ApJ **568**, 377
2. Vidal-Madjar, A., A. Lecavelier des Etangs, J.-M. Désert, G. E. Ballester, R. Ferlet, G. Hébrard, and M. Mayor 2003, Nature **422**, 143-146.
3. Vidal-Madjar A., Desert J.-M., Lecavelier des Etangs A., Hebrard G., Ballester G., Ehrenreich D., Ferlet R., Mcconnell J., Mayor M., & Parkinson C, ApJ accepted. 2004.
4. Brown, T. M., K. G. Libbrecht, and D. Charbonneau 2002, PASP **114**, 826.
5. Richardson, L. J., D. Deming, and S. Seager 2003. ApJ **597**, 581-589.
6. Richardson, L. J., D. Deming, G. Wiedemann, C. Goukenleuque, D. Steyert, J. Harrington, and L. W. Esposito 2003. ApJ **584**, 1053-1062.
7. Fortney, J. J., D. Sudarsky, I. Hubeny, C. S. Cooper, W. B. Hubbard, A. Burrows, and J. I. Lunine 2003. ApJ **589**, 615.
8. Iro, N., B. Bèzard, and T. Guillot, 2002, *A Radiative Equilibrium Model of HD209458b*, AAS/DPS Meeting 34.
9. Seager, S., Whitnet, B. and Sasselov, D., 2000, ApJ **537**, 916
10. Sudarsky, D., A. Burrows, and I. Hubeny 2003, ApJ **588**, 1121.
11. Partridge, H. and Schwenke, D. 1997, J. Chem. Phys. **106**, 4618
12. Horne, K., 1986, PASP **98**, 609

Klio: A 5-micron Camera for the Detection of Giant Exoplanets

Melanie Freed*, Philip M. Hinz* and Michael R. Meyer*

*Steward Observatory, 933 N. Cherry Ave, Tucson, AZ 85721
freed@as.arizona.edu

Abstract. We plan to take advantage of the unprecedented combination of low thermal background and high resolution provided by the 6.5m MMT's adaptive secondary mirror, to target the 3-5 micron atmospheric window where giant planets are expected to be anomalously bright. We are in the process of building a 3-5 micron coronograph that is predicted to be sensitive to planets as close as 0.4 arcsec to the parent star. We expect to be able to detect giant planets down to 5 times Jupiter's mass for a 1 Gyr old system at 10 pc. We plan to carry out a survey which is complementary to the radial velocity detections of planets and constructed to characterize the prevalence and distribution of giant planets around nearby, Sun-like stars.

WHY A 5 MICRON CORONOGRAPH?

With the number of known extrasolar planets now exceeding 100 (http://www.obspm.fr/encycl/catalog.html), we have reached the point where we can begin to study the statistical properties of these planets (i.e. [1]). However, while radial velocity surveys have provided a wealth of information about exoplanets, they are unable to determine the inclination of the systems or any spectral information. In addition, they are limited to relatively small star-planet separations over a practical baseline. Other indirect methods such as astrometry and microlensing are also valuable tools for the detection of exoplanets. However, only by direct detection can any spectral information about the exoplanets be determined, which is crucial to studying their physical properties (e.g. temperature, surface gravity, composition).

So far, the transit method is the only direct technique that has successfully detected an exoplanet [2] [3]. While transit techniques clearly have the potential to provide a wealth of information about exoplanets, they are also limited to close in planets, where multiple transits can quickly be observed and exoplanets are more likely to transit their primary star.

In an effort to directly detect exoplanets at separations left largely unexplored by current measurements, we are building a coronographic imager to be used in the thermal infrared with ground-based adaptive optics (AO). The camera will be optimized to target the thermal IR (3-5 microns), where exoplanets are thought to peak in brightness[4]. Until now, it has been impossible to use the thermal IR since no appropriate space-based telescope exists and the sky background is so large from the ground. Additional emissivity introduced by conventional AO systems overwhelms any astronomical signal.

CP713, *The Search for Other Worlds: Fourteenth Astrophysics Conference*,
edited by S. S. Holt and D. Deming

THE ADAPTIVE SECONDARY AT THE MMT

The adaptive secondary mirror at the 6.5m MMT provides a solution to the problem of increased emissivity from conventional adaptive optics systems. By making the secondary mirror of the telescope itself the deformable surface, approximately eight additional surfaces are eliminated compared to typical AO systems [5]. This results in both an emissivity and throughput that are similar to a non-AO equipped telescope. For faint objects, a conventional AO system at the MMT would need to integrate 2-3 times longer than the adaptive secondary to achieve the same signal-to-noise in L and M bands [5].

The adaptive secondary mirror is now operational at the MMT telescope [6] [7]. Measurements taken with the BLINC/MIRAC mid-IR camera give an emissivity of 6.5% of the telescope plus adaptive optics system, which is typical of a telescope with no adaptive optics. In January 2003, BLINC/MIRAC also recorded the first high order AO images taken in the thermal IR. These were taken at 10.3 microns and show an achieve Strehl ratio (SR) of 0.96 with the AO system operational. Once vibration-related issues are resolved, the adaptive secondary should be able to achieve SR=0.9 at 5 microns. This number is based on actual observations at H band with the vibrations removed in post-processing.

THE 5 MICRON CORONOGRAPH

Klio will be a three channel system that consists of f/20, f/35, and pupil imaging modes. The primary design driver for Klio is use in the high background regime (i.e. L and M bands). The f/20 channel has been optimized for the L and M bands with Nyquist sampling at L band (0.048"/pixel) and a field of view of 15 x 12". The f/35 channel is optimized for the H and K bands with Nyquist sampling at K band (0.027"/pixel) and a FOV of 8.7 x 7".

Figure 1 shows the optical design of the f/20 channel. At the secondary focus of the telescope, a wheel will allow a selection of coronographic stops to be moved into the beam. Optics will then form an image of the secondary, where another wheel will allow a selection of Lyot stops to be introduced into the beam before the final image is formed

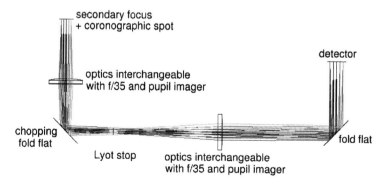

FIGURE 1. Optical drawing of the f/20 channel of Klio.

on the detector. The detector is an Indigo 320x256 pixel InSb chip with 30 micron pixels. We plan to have an engineering run without AO in the summer of 2004 and an engineering run with AO in the winter of 2004.

EXOPLANET SURVEY

So far, the only spectral information available for an exoplanet comes from observations of HD 209458b [8]. As a result, our understanding of exoplanets, especially those at Jupiter-like separations, would benefit greatly from even a single direct detection. With ages derived from their primary stars, IR fluxes would provide anchor points for models that are essentially untested. In addition, a carefully constructed survey has the ability to provide statistical information about the exoplanet population, even for a null result.

Radial velocity surveys provide a baseline from which we can make predictions about the distribution of exoplanets at further separations from their primaries. Currently, results indicate a companion mass distribution of $dn/dM \sim M^{-0.7}$ and a separation distribution of $dn/dloga \sim a^{>0}$ [1]. A fit to the data (for 0.2 AU < a < 2 AU) gives $dn/da \sim a^{0.7}$ as a starting point. Our survey will enable a quantitative test of the hypothesis that the mass and separation distribution of outer planets are consistent with those observed within 5 AU.

As a first order prediction, we can simply extend the observed distributions to larger separations. We performed a series of Monte Carlo simulations where the radial velocity mass distribution was taken to be valid from 1-15 $M_{Jupiter}$ and the separation distribution was extended from 0.01-50 AU. An outer boundary of 50 AU was chosen to be consistent with the outer edge of our Kuiper Belt [9] as well as the radius of the smallest silhouette disk in Orion [10]. Estimates of our sensitivities versus separation for L and M band are given in Heinze, Hinz, & McCarthy (2003) [11]. Here we consider the 0.45" gaussian

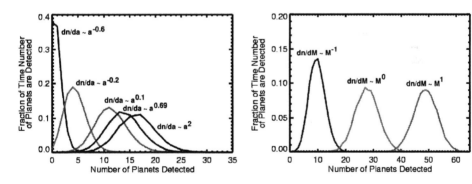

FIGURE 2. (left) Fractional chance of detecting various numbers of planets given a mass distribution of $dn/dM \sim M^{-0.7}$ and 5 different separation distribution functions. It appears unlikely that no planets will be detected with this survey. If, indeed, no planets are detected, we should be able to significantly rule out a large parameter space of distribution functions. (right) Fractional chance of detecting various numbers of planets given a separation distribution function of $dn/da \sim a^{0.7}$ and three different mass distribution functions. All three mass distributions are clearly distringuishable.

coronograph with 2 hour integrations at the MMT.

These simulations were used to define an optimal survey sample of 80 M0-F0 stars within 20 pc and less than 1 Gyr old. With this sample, and assuming $dn/da \sim a^{0.7}$ and $dn/dM \sim M^{-0.7}$, we expect to detect 14 ± 3 companions with 5-15 $M_{Jupiter}$ at 13-50 AU.

We do not necessarily expect our simple extrapolation to be valid given that at very small separations planets may have very different evolutionary histories than their larger separations counterparts (e.g. migration [12]). Our survey will provide an indication of the shape of the separation and mass distribution functions based on the number of companions detected.

Figure 2 indicates our chances of detecting various numbers of planets given $dn/da \sim a^{0.7}$ as well as three different mass distribution functions of the form $dn/dM \sim M^{\alpha}$ with α=[-1,0,1]. These three distribution functions are clearly distinguishable for the given sample. Assuming the given separation distribution function, we would be able to rule out $dn/dM \sim M^{-1}$ at 3σ and $dn/dM \sim M^{1}$ at 11σ in the event of no planet detections.

The same process can be repeated for various separation distributions. Figure 6 indicates our chances of detecting various numbers of planets, while this time keeping $dn/dM \sim M^{-0.7}$ and varying the separation distribution function from $dn/da \sim a^{\beta}$ with β=[-0.6,-0.2,0.1,0.7,2]. For a null result we can rule out β=-0.2 at 2σ and β=0.7 at 4σ.

These limits will provide valuable constraints on theories of planet formation and evolution as well as help focus the next generation of direct detection surveys.

ACKNOWLEDGMENTS

We would like to thank IRLabs for their expert help, especially Elliott Solheid and Ken Salvestrini for designing the instrument dewar and testing the detector. Thanks also to Michael Lesser for helping us integrate AZCam with Klio. M. Freed acknowledges support from the NASA Graduate Student Researchers Program (NGT5-50394). We also acknowledge support from AFOSR as well as the NASA Astrobiology Institute "Laplace Center" node at the UofA.

REFERENCES

1. Marcy, G., Butler, R.P., Fischer, D.A., & Vogt, S.S. 2003, in ASP Conf. Ser., Scientific Frontiers in Research on Extrasolar Planets, ed. D. Deming & S. Seager (San Francisco:ASP)
2. Charbonneau, D., Brown, T.M., Latham, D.W., & Mayor, M. 2000 ApJ, 529, L45.
3. Konacki, M., Torres, G., Jha, S., & Sasselov, D.D. 2003, Nature, 421, 507.
4. Sudarsky, D., Burrows, A., & Hubeny, I. 2003, ApJ, 588, 1121.
5. Lloyd-Hart, M. 2000, PASP, 112, 264.
6. Wildi, F.P., Brusa, G., & Lloyd-Hart, M. 2003, Proc. SPIE, 5169
7. Brusa-Zappellini et al. 2003, Proc. SPIE, 5169
8. Charbonneau, D., Brown, T.M., Noyes, R.W. & Gilliland, R.L. 2002, ApJ, 568, 377.
9. Trujillo, C.A. & Brown, M.E. 2001, ApJ, 554, L95
10. Bally, J., O'Dell, C.R., & McCaughrean, M.J. 2000, AJ, 119, 2919
11. Heinze, A. N., Hinz, P. M., & McCarthy, D. W. Jr. 2003, Proc. SPIE, 4839, 1154
12. Trilling, D.E., Lunine, J.I., & Benz, W. 2002, A&A, 394, 241

Effects of Helium Phase Separation on the Evolution of Giant Planets

Jonathan J. Fortney* and William B. Hubbard*

*Lunar and Planetary Laboratory, Department of Planetary Sciences,
The University of Arizona,
1629 E. University Blvd., Tucson, AZ 85721

Abstract. We present the first models of Saturn and Jupiter to couple their evolution to both a radiative-atmosphere grid and to high-pressure phase diagrams of hydrogen with helium. The purpose of these models is to quantify the evolutionary effects of helium phase separation in Saturn's deep interior. We find that prior calculated phase diagrams in which Saturn's interior reaches a region of predicted helium immiscibility do not allow enough energy release to prolong Saturn's cooling to its known age and effective temperature. We explore modifications to published phase diagrams that would lead to greater energy release, and find a modified H-He phase diagram that is physically reasonable, leads to the correct extension of Saturn's cooling, and predicts an atmospheric helium mass fraction Y_{atmos} in agreement with recent estimates. We then expand our inhomogeneous evolutionary models to show that hypothetical extrasolar giant planets in the 0.15 to 3.0 Jupiter mass range may have T_{eff}s 10-15 K greater than one would predict with models that do not incorporate helium phase separation.

INTRODUCTION

The interiors of Jupiter and Saturn, extrasolar giant planets (EGPs), and brown dwarfs (BDs) are all described by similar physics: these bodies are mainly composed of liquid metallic hydrogen, and their interior energy transport mainly occurs through efficient convection [1], leading to largely isentropic interiors. Jupiter and Saturn, whose radius, mass, luminosity, and age are known precisely, can serve as calibrators of thermal-history calculations for the entire class of objects. They can provide a test of the adequacy of the diverse physical models, including interior thermodynamics, heat transport mechanisms, and model-atmosphere grid, that enter into the general thermal-history theory for EGPs and BDs. However, at very low effective temperatures ($\sim$ 100 K), the corresponding interior temperatures may become low enough for phase separation of abundant interior components to occur, and this effect must be quantitatively evaluated before Jupiter and Saturn can be used as calibrators. This work provides a quantitative assessment of inhomogeneous evolution in Jupiter, Saturn, and low-mass EGPs. Relevant calculations and data for Jupiter and Saturn follow:

- Saturn is currently over 50% more luminous than one would predict with a homogeneous, adiabatic cooling model. Saturn models reach the planet's known T_{eff} of 95.0 K in only 2.0 to 2.5 Gyr [2].

CP713, *The Search for Other Worlds: Fourteenth Astrophysics Conference*,
edited by S. S. Holt and D. Deming

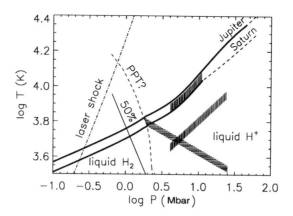

FIGURE 1. Temperature-pressure plot of the interiors of Jupiter and Saturn at $t = 4.56$ Gyr, superimposed on a hydrogen phase diagram (see text for details). The upper boundary of the horizontally-hatched region shows the minimum temperature at which He is fully miscible in metallic hydrogen with a mass fraction $Y = 0.27$, while the lower boundary shows the minimum temperature corresponding to $Y = 0.21$, according to HDW [6] theory. The lower vertically-hatched region shows the same He miscibility limits according to Pfaffenzeller, et al. [7], while the upper vertically-hatched region shows the modification to the Pfaffenzeller et al. theory that gives a realistic prolongation of Saturn's age. With this phase diagram, when He rains down, it falls all the way down to the core. Liquid metallic hydrogen (H^+), liquid molecular hydrogen (H_2), and the maximum pressures and temperatures reached by laser shock data are also shown.

- For Jupiter, homogeneous, adiabatic cooling models allow Jupiter to reach its known T_{eff} of 124.4 K in ~ 4.7 Gyr. This is a good match.
- The atmospheres of both Jupiter and Saturn are depleted in helium relative to the expected protosolar helium mass fraction of ~ 0.27. Y_{atmos} for Jupiter is 0.231 [3] and for Saturn is likely 0.18-0.25 [4], but is poorly constrained.

HYDROGEN AND HELIUM UNDER HIGH PRESSURE

Under extreme pressure dense molecular hydrogen dissociates and ionizes to become liquid metallic hydrogen. The transition is likely continuous over a pressure of 1-5 Mbar at the temperatures of interest ($\sim 10^4$ K) in giant planets. Also at these temperatures, helium, which makes up about 8% of the atoms in a solar composition mixture, likely becomes immiscible in liquid metallic hydrogen [5, 6, 7]. Figure 1 shows the interior temperature-pressure profile of Jupiter and Saturn superimposed on a hydrogen phase diagram.

Detailed calculations on the dynamics and distribution of helium in giant planets have been performed [8]. These authors found that when helium becomes immiscible in liquid metallic hydrogen, the composition that separates out is essentially pure helium, and this helium on fairly short timescales (relative to the convective timescale) will coalesce to form helium droplets. These droplets, being denser than the surrounding liquid metallic

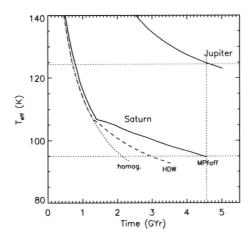

FIGURE 2. Evolutionary models including the phase separation of helium from liquid metallic hydrogen. The dotted curve for Saturn and the solid curve for Jupiter are for homogeneous evolution. The dashed curve includes the phase diagram of HDW[6], while the solid curve for Saturn uses the proposed modified phase diagram (MPfaff). The modified phase diagram allows both Jupiter and Saturn to reach their known ages and T_{eff}s. The interior of Jupiter begins He phase separation at $t \approx 5$ Gyr.

hydrogen, will fall through the planet's gravitational field. If the droplets reach a region where helium is again miscible at higher concentration, they will redissolve, enriching the deeper regions of the planet in helium. Helium would be lost from *all* regions with pressures lower than the pressures in the immiscibility region, since the planet is fully convective up to the atmosphere. Excess helium would be mixed down to the immiscibility region and be lost to deeper layers. This "helium rain" could be a substantial additional energy source for giant planets.

CALCULATIONS

We find that the calculated phase diagram of HDW [6], which is essentially equivalent to that of Stevenson [5], are inapplicable to the interiors of giant planets, if helium phase separation is Saturn's only additional energy source. As Figure 2 shows, this phase diagram prolongs Saturn's cooling 0.8 Gyr, even in the most favorable circumstance that all energy liberated is available to be radiated, and does not instead go into heating the planet's deep interior. As we show in our published work [2], we find that a modified version of the phase diagram of Pfaffenzeller et al. [7], with a higher temperature for the onset of helium immiscibility (see Figure 1), allows Saturn to reach its known T_{eff} and age, while Jupiter evolves homogeneously until $t \approx 5$ Gyr. Saturn's Y_{atmos} falls to 0.185 at 4.5 Gyr, which is at the low end of Saturn's derived value [4]. This modified phase diagram can be applied to various hypothetical giant planets [9]. We follow the evolution of planets with masses from 0.15 M_J (1/2 Saturn's mass) to 3.0 M_J. Figure 3 shows the

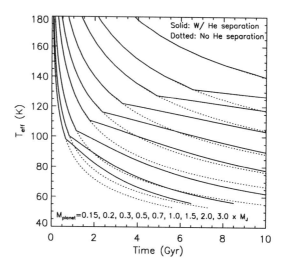

FIGURE 3. Evolution of the T_{eff} for planets of mass 0.15 to 3.0 M_J for our standard models with no stellar irradiation and 10 M_{Earth} cores. The dotted lines are models without helium phase separation, while the solid lines include the effects of helium phase separation on the planets' cooling. The top curve is for the highest mass planet while the bottom curve is for the lowest mass planet.

evolution of these planets with and without the effects of helium phase separation. These planets are in isolation and possess 10 M_{Earth} cores. At Gyr ages the model planets undergoing phase separation can have T_{eff}s 10-15 K higher than the homogeneous models, making the planets somewhat easier to detect.

ACKNOWLEDGMENTS

We thank Adam Burrows, Dave Sudarksy, & Tristan Guillot for interesing conversations. JJF is funded by a NASA GSRP Fellowship and WBH by a NASA PG&G grant.

REFERENCES

1. Hubbard, W. B., Burrows, A., & Lunine, J. I. 2002, Ann Rev of Astronomy & Astrophyics, 40, 103
2. Fortney, J. J. & Hubbard, W. B. 2003, Icarus, 164, 228
3. von Zahn, U., Hunten, D. M., & Lehmacher, G. 1998, Journal of Geophysical Research, 103, 22815
4. Conrath, B. J. & Gautier, D. 2000, Icarus, 144, 124
5. Stevenson, D. J. 1975, Physical Review B, 12, 3999
6. Hubbard, W. B. & Dewitt, H. E. 1985, Astrophysical Journal, 290, 388
7. Pfaffenzeller, O., Hohl, D., & Ballone, P. 1995, Physical Review Letters, 74, 2599
8. Stevenson, D. J. & Salpeter, E. E. 1977, Astrophysical Journal Supplements, 35, 239
9. Fortney, J. J. & Hubbard, W. B. 2004, Astrophysical Journal, submitted

Terrestrial Planets

Terrestrial Planet Formation

John E. Chambers

SETI Institute, 2035 Landings Drive, Mountain View CA 94043.

Abstract. Some of the astronomical and cosmochemical constraints on the formation of the Sun's terrestrial planets are reviewed, and the planetesimal theory of planetary accretion is described. Several difficulties remain for this model, especially in its earliest stage, but substantial progress has been made in recent years. While we have a fair understanding of terrestrial-planet formation in the Solar System, there are currently few available constraints on the abundance and characteristics of extrasolar Earth-like planets.

INTRODUCTION

Understanding the origin of terrestrial planets is a challenging problem, but also a fascinating one, especially for terrestrial-planet dwelling creatures like ourselves. As with many other aspects of planetary science, the study of terrestrial-planet formation is evolving rapidly thanks to new observations of extrasolar planetary systems and circumstellar disks, as well as advances in cosmochemistry and computer modelling. Here I will give a quick overview of some of these advances, together with a description of the theory which is struggling to keep up.

Planet formation in the Solar System took place in the Sun's protoplanetary nebula—a disk shaped region of gas and dust which orbited the young Sun. Several lines of evidence support the existence of such a nebula. Most stars less than a few million years old show signs of having an optically thick circumstellar disk [1, 2]. These disks typically extend for tens or hundreds of astronomical units (AU), and are thought to contain enough mass to form a planetary system similar to that in the Solar System. Few stars older than about 6 million years have optically thick disks [2], which suggests the Sun's nebula was removed on a comparable timescale. A second line of evidence for the nebula comes from chondrules and refractory inclusions, two components of primitive meteorites. These components appear to have formed from dust aggregates that melted while floating in a dust rich environment [3], presumed to be the solar nebula. Refractory inclusions predate chondrules by 1–4 million years [4, 5], which implies the nebula lasted for at least this length of time.

The Sun's protoplanetary nebula probably had a similar elemental composition to the Sun itself. Some 99.5% of the material in the terrestrial-planet region would have been gas, while the rest was solid grains of metal, silicates and sulfides. Temperatures at 1 AU from the Sun were probably too high for icy materials such as water. carbon monoxide and nitrogen to condense. This explains why each of the inner planets is highly depleted in elements such as hydrogen, carbon, nitrogen and the noble gases compared to the Sun. It is possible that temperatures in the inner nebula were high enough to vapourize

CP713, *The Search for Other Worlds: Fourteenth Astrophysics Conference*,
edited by S. S. Holt and D. Deming

silicates for a brief period of $\sim 10^5$ years at a very early stage. This may be the reason why the more volatile rock-forming elements are also depleted in the inner planets and in meteorites [6], although collisions may have played a role here too [7]. The nebula would have cooled rapidly as it grew older, and temperatures probably became low enough for water ice to condense in the asteroid belt while the asteroids were still forming.

This is the setting for the formation of the Sun's terrestrial planets. Some fraction of the dust in the inner Solar System (and possibly from further afield) was transformed into the inner planets and asteroids, while most of the gas was lost. The precise way in which this happened is still a matter of conjecture, although the evidence broadly supports a working model called the planetesimal hypothesis [8, 9]. In this model, the formation of the inner planets is divided into three stages. In stage one, dust grains coalesced to form bodies 0.1–10 km in size, called planetesimals, which were large enough to gravitationally perturb one another during close approaches. In stage two, the planetesimals interacted gravitationally and collided with one another, until most of the solid material was contained in a few tens of large bodies called planetary embryos. In the final stage, the embryos collided or were gravitationally scattered out of the Solar System until only the terrestrial planets and the Moon remained. It is likely that these stages overlapped in time to some degree, especially in different regions of the nebula, although I will describe them separately here.

THE ORIGIN OF PLANETESIMALS

The origin of planetesimals is the least understood stage of terrestrial planet formation. It is fair to say that we don't really understand how planetesimals formed in the Solar System, or under what circumstances they form in other protoplanetary disks. Two routes to planetesimal formation are being studied actively at present: (i) rapid production of large solid bodies by the gravitational collapse of dust-rich regions of the nebula, and (ii) the gradual accretion of planetesimals by collisions between dust grains and small solids such as chondrules. Both routes face severe obstacles, but substantial progress has been made recently in understanding the physics involved in each case.

In the gravitational instability model, dust grains sediment towards the midplane of the nebula due to the vertical component of the Sun's gravity, forming a solid-rich layer. If the concentration of mass in the midplane becomes high enough, portions of the dust-rich layer become gravitationally unstable, spontaneously collapsing to form planetesimals directly [10]. Unfortunately, matters are not quite so simple. As solid material collects near the midplane, it orbits the Sun with a velocity given by Kepler's law, dragging gas in the midplane along with it. However, gas above and below the midplane orbits the Sun slightly more slowly since the gas is partially supported by its radial pressure gradient. The velocity shear between the layers (about 50 ms^{-1} at 1 AU) generates turbulence which "puffs up" the solid rich layer again [11].

This balance between gravity and turbulence means that the solid to gas ratio in a vertical column of nebula material must exceed a critical value for gravitational instability to happen [12]. Calculations suggest that material with a solar composition is too gas rich by a factor of about 8 to permit instability, even when ices have condensed

- Ring instability due to local maximum in dust surface density (Ward 1976)

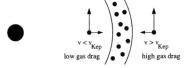

$v < v_{Kep}$
low gas drag

$v > v_{Kep}$
high gas drag

- Local maximum in gas pressure (Haghighipour & Boss 2003)

$dP/dr < 0$ $dP/dr > 0$ $dP/dr \ll 0$

- Collective particle drift (Goodman & Pindor 2000)

constant gas drag force

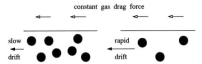

slow drift   rapid drift

FIGURE 1. Three ways to increase the local solid to gas ratio in the nebula: (i) a local maximum in the solid surface density leads to further concentration due to differential gas drag, (ii) a local maximum in the gas surface density causes different radial drift rates, (iii) overdense solid regions drift more slowly than less dense regions, leading to pile up of solids.

[13]. Thus, for gravitational instability to take place, solid material must collect in one region of the nebula, increasing the solid to gas column density. Several concentration mechanisms have been proposed—see figure 1 [14, 15, 16], but it is unclear whether these are mechanisms are effective enough to allow instability to occur [17].

The alternative to gravitational instability is coagulation of solid particles to form planetesimals. Experiments show that silicate dust grains typically stick together when they collide at speeds of up to 10 ms^{-1} [18]. Higher collision speeds are more likely to lead to rebound or fragmentation. The motion of dust grains is strongly coupled to the motion of the the nebula gas, so the grains are likely to experience gentle collisions allowing them to stick. Coagulation is aided by charge exchange during collisions, and the formation of electric dipoles [19]. It is unclear how accretion proceeds for metre-size bodies—these are too small to have an appreciable gravitational field, but surface forces between like-sized bodies will be quite weak. Possibly metre-sized bodies gain further mass by accreting much smaller particles.

Pairwise accretion must overcome potentially severe difficulties caused by gas drag [20]. Large bodies orbit the Sun slightly faster than the nebula gas so they experience a headwind. This gradually removes orbital angular momentum from the solid object, causing it to drift towards the Sun. Theoretical estimates for the drift rate are especially large for bodies with sizes of 0.1–10 metres—up to 1 AU in 500 years [21]. The actual drift rates are somewhat uncertain. If dust grains form a dense layer at the midplane, gas in tbis layer will be swept along at a similar speed to the solid material. In this case, the headwind felt by a solid body in the midplane is reduced, and it will drift inwards more slowly as a result. However, it seems clear that solid bodies must grow larger than a few metres in size relatively rapidly in order to survive.

Coagulation of solid particles can be enhanced if the nebula gas is turbulent. In this case, dust grains and aggregates are prevented from settling to a thin layer at the midplane. Instead, they tend to accumulate in stagnant regions in the turbulent flow [21]. Concentration factors of at least 100 can be achieved this way, possibly much higher. The degree of turbulent concentration depends on the aerodynamic properties of the particles involved. Chondrules appear to have the right size and density for effective concentration under plausible nebula conditions [22], and this may explain why primitive meteorites are largely composed of chondrules.

RUNAWAY GROWTH AND OLIGARCHIC GROWTH

The second stage of terrestrial-planet formation is better understood than the first. Interactions between planetesimals are dominated by gravitational forces and drag due to nebula gas, both processes that are well understood. In addition, the huge number of planetesimals present in the nebula means that their growth can be studied using statistical models like the molecules in a gas [23].

Close encounters between planetesimals play a major role in the second stage of planetary accretion. During an encounter between a pair of planetesimals, gravitational interactions alter their trajectories, and the net effect of many encounters determines each planetesimal's orbit around the Sun. Over time, encounters increase the relative velocities v_{rel} of planetesimals by making their orbits more eccentric and inclined. Conversely, gas drag decreases relative velocities. Low-mass planetesimals tend to acquire eccentric and inclined orbits with high relative velocities. High-mass planetesimals move on nearly circular orbits in the midplane due to dynamical friction with the smaller bodies [24]. The large bodies have low relative velocities during encounters which gives their mutual gravity enough time to "focus" their trajectories towards each other.

A planetesimal gains mass m at a rate given by

$$\frac{dm}{dt} = \frac{C\pi R^2 v_{rel} \sigma}{H} \left(1 + \frac{v_{esc}^2}{v_{rel}^2}\right) \tag{1}$$

where σ is the surface density of planetesimals, $H \propto v_{rel}$ is the scale height of the disk of planetesimals, R and $v_{esc} \propto R$ are the planetesimal's radius and escape velocity respectively, and C is a constant of order unity [25].

Equation 1 can give rise to several different growth modes for a population of planetesimals. At early times, v_{rel} and H remain small, so the growth rate is given by $1/m \times dm/dt \propto m^{1/3}$. A combination of dynamical friction and gravitational focussing means that large planetesimals grow more quickly than small ones, a state of affairs called runaway growth. The largest objects, called planetary embryos, grow rapidly, while most bodies remain small.

Runaway growth slows down when the embryos grow large enough that their gravitational perturbations become more important than interactions between the more numer-

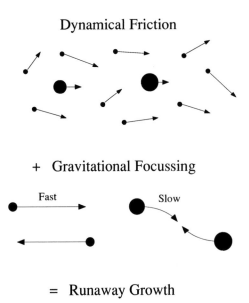

Dynamical Friction

+ Gravitational Focussing

Fast

Slow

= Runaway Growth

FIGURE 2. Numerous close encounters between planetesimals lead to dynamical friction, in which high mass bodies have low relative velocities and vice versa. Low encounter speeds lead to gravitational focussing, increasing the probability of a collision. Dynamical friction and gravitational focussing combine to give runaway growth of the largest planetesimals.

ous planetesimals. Gravitational focussing remains effective for planetary embryos, but large embryos perturb nearby planetesimals more than small embryos do. As a result, v_{rel} depends on the mass of the largest nearby embryo. The growth rate of an embryo is given by $1/m \times dm/dt \propto m^{-1/3}$. Accretion enters a new self-regulated regime called oligarchic growth, in which embryos grow faster than planetesimals, but large embryos grow more slowly than small ones [26].

Gravitational encounters between embryos and dynamical friction with the planetesimals tend to keep the embryos on widely separated orbits. Numerical simulations show that the embryos maintain a spacing of roughly 10–12 Hill radii [26], where the Hill radius is given by

$$ R_H = a \left(\frac{m}{3m_{sun}} \right)^{1/3} \qquad (2) $$

where a is the semi-major axis of the embryos' orbit. Each embryo mainly accretes planetesimals from its feeding zone—an annular region of the disk centred on its own orbit. As a result, embryos are likely to have distinct chemical compositions reflecting differencies in the chemical and elemental make up of the nebula at different distances from the Sun.

Calculations show that oligarchic growth supercedes runaway growth at 1 AU from the Sun when the embryos have masses about 10^5-10^6 times smaller than Earth [25]. Oligarchic growth continues for 10^5–10^6 years [23], until most of the solid mass is

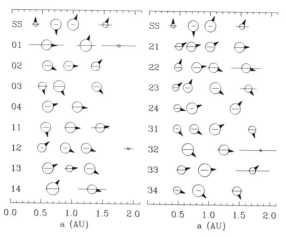

FIGURE 3. The results of 16 numerical simulations of the final stage of planetary accretion, reprinted from [28], ©2001, with permission from Elsevier. Each row of symbols shows the outcome of one simulation, with symbol radius proportional to planetary radius, and arrows indicating spin axis orientation. The horizontal bars indicate the non-circularity of each orbit. The top of row symbols shows the inner planets of the Solar System for comparison.

contained in a few tens of lunar-to-Mars size bodies, and the remaining mass is contained in a larger number of unaccreted planetesimals and collision fragments.

A third type of growth is possible if one or more of the giant planets formed quickly, before accretion had proceeded very far in the inner Solar System. In this case, gravitational perturbations from Jupiter and the other giant planets largely determine the relative velocities of planetesimals in the inner Solar System. Relative velocities are likely to be quite high, especially in the region that now contains the asteroid belt. Gravitational focussing is weak, and runaway growth is delayed or prevented altogether [27]. Instead, accretion takes place via orderly growth—all objects grow slowly and at similar rates. The rapid formation of Jupiter, and the prevention of runaway growth in the asteroid belt, is one reason why there may be no terrestrial planets beyond the orbit of Mars.

LATE STAGE ACCRETION

The final stage of planetary accretion is dominated by a few tens of lunar-to-Mars size planetary embryos. Most of the original planetesimals have been accreted or removed, so dynamical friction is weak. Close encounters between embryos rapidly increase their orbital eccentricities and inclinations. As a result, gravitational focussing becomes ineffective and the accretion rate slows substantially. Numerical simulations show that it took ∼ 20 million years for Earth to reach half its final mass and roughly 100 million years for accretion to be completed in the inner Solar System [28].

These timescales agree with cosmochemical estimates for the time of Earth's core formation. Each of the inner planets and some of the asteroids are differentiated, with iron-rich cores and silicate mantles. The time of core formation can be measured using the W-Hf isotope system, in which a radioactive lithophile (silicate-loving) isotope decays to a siderophile (iron-loving) isotope with a half life of 9 million years. Measurements of the daughter isotope in planetary mantles show that core formation on Earth or its precursors took place in 10–30 million years, while core formation on Mars took place somewhat earlier than this [29].

Late stage accretion necessarily involves a number of giant impacts between planet-size bodies. A certain amount of mass is lost in the form of fragments escaping from these collisions. Collision speeds are highest close to the Sun since orbital velocity increases with decreasing heliocentric distance. It is possible that Mercury underwent a high-speed collision late in its formation that removed much of the planet's silicate mantle, explaining the high density of Mercury today [30]. It is likely that the Moon formed as a result of another giant impact involving Earth and a Mars-sized planetary embryo. Mantle material from both bodies was thrown into a debris disk in orbit around Earth, and the Moon accreted from this material [31], which explains why the Moon is highly depleted in iron. The high temperatures generated by the impact prevented the condensation of volatiles in the debris disk, so that the Moon is essentially dry.

Despite the high eccentricities attained by embryos during the final stage of accretion, Earth and Venus now have orbits that are almost circular. This may be due to collisions or dynamical friction from the few remaining planetesimals, but each of these effects is probably small. Alternatively, the orbits of the inner planets may have been circularized by gravitational interaction with residual amounts of nebula gas, if this was still present [32]. Gravitational perturbations from the growing planets redistribute the nebula gas so it is no longer axisymmetric with respect to the Sun. The back reaction of the gas's gravity applies a torque which tends to circularize a planet's orbit. The presence of a large amount of gas may retard the final stage of accretion by preventing embryos from developing orbits that cross one another. As the gas begins to dissipate, embryos on neighbouring orbits begin to collide forming a system with a handful of planets whose orbits are then circularized again by the remaining gas [32].

Late stage accretion is highly stochastic affair (see figure 3) due to the small number of embryos involved, and the fact that the outcome of each close encounter depends very sensitively on the initial orbits of the bodies involved. For this reason, the modern characteristics of the inner planets are the result of a sequence of chance events. It is quite possible the Solar System could have ended with 2 or 3 planets instead of 4, and final masses and orbits of these planets could have turned out very differently as a result of minor events that occurred earlier in their accretion [28].

FORMATION OF THE ASTEROID BELT

Today, the total mass of the asteroid belt between 2 and 4 AU is about 5×10^{-4} Earth masses, while the largest asteroid, Ceres, is less than 2% of a lunar mass. This is in marked contrast to the 5 planetary-mass bodies that exist within 2 AU of the Sun. Unless there was a very sharp drop in surface density in the nebula between 2 and 4 AU, the asteroid belt would have initially contained up to several Earth masses of solid material.

Vesta, which is thought to be the parent body of the HED meteorites, required only a few million years to accrete and differentiate [33]. The formation of bodies this size in such a short amount of time means the asteroid region must originally have contained at least two orders of magnitude more mass than at present [34]. This excess mass has been lost since the asteroids formed.

The asteroid belt differs from the terresrial-planet region in that it contains several unstable mean-motion and secular resonances associated with the giant planets. An asteroid entering one of these resonances typically falls into the Sun or is ejected from the Solar System in $\sim$ 1 million years [35]. Models for the depletion of mass from the asteroid belt generally invoke perturbations from the giant planets and the presence of these resonances in particular. One possibility is that a combination of resonances and gas drag removed much of the original mass between Mars and Jupiter. Resonances tend to increase the eccentricities of nearby planetesimals, while gas drag is highly effective for bodies moving on eccentric orbits, causing rapid inward migration for objects up to $\sim$ 10 km in size. If most of the solid material in the asteroid belt remained in small bodies, a subtantial amount of mass would have drifted inwards into the terrestrial planet region [36], possibly preventing the formation of a planet in the asteroid region.

A second possibility is that planetary embryos formed in the asteroid belt, but these were lost subsequently. Planetary embryos are massive enough to cause large changes in one another's orbits during close encounters. Although unstable resonances occupy only a small fraction of the asteroid belt, close encounters between embryos may have deposited them one by one into unstable regions until all large bodies were removed from the asteroid belt. Numerical simulations show this is a likely outcome if planetary embryos larger than the Moon formed between 2 and 4 AU [37]. Many smaller asteroids would also have been scattered into unstable resonances and removed [38].

A natural byproduct of each of these models is that the inner planets accreted some planetesimals that formed beyond 2 AU from the Sun. Nebula temperatures in the asteroid belt were cooler than in the terrestrial-planet region when planetesimals were forming, allowing volatile materials such as water to condense. Carbonaceous chondrites contain up to 10% water by mass in the form of hydrated silicates, and this water generally has a similar D/H ratio to water in Earth's oceans. The asteroid belt may have been an important source of Earth's volatiles—a single carbonaceous embryo could supply all of the water present on Earth today [39].

EXTRASOLAR EARTHS

At present, we did not know how many stars possess terrestrial planets or how common analogues of Earth may be. Programmes to search for extrasolar Earths are in the planning and development stage, but results are at least several years away [40]. The existence of extrasolar giant planets does not prove that extrasolar terrestrial planets also exist since these objects may form in very different ways [41]. However, the observation that some stars to possess debris disks [42], perhaps produced by collisions between asteroids, or dust evaporating from comets, offers hope that planetesimal formation is a fairly widespread phenomenon.

Many of the known extrasolar systems are unlikely to contain an Earth-like planet in the star's habitable zone [43] since such a planet would have an unstable orbit. Even in systems where part of the habitable zone is stable, planets may not have formed due to gravitational perturbations from giant planets in the same system. This appears to be the case in the 47 UMa system, which is one of the best analogues of the Solar System known at present. Numerical simulations suggest that the habitable zone of this star is more likely to contain an asteroid belt than terrestrial planets [44]. It appears that giant planets play an important role in shaping the formation of terrestrial planets and asteroid belts in the same system. To date, we know about the characteristics of giant planets orbiting only a small fraction of stars, so it may be some time before we can say whether extrasolar Earths are common or rare indeed.

REFERENCES

1. Beckwith, S.V.W., Sargent, A.I., Chini, R.S. and Guesten, R. (1990) A survey for circumstellar disks around young stellar objects. *Astron. Journal.,* **99**, 924–945.
2. Haisch, K.E., Lada, E.A. and Lada, C.J. (2001) Disk frequencies and lifetimes in young clusters. *Astrophys. Journal,* **553**, L153–L156.
3. Jones, R.H., Lee, T., Connolly, H.C., Love, S.G. and Shang, H. (2000) Formation of chondrules and CAIs: theory versus observation. In Protostars and Planets IV, University of Arizona Press, Tucson AZ, eds: V. Mannings, A.P. Boss, and S.S. Russell, pp 927–961.
4. Wadhwa, M. and Russell, S.S. (2000) Timescales of accretion and differentiation in the early Solar System: the meteoritic evidence. In Protostars and Planets IV, University of Arizona, Tucson AZ, eds. V. Mannings, A.P. Boss, and S.S. Russell, p. 995.
5. Huss, G.R., MacPherson, G.J. Wasserburg, G.J., Russell, S.S. and Srinivasan, G. (2001) ^{26}Al in CAIs and chondrules from unequilibrated ordinary chondrites. *Meteoritics,* **36**, 975–997.
6. Cassen, P. (2001) Nebula thermal evolution and the properties of primitive planetary materials. *Meteoritics,* **36**, 671–700.
7. Halliday, A.N. and Porcelli, D. (2001) In search of lost planets—the paleocosmochemistry of the inner Solar System. *Earth Plan. Sci. Lett.,* **192**, 545–559.
8. Wetherill, G.W. (1990) Formation of the Earth. *Ann Rev. Earth Plan Sci.,* **18**, 205–256.
9. Lissauer, J.J. (1993) Planet Formation. *Ann Rev. Astron. Astrophys.,* **31**, 129–174.
10. Goldreich, P. and Ward, W.R. (1973) The formation of planetesimals. *Astrophys. Journal,* **183**, 1051–1062.
11. Weidenschilling, S.J. (1980) Dust to planetesimals—settling and coagulation in the solar nebula. *Icarus,* **44**, 172–189.
12. Sekiya, M. (1998) Quasi-equilibrium density distributions of small dust aggregations in the solar nebula. *Icarus,* **133**, 298–309.
13. Youdin, A.N. and Shu, F.H. (2002) Planetesimal formation by gravitational instability. *Astrophys. Journal,* **580**, 494–505.
14. Ward, W.R. (1976) The formation of the solar system. In Frontiers of Astrophysics, Harvard University Press, Cambridge MA, ed: E.H. Avrett, pp 1–40.
15. Haghighipour, N. and Boss, A.P. (2003) On pressure gradients and rapid migration of solids in a nonuniform solar nebula. *Astrophys. Journal,* **583**, 996–1003.
16. Goodman, J. and Pindor, B. (2000) Secular instability and planetesimal formation in the dust layer. *Icarus,* **148**, 537–549.
17. Weidenschilling, S.J. (2003) Radial drift of particles in the solar nebula: implications for planetesimal formation. *Icarus,* **165** 438–442.
18. Poppe, T., Blum, J. and Henning, T. (2000) Analogous experiments on the stickiness of micron sized preplanetary dust. *Astrophys. Journal,* **533**, 454–471.
19. Marshall, J. and Cuzzi, J. (2001) Electrostatic enhancement of coagulation in protoplanetary nebulae. 32nd Lunar Planetary Science Conference, Houston TX, abstract 1262.

20. Stepinski, T.F. and Valageas, P. (1997) Global evolution of solid matter in turbulent protoplanetary disks. *Astron. Astrophys., 319,* 1007–1019.
21. Cuzzi, J.N., Dobrovolskis, A.R. and Champney, J.M. (1993) Particle gas dynamics in the midplane of a protoplanetary nebula. *Icarus, 106,* 102–134.
22. Cuzzi, J.N., Hogan, R.C., Paque, J.M. and Dobrovolskis, A.R. (2001) Size selective concentration of chondrules and other small particles in protoplanetary nebula turbulence. *Astrophys. Journal, 546,* 496–508.
23. Wetherill, G.W. and Stewart, G.R. (1993) Formation of planetary embryos—effects of fragmentation, low relative velocity and independent variation of eccentricity and inclination. *Icarus, 106,* 190.
24. Wetherill, G.W. and Stewart, G.R. (1989) Accumulation of a swarm of small planetesimals. *Icarus, 77,* 330–357.
25. Thommes, E.W., Duncan, M.J. and Levison, H.F. (2003) Oligarchic growth of giant planets. *Icarus, 161,* 431–455.
26. Kokubo, E. and Ida, S. Oligarchic growth of protoplanets. *Icarus, 131,* 171–178.
27. Kortenkamp, S.J. and Wetherill, G.W. (2001) Runaway growth of planetary embryos facilitated by massive bodies in a protoplanetary disk. *Science, 293,* 1127–1129.
28. Chambers, J.E. (2001) Making more terrestrial planets. *Icarus, 152,* 205–224.
29. Kleine, T., Munker, C., Mezger, K. and Palme, H. (2002) Rapid accretion and early core formation on asteroids and the terrestrial planets from Hf-W chronology. *Nature, 418,* 952–955.
30. Benz, W., Slattery, W.L. and Cameron, A.G.W. (1988) Collisional stripping of Mercury's mantle. *Icarus, 74,* 516–528.
31. Canup, R.M. and Asphaug, E. (2001) Origin of the Moon in a giant impact near the end of Earth's formation. *Nature, 412,* 708–712.
32. Kominami, J. and Ida, S. (2002) The effect of tidal interaction with a gas disk on formation of terrestrial planets. *Icarus, 157,* 43–56.
33. Srinivasan, G., Goswami, J.N. and Bhandari, N. (1999) [26]Al in eucrite Piplia Kalan: plausible heat source and formation chronology. *Science, 284,* 1348–1350.
34. Wetherill, G.W. (1992) An alternative model for the formation of the asteroids. *Icarus, 100,* 307–325.
35. Gladman, B.J., Migliorini, F., Morbidelli, A. Zappala, V., Michel, P., Cellino, A., Froeschle, C., Levison, H.F., Bailey, M. and Duncan, M. (1997) Dynamical lifetimes of objects injected into asteroid belt resonances. *Science, 277* 197–201.
36. Franklin, F. and Lecar, M. (2000) On the transport of bodies within and from the asteroid belt. *Meteoritics, 35,* 331–340.
37. Chambers, J.E. and Wetherill, G.W. (2001) Planets in the asteroid belt. *Meteoritics, 36,* 381–399.
38. Petit, J.M., Morbidelli, A. and Chambers, J.E. Primordial excitation and clearing of the asteroid belt. *Icarus, 153,* 338–347.
39. Morbidelli, A., Chambers, J., Lunine, J., Petit, J.M., Robert, F., Valsechhi, G.B. and Cyr, K.E. (2000) Source regions and timescales for the delivery of water to Earth. *Meteoritics, 35,* 1309–1320.
40. Seager, S. (2003) The search for extrasolar Earth-like planets. *Earth Plan. Sci. Lett., 208,* 113–124.
41. Boss, A.P. (2002) Formation of gas and ice giant planets. *Earth Plan. Sci. Lett., 202,* 513–523.
42. Spangler, C., Sargent, A.I., Silverstone, M.D., Becklin, E.E. and Zuckerman, B. (2001) Dusty debris around solar-type stars: temporal disk evolution. *Astrophys. Journal, 555,* 932–944.
43. Kasting, J.F., Whitmire, D.P. and Reynolds, R.T. (1993) Habitable zones around main sequence stars. *Icarus, 101,* 108–128.
44. Laughlin, G., Chambers, J. and Fischer, D. (2002) A dynamical analysis of the 47 Ursae Majoris planetary system. *Astrophys. Journal, 579,* 455–467.

Characterizing Extrasolar Earths

Sara Seager

Department of Terrestrial Magnetism, Carnegie Institution of Washington,
5241 Broad Branch Rd. NW, Washington, DC, 20015

Abstract. For thousands of years people have wondered, are we alone? Is there life elsewhere in the Universe? For the first time in human history we are on the way to being able to answer this question by plans to search for Earth-like extrasolar planets. This review summarizes the current state of the scientific motivation for Earth-like planet detection and characterization. The relevance of the June 2004 transit of Venus to the the characterization of extrasolar planets is also discussed.

INTRODUCTION

For the first time in human history the possibility of detecting and studying Earth-like planets is on the horizon. Terrestrial Planet Finder (TPF) [1], with a launch date in the 2015 timeframe, is being planned by NASA to find and characterize planets in the habitable zones of nearby stars. The mission Darwin from ESA [2] has similar goals. The motivation for both of these space missions is the detection and spectroscopic characterization of extrasolar terrestrial planet atmospheres.

Extrasolar planets are unique among astronomical objects in that they have local counterparts available for study—the solar system planets (Figure 1). Indeed spacecraft have visited each of the solar system planets (except Pluto, which is widely considered to be a Kuiper Belt object rather than a planet). From the solar system we know that the terrestrial planets are very different from the gas giant planets. The giant planets are composed almost entirely of hydrogen and helium with massive atmospheres and liquid interiors. The giant planet atmospheres are primitive—little atmospheric evolution has taken place so that they contain roughly the same atmospheric gases as at their formation. The terrestrial planets are predominantly composed of rock and metals and have thin atmospheres. In contrast to the giant planets, atmospheric evolution from both atmospheric escape of light gases and gas-surface reactions has substantially changed each of the terrestrial planet atmospheres from their primitive states. Further complicating the composition and spectral signature of terrestrial planet atmospheres are such varied factors as the energy distribution of the host star, atmospheric circulation and planetary rotation rate, geological activity, the initial volatile inventory, and active biology. With these factors in mind we note that there are much larger challenges to predicting and interpreting spectral and photometric signatures of extrasolar terrestrial planets compared to extrasolar giant planets.

Direct detection of extrasolar planets is one of the biggest technological challenges for astronomers of our generation. Earth-like planets in our "local neighborhood" of 20 pc are not intrinsically faint from an astronomical perspective—they are brighter than the

CP713, *The Search for Other Worlds: Fourteenth Astrophysics Conference,*
edited by S. S. Holt and D. Deming

FIGURE 1. A collage of the solar system planets. Images from NASA. The planet sizes are approximately to scale but the planet separations are not. The four inner terrestrial planets are much smaller, less massive and therefore more difficult to detect in an extrasolar planetary system than giant planets like Jupiter and Saturn. Many decades will pass before we are able to obtain similar, spatially resolved images of extrasolar planets.

faint galaxies in the Hubble Deep Field. Earth-like planets, however, pose an immense challenge for observational detection and study due to the close proximity of the parent stars, which are orders of magnitude brighter and more massive than the planets. Two different TPF designs are being considered by NASA, a visible-wavelength, single aperture "coronagraphic" telescope as well as a mid-infrared nulling interferometer consisting of several telescopes either in free flying formation or connected on a long truss. A downselect to one architecture will occur in 2006 and a launch date is planned for the 2015 time frame. Darwin has a design similar to the TPF free-flying nulling interferometer.

DETECTING AND CHARACTERIZING TRANSITING PLANETS

Until the direct detection technological challenge is met, detection and study of transiting planets will be the best way to detect and study extrasolar planets, including Earth-sized planets. NASA's Kepler space mission (launch date 2007) is planned to detect Earth-sized planets. Kepler will monitor one field of the sky with approximately 100,000 sun-like stars ranging in brightness from 9th to 15th magnitude for three full years. In addition, COROT (CNES and European partners, launch date 2006) is capable of detecting short-period planets slightly larger than Earth. On the characterization front, it may be possible to study Earth-like planet atmospheres via transits in the not-too-distant future (R. Gilliland in Seager et al. 2004 "JWST and Astrobioloby" report to NASA, in preparation). If Earths are common, a transiting Earth could exist around a V=6 magnitude or brighter star. Atmospheric features with a slightly modified JWST could distinguish between a hostile-to-life Venus and an oxygen-bearing Earth transiting a bright star. For transit transmission spectra of Earth-sized planets with JWST to be measurable, however, we need to first detect transiting Earths around bright stars—much brighter than the Kepler targets. New ideas to detect such transiting Earths are required.

THE JUNE 2004 TRANSIT OF VENUS

Transiting planets have been and will be so important for understanding extrasolar planets that it is worth considering what can be learned from the upcoming transit of Venus. Note that although Venus will also transit the sun in June 2012, the next pair of Venus transits will be over 100 years from now in December 2117 and December 2125.

The past includes many exciting firsts from transiting planets. The first "exoEarth" atmosphere was detected during a transit—the 1761 transit of Venus. Lomonosov noticed a luminous ring surrounding Venus during ingress and egress (due to refraction of sunlight) and postulated the existence of a massive atmosphere in Venus. Nearly 150 years later, the first extrasolar planet atmosphere was discovered [3] using transit transmission spectra during a planetary transit of the 3.5-day period planet HD209458b.

The upcoming June 8, 2004 transit of Venus will help us with our understanding of transit transmission spectra. We plan to obtain spectra of the Venusian atmosphere during the June 8 2004 transit of Venus. We will analyze these spectra and use transmission spectra models (e.g., Figure 2) to "interpret" the data to determine Venus' composition, abundances, effective temperature, scale height, projected oblateness [4], and size. We will compare these "results" to what is known about Venus. If Venus were an extrasolar transiting planet, what would we learn about its atmosphere and physical characteristics? What would we miss? What would we misinterpret? More importantly, we will be able to test our transmission spectra models—the same models that we plan to use to interpret extrasolar planet transmission spectra during the next decade—to see if they are both complete and accurate.

In order to design the most useful experiment, we need to answer the following questions: *Venus is so well known; why not just compute a transmission spectra model and not bother observing the transit?* It is important to realize that although the Venusian atmosphere has been well measured, Venus has never been observed spectroscopically as a transiting planet. Furthermore, model transmission spectra have not been thoroughly calculated for Venus since the only situations in which transmission spectra would be used are Venus occultations of background stars—very rare events (e.g., [5]) *Are there really any parts of the Venus atmosphere model that are complex enough to seriously warrant a transit measurement (beyond general model testing)?* Refraction of starlight through the planet atmosphere is strong during the Venus transit. This refraction effect has been seen with the naked eye during ingress and egress when a luminous ring surrounded Venus (and was used by Lomonosov in 1761 to postulate the existence of a massive atmosphere on Venus). Refraction during the transit itself may in fact complicate abundance determination from transmission spectra because refraction affects continuum wavelengths, but not the absorption lines, and hence the continuum could be lowered causing line strengths to be overestimated. In addition, the magnitude of refraction with time during ingress and egress can tell us about the oblateness of the planet.

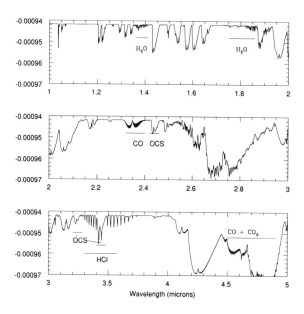

FIGURE 2. Model of the near-IR transmission spectrum of Venus at a spectral resolution of 10,000. The y axis is the normalized in-transit minus out-of-transit flux, thus cancelling out the stellar lines. Note that Venus' projected size on the sun is an order of magnitude larger than a distant extrasolar terrestrial planet projected onto its parent star. Major absorption features are identified, unless they are CO_2 features. At visible wavelengths, only very weak features of CO_2 are present and extremely weak lines of H_2O and H_2S. The transmission spectrum strength (y axis) is shown for the entire sun. In the case where we observe Venus and a small subsection of the sun, the signal will be at the few percent level and even up to tens of percent.

These two refraction effects—continuum lowering and oblateness determination—are complex effects that require further modeling and testing [6]. We expect transiting extrasolar planets (with the exception of the short-period planets) to have refraction effects [6]. With very high precision photometry with Kepler and potentially JWST, we may be able to use refraction effects during ingress and egress to determine the planet's oblateness and atmospheric density.

EARTH AS AN EXTRASOLAR PLANET

What information could be extracted if Earth were a distant extrasolar planet observable only as a point source of light? An emerging field of research aims to understand Earth as an extrasolar planet both with models and with observations.

Recent modeling work [7] has shown that physical properties of Earth can be determined by monitoring the time variation of the visible flux of Earth. Cloud patterns together with water vapor in the atmosphere would be indicative of large bodies of liquid water. As Earth rotates and continents come in and out of view, the total amount of reflected sunlight will change due to the high albedo contrast of different components of Earth's surface (<10% for ocean, >30-40% for land, >60% for snow and some types of ice). Without clouds this variation could be an easily detectable factor of a few. With clouds the variation is lower—10 to 20%. The rotational period, however, could still be extracted. From a planet with much less cloud cover than Earth, much more surface information may be extracted.

Real data of the spatially unresolved Earth are available. Global, instantaneous spectra and photometry can be obtained using Earthshine measurements taken from observations from Earth itself. Earthshine is sunlight scattered from Earth that scatters off of the moon and travels back to Earth. It is easily seen with the naked eye during crescent moon phase. Earthshine data are more relevant than remote sensing satellite data for studying Earth as an extrasolar planet. The latter are highly spatially resolved and limited to narrow spectral regions. Furthermore because remote sensing involves looking straight down at specific regions of Earth, hemispherical flux integration with lines-of-sight through different atmospheric path lengths is not available.

Several Earthshine studies are ongoing including one [8] to monitor changes in Earth's global albedo as a proxy for climate change. Other efforts have focused on a broad wavelength coverage of spectra ([9], [10], [11]). These efforts have shown Earth's blue color from Rayleigh scattering, the major absorption features at visible wavelengths, and Earth's vegetation red edge signature. Note that Earth's surface can be studied only where the atmosphere is transparent; that is, at visible wavelengths and at mid-IR wavelengths in the 8–12 micron window.

Turning satellites to look at Earth is also very useful because ultimately we need to consider viewing geometries other than the equatorial viewpoint probed with Earthshine measurements. In a landmark study the Galileo spacecraft took some visible and near-IR spectra of Earth during two gravitational assist flybys of Earth. Looking at small areas of Earth, Sagan et al. [12] concluded that the widespread presence of the vegetation red edge combined with abundant atmospheric oxygen, other molecules out of thermodynamic equilibrium, and radio signals "constituted evidence of life on Earth without any a priori assumptions about its chemistry." Later a spatially unresolved mid-IR spectrum of Earth was taken with Thermal Emission Spectrometer experiment on the Mars Global Surveyor spacecraft en route to Mars (Christensen and Pearl [13]; Figure 3). Even more useful would be a time series of spatially unresolved spectra of Earth (see [11]) at different viewing geometries including polar views. A small satellite for this purpose has been studied by the Canadian Space Agency [14].

BIOMARKERS

The search for Earth-like planets includes the goal of studying the planet atmosphere for evidence of habitability and life. Lederberg [15] and Lovelock [16] first addressed the idea of looking at an atmosphere for indications of life. They suggested searching for signs of severe chemical disequilibrium. Since Earth is our only example of a planet whose atmosphere is modified by life, it is clearly the best place to start for specific

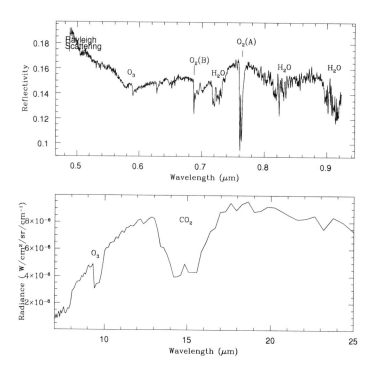

FIGURE 3. The spectrum of the spatially unresolved Earth. Top: A visible wavelength spectrum of the spatially unresolved Earth, as seen with Earthshine [10]. The viewpoint is largely centered equatorially on the Pacific ocean. The major atmospheric features are identified. The reflectivity scale is arbitrary. Data courtesy of N. Woolf and W. Traub from [10]. Bottom: A mid-infrared wavelength spectrum of the spatially unresolved Earth taken by the Thermal Emission Spectrometer experiment on the Mars Global Surveyor spacecraft en route to Mars in 1996 [13]. The viewpoint is over Hawaii. The structure in the water vapor bands is due to many strong, widely spaced spectral lines. Note also the emission at the center of the CO_2 band. Radiance is in units $W \, cm^{-2} \, sr^{-1} \, cm^{-1}$.

examples of atmospheric biomarkers.

There are several strong spectral diagnostics in Earth's atmosphere for habitability and life as we know it (Figure 3). These spectral features are described in detail in [17]. O_2 and its photolytic product O_3 are the most reliable biomarker gas indicators for life as we know it. O_2 is highly reactive and therefore must be continuously produced to remain in significant quantities in the atmosphere. There seem to be no non- biological sources that can continually produce large quantities of O_2 (Venus and Mars both have very small amounts of O_2) and only rare false positives that in most cases could likely be ruled out by other planetary characteristics. O_3 is a non-linear indicator of O_2; only a small amount of O_2 is needed to produce a relatively large quantity of O_3. N_2O is a second gas produced by life—albeit in small quantities—during microbial oxidation-reduction reactions. N_2O has a very weak spectroscopic signature. Furthermore, the N_2O

spectroscopic signature overlaps with H_2O absorption bands. Because of this, it may be more easily detectable on a planet with less water vapor than Earth. H_2O vapor itself is a very important spectral feature. H_2O, while not a biomarker, is indicative of habitability because all known life needs liquid water. Other spectral features, while not necessarily biomarkers or habitability indicators, can provide useful planet characterization information. CO_2 indicates a terrestrial planet atmosphere; Venus, Earth, and Mars all have CO_2, which is easily detectable due to a very strong mid-IR absorption feature. CO_2 originates from outgassing and is expected to be present on all small, rocky planets. High concentrations of CH_4, such as may have been present on early Earth, could indicate the presence of methanogenic bacteria, but could also be from midocean ridge volcanism. Other gases do not have strong enough absorption features (even at higher abundance than Earth levels) to be detectable with the expected signal-to-noise and spectral resolution. The simultaneous detection of combinations of the above features may be a more robust indicator of life or habitibility than detection of only a single unusual feature. Because of this there is some debate as to which wavelength regime is most suitable for life and habitability indicators.

In addition to atmospheric biomarkers, Earth has one very strong and intriguing surface biomarker: vegetation. The reflection spectrum of photosynthetic vegetation has a dramatic sudden rise in albedo around 750 nm (Figure 4). This rise is a factor of five or more. The "red-edge" feature is caused both by strong chlorophyll absorption to the blue of 700 nm and high reflectance due to plant cell structure to the red of 700 nm. If plants did not have this window of strong reflection (and transmission), they would overheat and their chlorophyll would degrade. On Earth, this red-edge signature is probably reduced to several percent ([9], [10], [11]) due largely to cloud cover and noncontinuous vegetation cover across the planet. Such a spectral surface feature could be much stronger on a planet with a lower cloud cover fraction. Recall that any observations of extrasolar Earth-like planets will not be able to spatially resolve the surface. A surface biomarker could be distinguished from an atmospheric signature by time variation. As the continents, or different concentrations of the surface biomarker, rotate in and out of view the spectral signal will change correspondingly.

BEYOND PRESENT-DAY EARTH

To date the study of Earth's biomarker signatures detectable in an Earth-like planet's atmosphere have been largely focused on the present Earth and small deviations from it. Yet there are almost certainly other potential biological signatures. One example is organo-sulfur compounds such as methanethiol and DMS [18], which are important in Earth's modern ecosystem and may have been important in Earth's distant past as well. If present in a high enough quantity, methanethiol or DMS would overlap with the atmospheric ozone signature in the mid-infrared, confusing any low-resolution spectral detection.

Vegetation's red edge (described in the above section) is an intriguing surface biomarker. If we accept that extraterrestrial light-harvesting organisms should be ubiquitous, would they likely have a strong spectroscopic signature? Light-harvesting

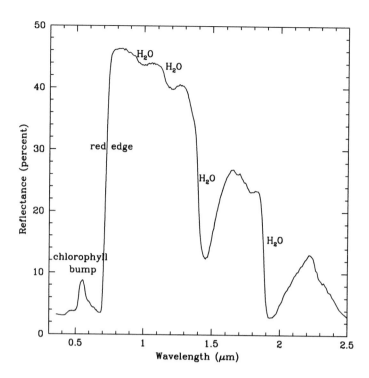

FIGURE 4. Reflection spectrum of a deciduous leaf (from [19]). The small bump near 0.5 μm is a result of chlorophyll absorption (at 0.45 μm and 0.68 μm) and gives plants their green color. The much larger sharp rise (between 0.7 and 0.8 μm) is known as the red edge and is due to the contrast between the strong absorption of chlorophyll and the otherwise reflective leaf.

organisms may have similar properties to vegetation in order to absorb the correct frequency energy for molecular transitions but not absorb all available energy. Light-harvesting pigments in vegetation cover the full range of the visible-light spectrum; most of these pigments have sharp spectral features at the red edge of the pigment's absorbing range. In addition many photosynthetic organisms have pigments at different wavelengths than vegetation's chlorophyll A, as shown in Figure 5. Thus, extrasolar light-harvesting organisms may have sharp spectral features similar to terrestrial vegetation's red-edge spectral feature but at different wavelengths.

Spectral signatures (i.e., atmospheric conditions and climate) have also been largely focused on the present Earth and slight variations [17]. However, physical characteristics (e.g., geography, obliquity, and climates) of those planets are likely to be very different than Earth. In order to realistically investigate global properties of planets with extremely different parameters from Earth, 3-D atmospheric transport is required, in contrast to theoretical spectra so far presented for the spatially unresolved Earth. For example, a 1-D interpretation of a planet with very slow rotation will be that this planet is

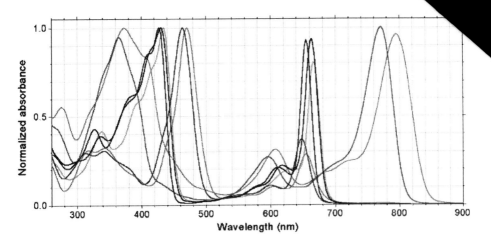

FIGURE 5. Some of the pigments involved in photosynthesis. Wavelength dependence of chlorophyll pigments. Chlorophyll a (black), chlorophyll b (red) and the bacteriochlorophylls which are found in photosynthetic bacteria: bacteriochlorophyll a (magenta), b (orange), c (cyan), d (blue), and e (green). Image from http://www.personal.psu.edu/faculty/n/x/nxf10/scitab/chlabs/, used by permission. Also, see [20].

not habitable: the side facing the star would be too hot for liquid water but water would be frozen on the dark side. A 3-D model is essential for heat transport and would likely show otherwise (e.g., Venus). For similar reasons, 3-D heat transport is necessary to understand the global surface temperature and cloud cover of Earth-like planets with other rotation rates, eccentricities, obliquities, etc. We have begun an investigation of Earth's climate under variable rotation rates using a general circulation model and have found that the cloud cover pattern and fraction is very sensitive to different rotation rates (Seager & Cho, in preparation). Others have begun to explore variation of Earth's obliquity and orbital eccentricity [21].

SUMMARY

Future prospects for detection and study of Earth-like planets are exciting. Spectral and surface biomarkers will allow us to assess the habitability of a given planet. The immense challenge of nulling out the parent star light, however, must be overcome for direct detection and detailed study of extrasolar planets. In the meantime transits will provide a less challenging, though still difficult, way to detect and study extrasolar planets. It may be possible to observe Earth-like planet atmospheres with transit transmission spectra should Earth-like planets be common and one be detected around a bright (V=6 mag) star. To thoroughly understand the transit transmission spectra method we can use the transit of Venus in June 2004 (and June 2012) to test our models. While direct detection techniques are being developed we need to go beyond present-day Earth to

ariations such as different rotation rates and continental distributions
n cloud cover and spectral signatures. Ultimately we will need to move
; present-day Earth to be prepared for the unexpected among the vast
olar terrestrial planets.

REFERENCES

1. Beichman, C. A., Woolf, N. J., & Lindensmith, C. A., *NASA/JPL Publication*, **99-3** (1999).
2. Friedlund et al. *ESA-SCI*, **12** (2000).
3. Charbonneau, D., Brown, T. M., Noyes, R. W., & Gilliland, R. L., *ApJ*, **568**, 377–384 (2002).
4. Seager, S., & Hui, L., *ApJ*, **574**, 1004 (2002).
5. Menzel, D. H., & de Vaucouleurs, G., *AJ*, **65**, 351 (1960).
6. Hui, L., & Seager, S., *ApJ*, **572**, 540–555 (2002).
7. Ford, E. B., Seager, S., & Turner, E. L. *Nature*, 412, 885-887 (2001).
8. Goode, P. R., Qiu, J., Yurchyshyn, V., Hickey, J., Chu, M.-C., Kolbe, E., Brown, C. T. & Koonin, S. E., *J. Geophys. Res. Lett.*, **28**, 1671–1674 (2001).
9. Arnold, L., Gillet, S., Lardiere, O., Riaud, P., & Schneider, J., *A&A*, **392**, 231–237 (2002).
10. Woolf, N. J., Smith, P. S., Traub, W. A., & Jucks, K. W., *ApJ*, **574**, 430 (2002).
11. Seager, S., Turner, E. L., Schafer, J., & Ford, E. B., submitted to *Astrobiology* (2004).
12. Sagan, C., Thompson, W. R., Carlson, R., Gurnett, D., & Hord, C. 1993, *Nature*, **365**, 715 (1993).
13. Christensen, P. R., & Pearl, J. C. *JGR*, **102**, 10875-10880 (1997).
14. Davis, G. R., Calcutt, S. B., Drummond, J. R., Naylor, D. A., Penny, A. J., & Seager, S. Measurements of the Unresolved Spectrum of Earth (MUSE), Final Report of the Concept Study for the Canadian Space Agency Space Science Program, April 2002.
15. Lederberg, J., *Nature*, **207**, 9-13 (1965).
16. Lovelock, J. E., *Nature*, **207**, 568-570 (1965).
17. Des Marais, D. J., Harwit, M., Jucks, K., Kasting, J. F., Lunine, J. I., Lin, D., Seager, S., Schneider, J., Traub, W., & Woolf, N. *Astrobiology*, **2**, 153-181 (2002).
18. Pilcher, C. P. *Astrobiology*, **3**, 471 (2003).
19. Clark, R. N., Swayze, G. A., Gallagher, A. J., King, T. V. V., & Calvin, W. M. 1993, *The U.S. Geological Survey, Digital Spectral Library: Version 1: 0.2 to 3 microns, U.S. Geological Survey Open File Report 93-592*, 1340 pages, URL http://speclab.cr.usgs.gov.
20. Frigaard, N.-U., Larsen, K. L., & Cox, R. P., *FEMS Microbiology Ecology*, **20**, 69-77 (1996).
21. Williams, D. M. & Pollard, D., *BAAS*, **32**, 1050.

Imaging Terrestrial Planets

M. Kochte[*], A. B. Schultz[*], D. Fraquelli[*], I. J. E. Jordan[*], R. G. Lyon[†],
K. G. Carpenter[†], H. M. Hart[**], M. DiSanti[†], F. Bruhweiler[‡], C. Miskey[‡],
M. Rodrigue[§], M. S. Fadali[§], D. Skelton[¶] and K-P. Cheng[||]

Computer Sciences Corporation, Space Telescope Science Institute
†Goddard Space Flight Center/NASA
**Computer Sciences Corporation, FUSE Project, JHU
‡Catholic University of America
§University of Nevada
¶Optical Sciences Corporation
||California State University, Fullerton

Abstract. We present optical simulations of a new approach to directly image terrestrial planets. Terrestrial planets typically are 10 orders of magnitude fainter than the central star, a difficult challenge for any optical system. Our studies show that the combination of an external occulter and an apodizer yields the required contrast, with significantly reduced requirements on stray light and diffraction. This mitigates the very high mirror tolerances required of other coronagraphic methods and makes exo-planet detection feasible with current technology.

SPACECRAFT OCCULTERS

Direct images of terrestrial planets around nearby stars will be spectacular and profound - one of the great scientific accomplishments of the 21st Century. To date, radial velocity studies of nearby solar-type stars have discovered over 100 Jupiter-sized planets. Unlike our own Solar System, these giant, Jupiter-like planets dominate their systems, orbiting

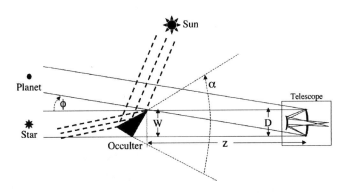

FIGURE 1. Constellation of spacecraft.

CP713, *The Search for Other Worlds: Fourteenth Astrophysics Conference,*
edited by S. S. Holt and D. Deming
© 2004 American Institute of Physics 0-7354-0190-X/04/$22.00

within a fraction of an AU of the primary. Currently, direct imaging capabilities are limited to ground-based coronagraphic and adaptive optics (AO) imaging with 8-m or larger telescopes and with the instruments onboard the Hubble Space Telescope (HST). None of these platforms has imaged any verifiable exo-planet. The task is more challenging for a terrestrial planet, which, in reflected light, is typically 10 orders of magnitude dimmer than the central star, *i.e.*, the luminosity ratio of the planet to star is 10^{-10}. NASA's proposed Terrestrial Planet Finder (TPF) program is investigating only two imaging detection concepts: interferometry and coronagraphy.

We have explored combining an apodized square aperture space telescope with an external occulter [1], which we call the Umbral Missions Blocking Radiating Astronomical Sources (UMBRAS) project, for the purpose of observing extrasolar terrestrial planets [2, 3, 4]. Relative configuration of an external occulter operating with a space telescope is depicted in Figure 1. The telescope points at the target as the occulter craft interposes itself between the target and the telescope. The occulter, a Solar Powered Ion-Driven Eclipsing Rover (SPIDER), is autonomous and coordinates its activities with the telescope. Imaging of the field is used to map the star field, locate the target, locate the occulter, and support formation control once the occulter is in the target-telescope line-of-sight (TTLOS).

The physical diameter of the occulter (W), its distance from the telescope (z), observing wavelength (λ), and ability to maintain alignment are major factors in the resulting on-axis starlight reduction. Velocity matching of the SPIDER with the telescope and formation control are achieved with small impulse thrusters on the SPIDER and with communication between the telescope and the SPIDER [5]. The automated SPIDER formation control tasks are implemented by the Attitude and Translational Control System (ATCS) [6].

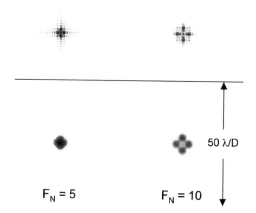

FIGURE 2. Focal plane simulation of PSF for a square occulter with a square aperture telescope for two Fresnel numbers F_N. Higher F_N is equivalent to smaller separation between telescope and occulter. *Upper pane*: unapodized. *Lower pane*: crossed 4th order Sonine apodization.

Apodized square aperture telescopes have the advantage of redirecting much of the diffracted starlight along preferred orthogonal spikes. Apodization reduces the resolution, but increases the depth of the nulls between diffraction spikes [7, 8, 9]. We combine these techniques with an external occulter.

OPTICAL ANALYSIS AND SIMULATIONS

Optical analysis of a square aperture telescope with a square external occulter show that the diffraction pattern can be characterized by two parameters: the Fresnel number ($F_N = W^2/\lambda z$) and the ratio of the telescope diameter to the occulter width (D/W). Figure 2 illustrates model PSFs expected at the focal plane, with and without telescope apodization, for D/W=0.5 and two Fresnel numbers. For constant occulter width and a given wavelength, increasing the Fresnel number is the mathematical equivalent of moving the occulter closer to the telescope.

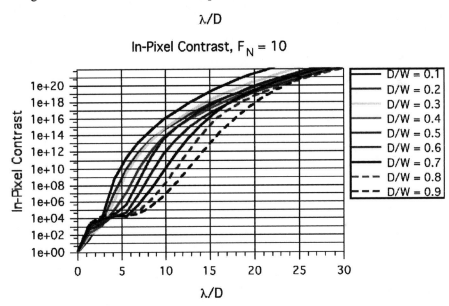

FIGURE 3. Contrast ratio in the focal plane PSF wavelength normalized to aperture.

Figure 3 shows the "in-pixel contrast", $i.e.$, the ratio of the star flux to the planetary flux in a $(\lambda/D)^2$ pixel, as a function of angular separation between the planet and star. In reflected visible light, the luminosity ratio of a terrestrial planet to the host star is typically in the range of 10^{-10} (corresponding to the label 1e+10, which is 10^{10}, on the y-axis in Figure 3), while Jupiter-like planets have contrast ratios of 10^{-8} (1e+08, or 10^8, on the y-axis in Figure 3).

The actual measured contrast is obtained by multiplying the luminosity ratio by the in-pixel contrast that is plotted in Figure 3. By intersecting the expected terrestrial planet contrast (a horizontal line on the y-axis) with one of the plotted contrast ratio curves (D/W), one can determine the angular separation needed to detect the planet. For a 4-meter telescope at 5500A, $\lambda/D=0.026$". Using D/W=0.5 (8-meter occulter) and a Fresnel number of 10 (which places the occulter at 12,800 kilometers from the telescope), the in-pixel contrast ratio of 10^{10} is achievable at ~ 7.5 λ/D (0.19"). If instead a 10-meter occulter (D/W=0.4) is used, the telescope-occulter distance becomes 20,000 kilometers, and the in-pixel contrast ratio of 10^{10} is reached at $\lambda/D=7$; i.e., 0.18" from the target star.

An external occulter has a number of advantages over other coronagraphic and interferometric techniques:

- Approximately 2 orders of magnitude less on-axis starlight enters the telescope for the same mirror tolerances.
- Mirror polishing tolerances can be significantly lower, and can be achieved with existing technology.
- Higher technology readiness levels than other coronagraphic techniques.
- Formation flying requirements are on the order centimeters not nanometers.
- It can be used to improve any other direct exo-planet imaging technique.
- It can be used with any instrument on the telescope – imaging, spectroscopic, or interferometric.

REFERENCES

1. Kochte, M, et al., 2002, 199th AAS Conference, Washington, D.C., *Free-flying Occulters for Use with Space Telescopes*
2. Schultz, A., et al., 2003 SPIE Meeting, San Diego, CA, *Imaging Terrestrial Planets with a Free Flying Occulter and Space Telescope: An Optical Simulation*
3. Schultz, A, et al., 2002, SPIE Conference, Waikoloa, HI, *UMBRAS: A Matched Occulter and Telescope for Imaging Extrasolar Planets*
4. Schultz, A. et al., 2000, SPIE Meeting, San Diego, CA, *Imaging Planets About Other Stars with UMBRAS II*
5. Hart, H.M, et al., 2000, SPIE Meeting Quebec City, Canada, *Imaging Planets About Other Stars with UMBRAS: Target Acquisition and Station Keeping*
6. Jordan, I.J.E., et al., 2000, AIAA Space 2000 Conference & Exposition, Long Beach, CA, *UMBRAS: Design of a Free-Flying Occulter for Space Telescopes*
7. Schultz, A., Frazier, T.V., Kosso, E., 1984, Applied Optics 23, 1914, *Sonine's Bessel Identity Applied to Apodization*
8. Nisenson, P., Papaliolios, C., 2001, ApJ, 548, L201, *Detection of Earth-like Planets Using Apodized Telescopes*
9. Gezari, D.Y., et al., 2002, SPIE Conference, Waikoloa, HI, *ExPO: A Discovery-class Apodized Square Aperture (ASA) Exo-Planet Imaging Space Telescope Concept*

Prospects for 'Earths' in the Habitable Zones of Known Exoplanetary Systems

Barrie W. Jones*, David R. Underwood* and P. Nick Sleep*

*The Open University, Walton Hall, Milton Keynes MK7 6AA, UK

Abstract. We have shown that Earth-mass planets could survive in variously restricted regions of the habitable zones (HZs) of most of a sample of nine of the 104 main-sequence exoplanetary systems confirmed by mid-November 2003. In a preliminary extrapolation of our results to the other systems, we estimate that roughly a half of these systems could have had an Earth-mass planet confined to the HZ for at least the most recent 1000 Ma. The HZ migrates outwards during the main-sequence lifetime, and so this proportion varies with stellar age – about two thirds of the systems could have such a planet confined to the HZ for at least 1000 Ma at *sometime* during the main-sequence lifetime. Clearly, these systems should be targeted for exploration for terrestrial planets. We have reached this conclusion by launching putative Earth-mass planets in various orbits and following their fate with mixed-variable symplectic and hybrid integrators. Whether the Earth-mass planets could *form* in the HZs of the exoplanetary systems is an urgent question that needs further study.

INTRODUCTION

It might be some years before we know for certain whether any exoplanetary systems have planets with masses the order of that of the Earth. We have therefore used a computer model to investigate a representative sample of the known exoplanetary systems, to see whether such planets could be present, and in particular whether they could remain confined to the habitable zones (HZs). If so, then it is possible that life is present on any such planets.

The HZ is that range of distances from a star where water at the surface of an Earth-like planet would be in the liquid phase. We have used boundaries for the HZ originating with Kasting, Whitmire, & Reynolds [1]. We have revised the boundaries of Kasting et al by using a more recent model of stellar evolution [2].

To test for confinement to the HZ of interest, we launch putative Earth-mass planets (labelled EM) into various orbits in or near the HZ and use a mixed-variable symplectic integrator to calculate the evolution of the orbit. The integration is halted when an Earth-mass planet comes within three Hill radii $(3R_H)$ of the giant planet. This is the distance at which a symplectic integrator becomes inaccurate, and is also the distance by which severe orbital perturbation of the Earth-mass planet will have occurred.

Some of the integrations halted in this way were re-calculated using a hybrid integrator, where there is a switch to a Bulirsch-Stoer integrator at $3R_H$ [3]. This enables the evolution of the orbit to be followed for longer.

Confinement to the HZ over the interval of interest requires that there is no early termination of the integration by an approach within $3R_H$, *and* that the semimajor axis of the Earth-mass planet remains within the HZ. We mostly used 1000 Ma (1000 million

CP713, *The Search for Other Worlds: Fourteenth Astrophysics Conference,*
edited by S. S. Holt and D. Deming
© 2004 American Institute of Physics 0-7354-0190-X/04/$22.00

years) as the standard pre-set time.

Based on the Earth, 1000 Ma is the order of time required for a biosphere to begin to have an effect on a planet's surface or atmosphere that could be detected from afar, provided that we exclude any early heavy bombardment from space such as might have frustrated the emergence of life on Earth for the first 700 Ma, or possibly to several 100 Ma later [4].

We selected nine systems that between them represent a large proportion of the 104 main-sequence exoplanetary systems confirmed by mid-November 2003 [5]. Details of our work on Ups And, Gl876, Rho CrB, 47 UMa, and ε Eridani are in Jones et al [6], Jones and Sleep [7], Jones and Sleep [8].

RESULTS

We have found that for EM *interior* to the giant planet(s), symplectic integrations are typically halted by secular increases in the eccentricity of EM orbits launched near to $3R_H$ from the giant. The semimajor axis of EM changes little up to this point. The hybrid integrator has been used to explore the fate of some of the halted orbits within $3R_H$. In almost all cases the outcome is ejection of EM to an astrocentric distance beyond 100 AU.

For EM *exterior* to the giant planet, secular increases in the eccentricity of EM occur when it is launched within about nR_H of the giant, where n depends on the orbital eccentricity and whether the orbit is interior to or exterior to the giant planet's orbit. A rough rule (that we are in the process of refining), is that $n \sim 3$ if the eccentricity of the giant planet's orbit is less than about 0.1, rising to 7-8 if it is about 0.3.

In order to avoid the extensive integrations necessary to establish the confinement or otherwise in other exoplanetary systems, we have applied the nR_H criteria at the HZ boundaries in each of these systems.

Figure 1 shows a notional HZ migrating outwards during the main sequence. Consider first a single giant planet closer to the star than the HZ, and suppose that it has the apastron and periastron distances shown, with $7R_H$ extending outwards from apastron, corresponding to high eccentricity. The whole HZ lies beyond $7R_H$ for the whole of the main-sequence, so we conclude that confined EM orbits are likely anywhere in the HZ at any time during the main-sequence. Of particular interest from a biological perspective is whether there could have been such orbits throughout at least the most recent 1000 Ma, excluding the first 700 Ma of the main sequence, as discussed in the 'Introduction'. It is also of interest whether such a span of 1000 Ma can be found at any time during the main-sequence, and not just recently.

Figure 1 also shows a single giant further from the star, with $3R_H$ extending inwards from periastron. The whole HZ lies beyond $3R_H$ for the whole of the main-sequence, so we again conclude that confined EM orbits are likely anywhere in the HZ at any time during the main-sequence. In this case too the most recent 1000 Ma and any span of 1000 Ma are of biological interest.

TABLE 1. Habitability of extrasolar planetary systems; see "Results" for explanation.

Star	now	sometime
OG-TR-56	yes	yes
HD73256	yes	yes
HD83443	yes	yes
HD46375	yes	yes
HD179949	yes	yes
HD187123	yes	yes
Tau Boo	yes	yes
BD103166	yes	yes
HD75289	yes	yes
HD209458	yes	yes
HD76700	yes	yes
51 Peg	yes	yes
Ups And	NO	NO
HD49674	yes	yes
HD68988	yes	yes
HD168746	yes	yes
HD217107	yes	yes
HD162020	yes	yes
HD130322	yes	yes
HD108147	yes	yes
HD38529	NO	NO
55 Cancri	yes	yes
Gliese 86	yes	yes
HD195019	yes	yes
HD6434	yes	yes
HD192263	yes	yes
Gliese 876	NO	NO
Rho CrB	yes	yes
HD74156	NO	NO
HD168443	NO	NO
HD3651	?	yes
HD121504	yes	yes
HD178911B	?	yes
HD16141	yes	yes
HD114762	part	yes
HD80606	NO	part
HD219542B	yes	yes
70 Vir	?	yes
HD216770	?	yes
HD52265	yes	yes
GJ3021	NO	part
HD37124	NO	NO
HD 219449	NO	part
HD73526	part	part
HD104985	?	yes
HD82943	NO	part
HD169830	part	part
HD8574	part	yes
HD89744	part	yes
HD134987	part	yes
HD40979	part	part
HD12661	NO	NO

Star	now	sometime
HD150706	?	part
HD59686	NO	yes
HR810	part	yes
HD142	?	part
HD92788	NO	part
HD28185	NO	part
HD142415	?	part
HD177830	part	part
HD108874	NO	part
HD4203	part	part
HD128311	NO	NO
HD27442	part	yes
HD210277	NO	NO
HD19994	part	part
HD20367	?	part
HD114783	part	part
HD147513	NO	NO
HIP75458	NO	NO
HD222582	NO	NO
HD65216	?	NO
HD160691	NO	NO
HD141937	NO	NO
HD41004A	part	part
HD47536	NO	NO
HD23079	NO	part
16 CygB	NO	NO
HD4208	part	part
HD114386	yes**	yes**
gam Ceph	part	part
HD213240	NO	NO
HD10647	NO	part
HD10697	NO	NO
47 UMa	part	part
HD190228	NO	NO
HD114729	part	part
HD111232	?	part
HD2039	NO	NO
HD136118	NO	NO
HD50554	NO	NO
HD196050	part	part
HD216437	?	part
HD216435	NO	NO
HD106252	NO	NO
HD23596	NO	NO
14 Her	NO	NO
HD39091	?	part
HD72659	yes**	yes**
HD70642	yes**	yes**
HD33636	NO	NO
eps Eridani	NO	part
HD30177	part	part
Gliese 777A	yes**	yes**

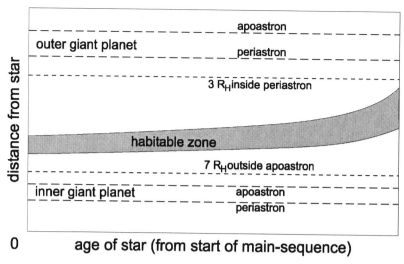

FIGURE 1. A notional habitable zone (HZ) migrating outwards during the main-sequence, with a giant planet interior to the HZ, and, alternatively, a giant planet exterior to the HZ. R_H is the Hill radius of the giant planet.

The Table summarises the results of this kind of analysis applied to all of the main-sequence exoplanetary systems in Schneider [5]. The following conventions are adopted.

1. The systems are in the order listed by Schneider, by increasing period of the planet with the shortest period.

2. The nR_H are calculated using the minimum giant masses, though R_H varies slowly, as $m^{1/3}$.

3. Systems with more than one planet are shown italicised.

4. The column 'now' shows whether an EM could be confined to the HZ within at least the past 1000 Ma (excluding the first 700 Ma of the main-sequence). If the entry is 'yes' then it could do so almost anywhere in the HZ. If the entry is 'NO' then nowhere in the HZ should offer confinement. If the entry is 'part' then some small proportion of the HZ should offer confinement, for example near its outer boundary for a giant planet not much closer to the star than the inner boundary.

5. The column 'sometime' refers to whether an EM could be confined to the HZ for at least 1000 Ma at any time in the main-sequence (again excluding the first 700 Ma).

6. A '?' denotes a star of unknown age, where this is crucial to the evaluation.

7. A '**' denotes those very few cases where the periastron of the giant lies beyond the HZ even at the end of the main-sequence. These are the systems most like the Solar System.

Overall, we estimate that roughly half of the systems in the Table could have had an Earth-mass planet confined to the HZ for at least the most recent 1000 Ma ('now' = 'yes' or 'part'), and that about two thirds of the systems could have such a planet confined to the HZ for at least a billion years *sometime* during the main-sequence lifetime ('sometime' = 'yes' or 'part').

Our results are broadly in accord with those of Menou and Tabachnik [9], who performed short numerical integrations (1 Ma) for mass-less particles in many of the systems in the Table.

Whether the Earth-mass planets could form in the HZs of the exoplanetary systems is an urgent question that needs further study, though work by Mandell and Sigurdsson [10] is encouraging.

REFERENCES

1. Kasting, J.F., Whitmire, D.P., & Reynolds, R.T. 1993, Icarus, 101, 108.
2. Mazzitelli, I. 1989, ApJ, 340, 249.
3. Chambers, J.E. 1999, MNRAS, 305, 793.
4. Brasier, M.D., et al. 2002, Nature, 416, 76.
5. Schneider, J. 2003, *http://www.obspm.fr/encycl/cat1.html*.
6. Jones, B.W., Sleep, P.N., & Chambers, J.E. 2001, A&A, 366, 254.
7. Jones, B.W., & Sleep, P.N. 2002, A&A, 393, 1015.
8. Jones, B.W., & Sleep, P.N. 2003, ASP Conf Ser, 294, 225.
9. Menou, K., & Tabachnik, S. 2003, ApJ, 583, 473.
10. Mandell, A.M., & Sigurdsson, S. 2003, astro-ph/0307512.

Dynamics and Dynamical Evolution
of Planetary Systems

Planet–disk Interactions and Orbital Migration

Caroline E. J. M. L. J. Terquem

Institut d'Astrophysique de Paris, 98 bis, bd Arago, 75014 Paris, France
Université Denis Diderot–Paris VII, 2 Place Jussieu, 75251 Paris Cedex 5, France

Abstract. The recent observations of extrasolar giant planets orbiting close to their companion stars suggest that tidal interactions between the disk and an embedded planet can play a fundamental role in determining the latter's mass and orbital properties. We describe briefly these interactions and review recent work on orbital migration, focusing on runaway migration, as well as migration in magnetized and turbulent disks.

INTRODUCTION

Almost 20% of the ~ 120 extrasolar planets detected so far orbit at a distance between 0.038 and 0.1 astronomical unit (au) from their host star.

It is very unlikely that these so–called 'hot Jupiters' have formed *in situ*. In most of the standard disk models, temperatures at around 0.05 au are larger than 1500 K ([1], [2]), preventing the condensation of rocky material and therefore the accretion of a solid core there. Even if models with lower temperatures are considered, giant planets may form in close orbits according to the standard core accretion model only if the disk surface density is rather large and the accretion process very efficient ([3], [4]). This is because at 0.05 au a solid core of about 40 earth masses has to be built–up before a massive gaseous envelope can be accreted ([2], [3]). More likely, the hot Jupiters have formed further away in the protoplanetary nebula and have migrated down to small orbital distances. It is also possible that migration and formation were concurrent ([2]). In general, orbital decay of a protoplanet is believed to result from the tidal interaction between the protoplanet and the gas in the surrounding protoplanetary nebula ([5], [6], [7], [8] and references therein, [9], [10], [11]). Other mechanisms have been proposed but they are not as efficient (see [12] and references therein).

We now review the different types of migration that tidal interactions between a disk and a planet may lead to, and the recent advances that have been made in the study of orbital migration.

DIFFERENT TYPES OF MIGRATION

Depending on the mass of the migrating protoplanet, three regimes can be distinguished, leading to three markedly different migration timescales.

CP713, *The Search for Other Worlds: Fourteenth Astrophysics Conference,*
edited by S. S. Holt and D. Deming

Type–I migration

Cores with masses below about 10 $M_\oplus$ interact linearly with the surrounding nebula and migrate inward *relative* to the gas. The drift timescale for a planet of mass M_{pl} undergoing type I migration in a uniform disk is ([10], [11]):

$$\tau_I(\text{yr}) \sim 10^8 \left(\frac{M_{pl}}{M_\oplus}\right)^{-1} \left(\frac{\Sigma}{\text{g cm}^{-2}}\right)^{-1} \left(\frac{r}{\text{au}}\right)^{-1/2} \times 10^2 \left(\frac{H}{r}\right)^2 \tag{1}$$

where Σ is the disk surface density, r is the distance to the central star and H is the disk semithickness. It has been assumed here that the torque exerted by the material which corotates with the perturbation is small compared to the torque exerted at the Lindblad resonances.

The migration timescale given by equation (1) applies to a planet on a circular orbit. It has been shown ([13], [14]) that a planetary core on an eccentric orbit (in an axisymetric disk) with an eccentricity e significantly larger than the disk aspect ratio H/r may undergo outward migration.

Type–II migration

Planets with masses larger than or comparable to that of Jupiter interact nonlinearly with the disk and may open up a gap ([6], [7], [8] and references therein). The planet is then locked into the angular momentum transport process of the disk, and migrates *with* the gas. The direction of type II migration is that of the viscous diffusion of the disk. Therefore it is inward except in the outer parts of the disk which diffuse outward.

For type II migration, the characteristic orbital decay timescale is the disk viscous timescale:

$$\tau_{II}(\text{yr}) = \frac{1}{3\alpha} \left(\frac{r}{H}\right)^2 \Omega^{-1} = 0.05 \frac{1}{\alpha} \left(\frac{r}{H}\right)^2 \left(\frac{r}{\text{AU}}\right)^{3/2} \tag{2}$$

where α is the standard Shakura & Sunyaev's parameter ([15]) and Ω is the angular velocity at radius r.

The expression of τ_{II} is independent of the protoplanet's mass and the surface mass density of the disk, but it is implicitly assumed here that the mass of gas within the characteristic orbital radius of the planet is at least comparable to the mass of the planet itself. If the disk is significantly less massive, then there is not enough gas in the vicinity of the planet to absorb its orbital angular momentum, and migration is slown down ([16]).

Runaway migration

It has recently been shown ([17]) that a planet in the intermediate mass range embedded in a disk massive enough may undergo a runaway migration (also called *type–III*

236

migration). This happens when the planet's coorbital region is partially depleted. When the planet migrates inward, the fluid elements of the inner disk gain angular momentum as they pass to the outer disk. The material trapped in the coorbital region, and which drifts radially with the planet, loses angular momentum. For a partial gap, the net effect is a negative corotation torque on the planet which scales with its drift rate. If the planet migrates outward, the net corotation torque is positive. Thus the corotation torque provides a positive feedback that can lead to a runaway when the disk is massive enough and vary the planet semimajor axis by 50% over a few tens of orbits. The timescale for runaway migration can then be much shorter than that for type I or type II migration.

This typically happens for Saturn–sized giant planets embedded in disks with a mass several times the minimum mass of the solar nebula.

MIGRATION IN A MAGNETIZED DISK

The effect of a toroidal magnetic field on type I migration for a planet on a circular orbit has recently been investigated ([18]). When a field is present, in contrast to the nonmagnetic case, there is no singularity at the corotation radius, where the frequency of the perturbation matches the orbital frequency. However, all fluid perturbations are singular at the so–called *magnetic resonances*, where the Doppler shifted frequency of the perturbation matches that of a slow MHD wave propagating along the field line. There are two such resonances, located on each side of the planet's orbit and within the Lindblad resonances. Like in the nonmagnetic case, waves propagate outside the Lindblad resonances. But they also propagate in a restricted region around the magnetic resonances.

The magnetic resonances contribute to a significant global torque on the disk which, like the Lindblad torque, is positive (negative) outside (inside) the planet's orbit. Since these resonances are closer to the planet than the Lindblad resonances, they couple more strongly to the tidal potential and the torque they contribute dominates over the Lindblad torque if the magnetic field is large enough. In addition, if $\beta \equiv c^2/v_A^2$, where c is the sound speed and v_A the Alfven velocity, increases fast enough with radius, the outer magnetic resonance becomes less important (it disappears altogether when there is no magnetic field outside the planet's orbit) and the total torque is then negative, dominated by the inner magnetic resonance. This leads to outward migration of the planet.

Figure 1 shows the tidal torque $\tilde{T}_m(r_p, r)$, which is the torque is units of $\Sigma_p r_p^4 \Omega_p^2$, with the subscript p denoting values at the corotation radius r_p, exerted between the radii r_p and r for $m = 10$ and for both $B = 0$ and B non zero. The effect of the magnetic resonances can clearly be seen in the magnetic case.

The amount by which β has to increase outward for the total torque exerted on the disk to be negative depends mainly on the magnitude of β. It was found that, for $\beta = 1$ or 100 at corotation, the torque exerted on the disk is negative when β increases at least as fast as r^2 or r^4, respectively.

Figure 2 shows the total torque $\tilde{T}_m$ exerted by the planet on the disk versus m for $d\ln\Sigma/d\ln r = 0$, $c/(r_p\Omega_p) = 0.03$, $d\ln\left(r^2\langle B^2\rangle\right)/d\ln r = -1$, which correspond to $\beta \propto r^3$, and different values of $\beta(r_p)$. As we can see, the cumulative torque $\tilde{T}$ is significantly

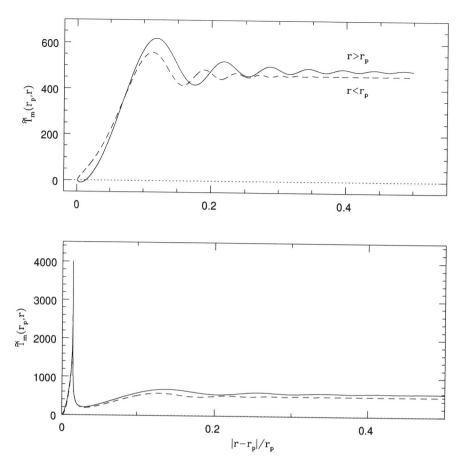

FIGURE 1. Tidal torque $\tilde{T}_m$ exerted by the planet on the disk between the radii r_p and r versus $|r - r_p|/r_p$ for $B = 0$ (*upper panel*) and $B \neq 0$ (*lower panel*). The two curves in each panel correspond to $r > r_p$ (*solid line*) and $r < r_p$ (*dashed line*). Here $m = 10$, $d\ln\Sigma/d\ln r = 0$ and $c/(r_p\Omega_p) = 0.03$. For the magnetic case, $\beta = \text{const} = 1$ and $d\ln\left(r^2\langle B^2 \rangle\right)/d\ln r = 2$. For these parameters, the total torque on the disk is positive in both the magnetic and non magnetic cases.

decreased for values of $\beta(r_p)$ as large as 10^2 and becomes negative for values of $\beta(r_p)$ between 10^2 and 10.

The migration timescales that correspond to the torques calculated when a magnetic field is present are rather short. The orbital decay timescale of a planet of mass M_{pl} at radius r_p is $\tau = M_{\text{pl}} r_p^2 \Omega_p / |T|$, where T is the torque exerted by the planet on the disk and Ω_p is the angular velocity at radius r_p. This gives:

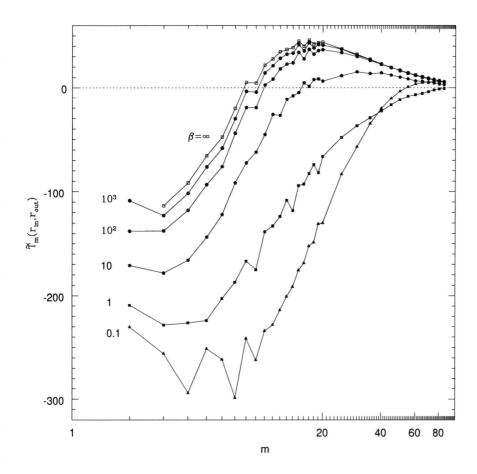

FIGURE 2. Total torque $\tilde{T}_m$ exerted on the disk versus m for $d\ln\Sigma/d\ln r = 0$ and $c/(r_p\Omega_p) = 0.03$. The open squares correspond to $\beta = \infty$, i.e. $B = 0$. The filled symbols correspond to $d\ln\left(r^2\langle B^2\rangle\right)/d\ln r = -1$ and $\beta(r_p) = 0.1$ (*triangles*), 1 (*squares*), 10 (*pentagons*) 10^2 (*hexagons*) and 10^3 (*septagons*). An estimate of the cumulative torque gives $\tilde{T} = 1362, 1129, 855, -508, -3975$ and -5483 for $\beta(r_p) = \infty, 10^3, 10^2, 10, 1$ and 0.1, respectively. This shows that $\tilde{T}$ is significantly reduced for values of β as large as 10^2. Note that a negative (positive) cumulative torque corresponds to the planet migrating outward (inward).

$$\tau(\text{yr}) = 4.3 \times 10^9 \left(\frac{M_{\text{pl}}}{M_\oplus}\right)^{-1} \left(\frac{\Sigma_p}{100\ \text{g cm}^{-2}}\right)^{-1} \left(\frac{r_p}{1\ \text{au}}\right)^{-1/2} \left(\frac{|T|}{\Sigma_p r_p^4 \Omega_p^2}\right)^{-1} \quad (3)$$

where Σ_p is the disk surface density at radius r_p. In a standard disk model, $\Sigma \sim 100$–10^3 g cm^{-2} at 1 au (see, for instance, [2]). Therefore, $\tau \sim 10^5$–10^6 yr for a one earth mass planet at 1 au in a nonmagnetic disk, as $|T| / \left(\Sigma_p r_p^4 \Omega_p^2 \right) \sim 10^3$ in that case (see fig. 2). This is in agreement with Ward ([10], [11], see eq. [1] above). In a magnetic disk, $|T| / \left(\Sigma_p r_p^4 \Omega_p^2 \right)$ may become larger (see fig. 2) leading to an even shorter migration timescale. However, it is important to keep in mind that these timescales are *local*. Once the planet migrates outward out of the region where β increases with radius, it may enter a region where β behaves differently and then resume inward migration for instance.

The calculations summarized here indicate that a planet migrating inward through a nonmagnetized region of a disk would stall when reaching a magnetized region. It would then be able to grow to become a terrestrial planet or the core of a giant planet. We are also led to speculate that in a turbulent magnetized disk in which the large scale field structure changes sufficiently slowly a planet may alternate between inward and outward migration, depending on the gradients of the field encountered. Its migration could then become diffusive, or be limited only to small scales.

MIGRATION IN A TURBULENT DISK

Numerical simulations of planet migration in a turbulent disk have been performed for giant planets ([19], [20]) and for embedded protoplanets ([21]). In these (MHD) simulations, the turbulence is self–consistently generated by the magnetorotational instability.

When the planet is massive enough to open a gap, it is found that inward migration occurs on the viscous timescale, as expected from previous work. However, the gap was found to be wider and cleaner than in laminar disk models with an equivalent α viscosity (i.e. an α parameter giving a stress equal to that produced by the turbulence).

For embedded protoplanets, the interaction with the disk is severely affected by the turbulence. The torque exerted on the planet has large fluctuations and oscillates between negative and positive values. Therefore migration is neither inward or outward, but occurs as a random walk.

Note that diffusive migration was predicted ([18]) on the basis of the effect of a magnetic field. However, fluctuations of the mass density in the vicinity of the planet certainly also play an important role in the diffusive behavior observed in the simulations.

REFERENCES

1. Bell, K. R., Lin, D. N. C., Hartmann, L. W., and Kenyon, S. J., *Astroph. J.*, **444**, 376–395 (1995).
2. Papaloizou, J. C. B., and Terquem, C., *Astroph. J.*, **521**, 823–838 (1999).
3. Bodenheimer, P., Hubickyj, O., and Lissauer, J. J., *Icarus*, **143**, 2–14 (2000).
4. Ikoma, M., Emori, H., and Nakazawa, K., *Astroph. J.*, **553**, 999–1005 (2001).
5. Goldreich, P., and Tremaine, S., *Astroph. J.*, **233**, 857–871 (1979).
6. Goldreich, P., and Tremaine, S., *Astroph. J.*, **241**, 425–441 (1980).
7. Lin, D. N. C., and Papaloizou, J. C. B., *M.N.R.A.S.*, **188**, 191–201 (1979).
8. Lin, D. N. C., and Papaloizou, J. C. B., "On the tidal interaction between protostellar disks and companions," in *Protostars and Planets III*, edited by E. H. Levy and J. I. Lunine, University of Arizona Press, Tucson, 1993, pp. 749–835.
9. Papaloizou, J. C. B., and Lin, D. N. C., *Astroph. J.*, **285**, 818–834 (1984).
10. Ward, W. R., *Icarus*, **67**, 164–180 (1986).
11. Ward, W. R., *Icarus*, **126**, 261–281 (1997).

12. Terquem, C., Papaloizou, J. C. B., and Nelson, R. P., "Disks, extrasolar planets and migration," in *From Dust to Terrestrial Planets*, edited by W. Benz, R. Kallenbach, G. Lugmair and F. Podosek, ISSI Space Sciences Series, 9 (reprinted from Space Science Reviews, 92), 2000, pp. 323–340.
13. Papaloizou, J. C. B., and Larwood, J. D., *M.N.R.A.S.*, **315**, 823–833 (2000).
14. Papaloizou, J. C. B., *Astron. Astrophys.*, **388**, 615–631 (2002).
15. Shakura, N. I. and Sunyaev, R. A., *Astron. Astroph.*, **24**, 337–355 (1973).
16. Ivanov, P. B., Papaloizou, J. C. B., and Polnarev, A. G., *M.N.R.A.S.*, **307**, 791–891 (1999).
17. Masset, F. and Papaloizou, J. C. B., *Astrophys. J.*, **588**, 494–508 (2003).
18. Terquem, C. E. J. M. L. J., *M.N.R.A.S.*, **341**, 1157–1173 (2003).
19. Winters, W. F., Balbus, S. A., and Hawley, J. F., *Astroph. J.*, **589**, 543–555 (2003).
20. Nelson, R. P., and Papaloizou, J. C. B., *M.N.R.A.S.*, **339**, 993–1005 (2003).
21. Nelson, R. P., and Papaloizou, J. C. B., *M.N.R.A.S.*, *in press* (2004), URL http://xxx.lanl.gov/abs/astro-ph/0308360.

Extrasolar Planet Orbits and Eccentricities

Scott Tremaine* and Nadia L. Zakamska*

*Department of Astrophysical Sciences, Princeton University, Princeton NJ 08544

Abstract. The known extrasolar planets exhibit many interesting and surprising features—extremely short-period orbits, high-eccentricity orbits, mean-motion and secular resonances, etc.—and have dramatically expanded our appreciation of the diversity of possible planetary systems. In this review we summarize the orbital properties of extrasolar planets. One of the most remarkable features of extrasolar planets is their high eccentricities, far larger than seen in the solar system. We review theoretical explanations for large eccentricities and point out the successes and shortcomings of existing theories.

Radial-velocity surveys have discovered ~ 120 extrasolar planets, and have provided us with accurate estimates of orbital period P, semimajor axis a, eccentricity e, and the combination $M \sin i$ of the planetary mass M and orbital inclination i (assuming the stellar masses are known). Understanding the properties and origin of the distribution of masses and orbital elements of extrasolar planets is important for at least two reasons. First, these data constitute essentially everything we know about extrasolar planets. Second, they contain several surprising features, which are inconsistent with the notions of planet formation that we had before extrasolar planets were discovered. The period distribution is surprising, because no one predicted that giant planets could have orbital periods less than 0.1% of Jupiter's; the mass distribution is surprising because no one predicted that planets could have masses as large as 10 Jupiters, or that there would be a sharp cutoff at this point; and the eccentricity distribution is surprising because we believed that planets forming from a disk would have nearly circular orbits.

1. MASS AND PERIOD DISTRIBUTIONS

The mass distribution of extrasolar planets is sharply cut off above $10M_J$, where M_J is the Jupiter mass [1, 2]. The absence of companions with masses in the range $10M_J < M < 100M_J$ (the "brown dwarf desert") provides the strongest reason to believe that the extrasolar planets are formed by a different mechanism than low-mass companion stars, and thus the criterion $M < 10M_J$ is probably the least bad definition of an extrasolar planet. The observed mass distribution for $M < 10M_J$ can be modeled as a power law $dn \propto M^{-\alpha}dM$ with $\alpha = 1.1 \pm 0.1$ [1, 3, 4, 5]; in other words the distribution is approximately flat in $\log M$, at least over the range $M_J \lesssim M \lesssim 10M_J$. Weak evidence that this distribution can be extrapolated to smaller masses comes from the solar system: fitting the distribution of the masses of the 9 planets to a power law over the interval from M_{Pluto} to M_J yields $\alpha = 1.0 \pm 0.1$, with a high Kolmogorov-Smirnov (KS) confidence level (over 99.9%).

CP713, *The Search for Other Worlds: Fourteenth Astrophysics Conference*,
edited by S. S. Holt and D. Deming
© 2004 American Institute of Physics 0-7354-0190-X/04/$22.00

The period distribution can also be described by a power law to a first approximation: $dn \propto P^{-\beta}dP$, where $\beta = 0.73 \pm 0.06$ [5]. Some analyses [3, 4, 6] find a steeper slope, $\beta \simeq 1$, probably because they do not correct for selection effects (the reflex motion of the star due to a planet of a given mass declines with increasing period). Stepinski & Black [4] and Mazeh & Zucker [6] stress that the distribution of periods (and eccentricities) is almost the same for extrasolar planets and spectroscopic binaries—only the mass distributions are different—and this striking coincidence demands explanation (see §2.7). Recall that the standard minimum solar nebula has surface density $\Sigma(r) \propto r^{-1.5}$ [7]. If the formation process assigned this mass to equal-mass planets without migration we would have $\beta = 0.67$, consistent with the exponent found above for extrasolar planets.

In Figure 1 we plot the masses and periods of extrasolar planets.[1] Although power laws are good first approximations to the mass and period distributions for $M < 10M_J$, there have been a number of suggestions of additional structure. (i) Zucker & Mazeh [8] and Udry et al. [9] suggest that there are too few massive planets ($M \gtrsim 3M_J$) on short-period orbits ($P \lesssim 50$ days), although (ii) Zucker & Mazeh argue that this shortfall is not present if the host star is a member of a wide binary system (denoted by star symbols in Figure 1). In addition, Udry et al. [9] suggest that there is (iii) a deficit of planets in the range $10\,\text{d} \lesssim P \lesssim 100\,\text{d}$; and (iv) a deficit of low-mass planets ($M \lesssim 0.8M_J$) on long-period orbits ($P \gtrsim 100\,\text{d}$). We do not find the latter two features persuasive, since their significance has not yet been verified by statistical tests, and of course low-mass planets on long-period orbits produce the smallest reflex velocity and so are most difficult to detect.

Whether or not these features exist, in our view the most impressive feature of the period distribution is that it is described so well by a power law. In particular, the mechanisms that might halt planetary migration often operate most effectively close to the star [10], so it is natural to expect a "pile-up" of planets at orbital periods of a few days; however, there is only weak evidence for an excess of this nature compared to a power law distribution.

2. ECCENTRICITY DISTRIBUTION

One of the remarkable features of extrasolar planets is their large eccentricities: the *median* eccentricity of 0.28 is larger than the *maximum* eccentricity of any planet in the solar system. The eccentricity distribution of extrasolar planets differs from that of solar system planets at the 99.9% KS confidence level.

Observational selection effects in radial-velocity surveys could either favor or disfavor detection of high-eccentricity planets: as the eccentricity grows at fixed period the periastron velocity grows, but the time spent near periastron declines. The limited simulations that have been done [12, 13] suggest that these two effects largely cancel, so the eccentricity-dependent selection effects are small. Thus, the observed distribution should accurately reflect the eccentricities of massive, short-period extrasolar planets.

---

[1] The data for figures are taken from http://www.obspm.fr/planets as of October 2003.

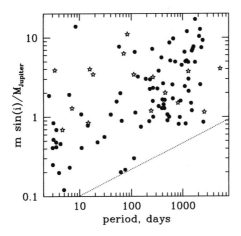

FIGURE 1. Minimum masses versus periods of extrasolar planets. Planets found in wide binary systems (from [11]) are denoted by stars. The absence of planets in the lower right part of the figure is due to the limited sensitivity of the radial-velocity surveys (the dotted line shows an amplitude of 10 m s^{-1} for a planet on a circular orbit around a solar-mass star).

Planets that form from a quiescent disk are expected to have nearly circular orbits, and thus the high eccentricities of extrasolar planets demand explanation. A wide variety of eccentricity excitation mechanisms has been suggested so far.

2.1. Interactions with the protoplanetary gas disk

Gravitational interactions between a planet and the surrounding protoplanetary disk can excite or damp the planet's eccentricity. These interactions are concentrated at discrete resonances, the most important of which are given by the relation [14, 15, 16, 17, 18]

$$\Omega + \frac{\varepsilon}{m}(\Omega - \dot{\varpi}) = \Omega_p + \frac{\varepsilon_p}{m}(\Omega_p - \dot{\varpi}_p); \qquad (1)$$

here Ω is the mean angular speed, $\dot{\varpi}$ is the apsidal precession rate, the integer $m > 0$ is the azimuthal wavenumber, $\varepsilon = 0, \pm 1$, and the unsubscripted variables refer to the disk while those with the subscript "p" refer to the planet. Resonances with $\varepsilon = 0$ are called corotation resonances and those with $\varepsilon = +1$ and -1 are called outer and inner Lindblad resonances, respectively. Resonances with $\varepsilon = \varepsilon_p$ are called "coorbital" resonances since the resonant condition (1) is satisfied when the gas orbits at the same angular speed as the planet; coorbital resonances are absent if the planet opens a gap in the disk. Resonances that are not coorbital are called external resonances. Apsidal or secular resonances have $m = 1$, $\varepsilon = \varepsilon_p = -1$ so the resonant condition is simply $\dot{\varpi} = \dot{\varpi}_p$; in this case the collective effects in the disk (mostly pressure) only have to compete with differential precession rather than with differential rotation, so the resonance is much broader.

Interactions at external Lindblad and corotation resonances excite and damp the planet's eccentricity, respectively [15]. In the absence of other effects, damping exceeds driving by a small margin ($\sim 5\%$). However, corotation resonances are easier to saturate, by trapping the orbital angular speed into libration around the pattern speed; if the corotation resonances are saturated and the Lindblad resonances are not, then the eccentricity can grow.

Interactions at coorbital resonances damp the planet's eccentricity [17, 18], but these only operate in a gap-free disk. Apsidal resonances can also damp eccentricity [19, 20], but at a rate that depends sensitively on the disk properties [21].

Given these complexities, about all we can conclude is that either eccentricity growth or damping may occur, depending on the properties of the planet and the disk. Numerical simulations have so far provided only limited insight: two-dimensional simulations [22, 23] show eccentricity growth if and only if the planet mass exceeds $10-20M_J$, but the relation of these results to the resonant behavior described above remains unclear.

2.2. Close encounters between planets

In the final stages of planet formation dynamical instability can develop, either as the masses of the planets increase due to accretion or as their orbital separation decreases due to differential migration. The instability usually leads one or more planets to be ejected from the system or to collide with one another or with the central star (this last outcome is relatively rare). The surviving planets are left on eccentric orbits [24, 25, 26, 27].

A great attraction of this mechanism is that it makes calculable predictions; unfortunately, the predictions have some difficulty matching the observations: (i) It has been suggested that this mechanism could produce the 'hot Jupiters'—planets such as 51 Peg B that are found on low-eccentricity, very short-period orbits (a few days) [24]—by close encounters of planets at distances of a few AU, which throw one planet onto a highly eccentric orbit that is later circularized by tidal dissipation. However, the frequency of hot Jupiters is far larger than this mechanism could produce [27]. (ii) The median eccentricity of the surviving planet following an ejection is about 0.6, compared to 0.3 in the observed systems [27]; on the other hand, so far the simulations have focused on equal-mass planets and the eccentricity is likely to be lower when a more massive planet ejects a less massive one. (iii) Collisions between planets lead to a population of collision remnants on nearly circular orbits, which is not observed [21, 27]. To avoid collisions would require that the planets are much more compact than Jupiter (i.e. the escape speed from the planet must be much larger than the orbital speed).

Another potential problem is the predicted mass-eccentricity relation: in this scattering process, low-mass planets are naturally expected to be excited to higher eccentricities. Unfortunately, so far no simulations have looked for mass-eccentricity correlations in a large sample of simulated planetary systems, so we do not know how strong the correlation should be. The eccentricities and masses of the known extrasolar planets show

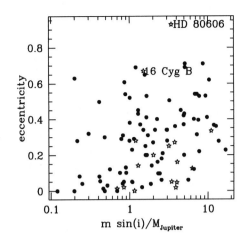

FIGURE 2. Eccentricity versus minimum mass for extrasolar planets. Planets found in wide binaries (from [11]) are shown with stars, and two examples with extreme eccentricities (16 Cyg B and HD 80606) are labeled.

a weak correlation (at about 90% confidence level) in the *opposite* sense: more massive planets seem to have higher eccentricities (Figure 2).

A related possibility is that the eccentricity is excited by interactions with massive planetesimals [28, 29]. Normally, close encounters with small bodies tend to damp the eccentricity and resonant interactions excite eccentricity. Because this process relies on interactions with solid bodies, rather than gas giant planets, it requires a rather massive planetesimal disk, comparable to the mass of the planet, which in turn requires a gaseous protoplanetary disk that is even more massive, $\gtrsim 0.1 M_\odot$.

2.3. Resonant interactions between planets

Many satellites in the solar system, as well as Neptune and Pluto, are locked in orbital resonances. The formation and evolution of these resonances has been thoroughly studied (see for example [30] and references therein).

Typically, resonance capture in the solar system occurs because of convergent outward migration, in which an inner satellite migrates outward faster than an outer satellite, thereby entering a mean-motion resonance. Continued migration after resonance capture excites the eccentricities of the resonant bodies; for example, this process is believed to be responsible for Pluto's eccentricity of 0.25 [31].

In contrast, the known extrasolar planets have presumably migrated *inward*. In this case, migration after resonance capture excites eccentricities even more efficiently. For example, a test particle that is captured into the 2:1 resonance with an exterior planet will be driven to an eccentricity of unity when the planet has migrated inward by a factor of four following capture [32]. Lee & Peale [33] have analyzed the dynamics of the planets in GJ 876, which are locked in a 2:1 resonance; they argue that to avoid exciting the eccentricities above the observed values requires either (i) extremely strong eccentricity damping, or (ii) fine-tuning the resonance capture to occur just before migration stops.

Chiang et al. [34] point out that eccentricities can also be excited by divergent migration, in which the ratio of the semimajor axes of two planets increases; the planets traverse a series of resonances, each of which excites additional eccentricity.

There are at least two concerns with models based on resonant interactions: (i) this mechanism requires at least two planets, whereas only a single planet has been discovered so far in most of the known extrasolar planetary systems; (ii) like close encounters, resonant interactions are expected to excite low-mass planets to higher eccentricities, but the observed mass-eccentricity correlation is small, and if anything in the opposite sense (Figure 2).

2.4. Secular interactions with a distant companion star

A planet's eccentricity can be excited by secular interactions with a distant companion that does not lie in the planet's orbital plane—this is sometimes called the Kozai mechanism [35, 36, 37, 38, 39, 40]. This mechanism has the interesting property that the separation and mass of the companion affect the period but not the amplitude of the eccentricity oscillation; thus even a weak tidal force from a distant stellar or planetary companion can excite a large eccentricity. However, the precession rate due to the companion must exceed the precession rate due to all other effects: other planets in coplanar orbits, any residual planetesimal disk, general relativity, etc.

There are several straightforward predictions for this mechanism: (i) There should be no correlation between the planet's mass and eccentricity, which is marginally consistent with the weak correlation seen in Figure 2. (ii) High-eccentricity planets should be found in binary star systems. This does not appear to be the case (Figure 2), although two of the highest eccentricities, 16 Cyg B and HD 80686, are found in binary systems. Perhaps some of the other high eccentricities are excited by unseen companions—planets or brown dwarfs. (iii) Coplanar multi-planet systems should have low eccentricities, since their mutually induced precession is far larger than the precession due to a distant companion, thereby suppressing the Kozai mechanism. This prediction is not supported by the dozen or so multi-planet systems, which have substantial eccentricities, although in most cases we have no direct evidence that these systems are coplanar (see §3).

2.5. Phenomenological diffusion

Given the limited success of the models we have described so far, it is useful to ask what constraints we can place on the eccentricity excitation mechanism from simple phenomenological models. Here we explore the possibility that the excitation can be modeled as a diffusion process in phase space, with an eccentricity-dependent diffusion coefficient $D(e)$.

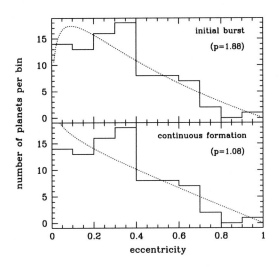

FIGURE 3. Eccentricity distribution: observed (histogram) and predicted by the diffusion models (dotted lines). Tidally circularized planets ($P < 6$ days) are excluded from the analysis. The diffusion coefficient in the eccentricity vector plane is $D(e) \propto e^p$.

We shall work with the eccentricity vector $\mathbf{e} = (e\cos\varpi, e\sin\varpi)$, since its components are approximately canonical variables for small eccentricities. Let $n(\mathbf{e})d\mathbf{e}$ be the number of planets with eccentricity vector in the range $d\mathbf{e}$. Then the diffusion equation in the eccentricity plane can be written as

$$\frac{\partial n(e,t)}{\partial t} = \nabla(D\nabla n) = \frac{1}{e}\frac{\partial}{\partial e}\left[eD(e)\frac{\partial n(e,t)}{\partial e}\right]. \qquad (2)$$

We solve this equation for two cases: (i) a single initial burst of formation at $t = 0$ ($n(e,0) \propto \delta(e)/e$); (ii) continuous formation at a constant rate ($n(0,t) =$const). Since planets are ejected from the system when they reach $e = 1$, there is also a boundary condition $n(1,t) = 0$. We assume that the diffusion coefficient is a power law, $D(e) \propto e^p$.

The best-fit ccentricity distributions that emerge from the diffusion models are plotted in Figure 3. They are shown at late times when the planets in the initial-burst model have diffused away from the center of the eccentricity plane ($e = 0$) and the continuous-formation model has reached its steady state $n(e) \propto e^{1-p} - e$. Both the initial-burst model and the continuous-formation model produce satisfactory fits (with KS confidence level about 60%), so we cannot distinguish between the two types of models by comparing to the current observations; however, the predicted behavior is very different near $e = 0$ so the models should be distinguishable with more data.

In both cases, the best-fit values of p ($p = 1.9$ for the initial-burst model and $p = 1.1$ for the continuous-formation model) suggest that eccentricity diffusion is produced by a mechanism that is ineffective for circular orbits.

2.6. Propagation of eccentricity disturbances from other planets

Stars in the solar neighborhood approach passing stars to within about 400 AU during a lifetime of 10^{10} y. Such encounters cannot excite the eccentricity of a single planet on a orbit like those of the known extrasolar or solar system planets. However, disks of gas and dust around young stars are observed to extend to radii of hundreds of AU, and it is plausible to assume that these disks form planetesimals or planets that survive to the present at radii $\sim 10^2$ AU. If so, then passing stars can efficiently excite the eccentricities of the outer bodies in the system, and the eccentricity disturbance can then propagate inward via secular interactions between the planets or planetesimals, much like a wave. This process is described more fully in the article "Propagation of eccentricity disturbances in planetary systems" [41] elsewhere in these proceedings. The most important disk properties that determine the efficiency of this excitation mechanism are the radial extent of the disk and the steepness of its smeared-out surface-density profile; the total mass of the disk is not a factor. If the surface density is a power law, $\Sigma(r) \propto r^{-q}$, efficient eccentricity excitation at small radii requires a rather flat profile, $q \approx 1$, compared to the minimum solar nebula profile $q \approx 1.5$.

2.7. Formation from a collapsing protostellar cloud

It is conventional to assume that the components of binary star systems form simultaneously from condensations in a collapsing protostellar cloud (and hence have high-eccentricity orbits), whereas planets form later from the protoplanetary disk (and hence initially have low-eccentricity orbits). The two populations of companions seem to be clearly separated by the brown-dwarf desert.

Thus it is remarkable that the eccentricity and period distributions of extrasolar planets and low-mass secondaries of spectroscopic binaries are almost indistinguishable [1, 4]. This observation suggests that perhaps the extrasolar planets, like binary stars, formed in the collapsing protostellar cloud [42, 43, 44].

Simulations by Papaloizou & Terquem [43] suggest that this process preferentially forms either "hot Jupiters" or planets with high eccentricities and large semimajor axes; although Papaloizou & Terquem did not carry out a detailed comparison, it seems likely that, with some tuning, their simulations could reproduce the distributions of periods and eccentricities of the known extrasolar planets. A possible concern is that the lower limit for opacity-limited fragmentation is a few Jupiter masses, and many planets are known to have smaller masses. Terquem & Papaloizou [44] postulate that the low-mass planets formed in a disk and their eccentricities were later excited by encounters with high-mass planets formed by fragmentation.

3. INCLINATION DISTRIBUTION

So far there is little direct evidence that the planets in multi-planet systems have small mutual inclinations, as we might expect if they formed from a disk. Chiang et al. [45]

argue that if the apsidal alignment between Ups And C and D ($\Delta\omega = 5° \pm 5°$) is not a coincidence, then these two planets must have mutual inclination $\lesssim 20°$. Laughlin et al. [46] conclude that stability considerations restrict the mutual inclination between the two planets in 47 UMa to $\lesssim 40°$.

In multi-planet systems with large mutual inclinations the Kozai mechanism can lead to large inclination and eccentricity oscillations, which promote instability by enabling close encounters between planets or with the central star. Thus, even if planetary systems are formed from collapsing clouds, as suggested in §2.7, multi-planet systems may be restricted to a disk-like configuration, containing planets on both prograde and retrograde orbits.

4. CONCLUSIONS

Planetary eccentricities can be excited by a variety of mechanisms, and it is likely that more than one of these plays a role in determining the eccentricity distribution of extrasolar planets. As the planet search programs continue to find planets of lower and lower masses, and longer and longer periods, which resemble more closely the giant planets in the solar system, the issue of why our planets have much lower eccentricities than their extrasolar analogs becomes more acute.

REFERENCES

1. Zucker, S. & Mazeh T. 2001, ApJ, 562, 1038
2. Jorissen, A., Mayor, M. & Udry, S. 2001, A&A, 379, 992
3. Heacox, W.D. 1999, ApJ, 526, 928
4. Stepinski, T.F. & Black, D.C. 2001, A&A, 371, 250
5. Tabachnik, S. & Tremaine, S. 2002, MNRAS, 335, 151
6. Mazeh, T. & Zucker, S. 2002, in JENAM 2001: Astronomy with Large Telescopes from Ground and Space, Reviews in Modern Astronomy 15, ed. R. E. Schielicke, 133 (astro-ph/0201337)
7. Hayashi, C. 1981, Prog. Theor. Phys. Suppl., 70, 35
8. Zucker, S. & Mazeh, T. 2002, ApJ, 568, L113
9. Udry, S., Mayor, M. & Santos, N.C. 2003, A&A, 407, 369
10. Terquem, C. 2003, in Planetary Systems and Planets in Systems, eds. S. Udry, W. Benz & R. von Steiger (Dordrecht: Kluwer), in press (astro-ph/0309175)
11. Eggenberger, A., Udry, S. & Mayor, M. 2003, in Scientific Frontiers in Research on Extrasolar Planets, ASP Conf. Series, 294, eds. D. Deming & S. Seager (San Francisco: Astronomical Society of the Pacific), 43
12. Fischer, D.A. & Marcy, G.W. 1992, ApJ, 396, 178
13. Mazeh, T., Latham, D.W. & Stefanik, R.P. 1996, ApJ, 466, 415
14. Goldreich, P. & Tremaine, S. 1979, Icarus, 34, 240
15. Goldreich, P. & Tremaine, S. 1980, ApJ, 241, 425
16. Goldreich, P. & Tremaine, S. 1981, ApJ, 243, 1062
17. Ward, W.R. 1986, Icarus, 67, 164
18. Artymowicz, P. 1993, ApJ, 419, 166
19. Ward, W.R. & Hahn, J.M. 1998, AJ, 116, 489
20. Ward, W.R. & Hahn, J.M. 2000, in Protostars and Planets IV, eds. V. Mannings, A.P. Boss, & S.S. Russell (Tucson: University of Arizona Press), 1135
21. Goldreich, P. & Sari, R. 2003, ApJ, 585, 1024

22. Nelson, R.P., Papaloizou, J.C.B., Masset, F. & Kley, W. 2000, MNRAS, 318, 18
23. Papaloizou, J.C.B., Nelson, R.P. & Masset, F. 2001, A&A, 366, 263
24. Rasio, F.A. & Ford, E.B. 1996, Science, 274, 954
25. Weidenschilling, S.J. & Marzari, F. 1996, Nature, 384, 619
26. Lin, D.N.C. & Ida, S. 1997, ApJ, 477, 781
27. Ford, E.B., Havlickova, M. & Rasio, F.A. 2001, Icarus, 150, 303
28. Murray, N., Hansen, B., Holman, M. & Tremaine, S. 1998, Science, 279, 69
29. Murray, N., Paskowitz, M. & Holman, M. 2002, ApJ, 565, 608
30. Peale, S.J. 1999, ARAA, 37, 533
31. Malhotra, R. 1993, Icarus, 106, 264
32. Yu, Q. & Tremaine, S. 2001, AJ, 121, 1736
33. Lee, M.H. & Peale, S.J. 2002, ApJ, 567, 596
34. Chiang, E.I., Fischer, D. & Thommes, E. 2002, ApJ, 564, L105
35. Kozai, Y. 1962, AJ, 67, 591
36. Holman, M., Touma, J. & Tremaine, S. 1997, Nature, 386, 254
37. Innanen, K. A., Zheng, J.Q., Mikkola, S. & Valtonen, M.J. 1993, AJ, 113, 1915
38. Mazeh, T., Krymolowski, Y. & Rosenfeld, G. 1997, ApJ, 477, L103
39. Krymolowski, Y. & Mazeh, T. 1999, MNRAS, 304, 720
40. Ford, E.B., Kozinsky, B. & Rasio, F.A. 2000, ApJ, 535, 385
41. Zakamska, N.L. & Tremaine, S. 2003, these proceedings
42. Black, D.C. 1997, ApJ, 490, 171
43. Papaloizou, J.C.B. & Terquem, C. 2001, MNRAS, 325, 221
44. Terquem, C. & Papaloizou, J.C.B. 2002, MNRAS, 332, 39
45. Chiang, E.I., Tabachnik, S. & Tremaine, S. 2001, AJ, 122, 1607
46. Laughlin, G., Chambers, J. & Fischer, D. 2002, ApJ, 579, 455

Disk-Planet Interaction: Triggered Formation and Migration

Graeme Lufkin*, Thomas Quinn* and Fabio Governato*

*Department of Astronomy, University of Washington, Box 351580, Seattle, WA 98195-1580

Abstract. We present three-dimensional SPH calculations of giant planets embedded in gaseous disks. Our findings are collected into a map of parameter space, exhibiting four distinct regions: Type I migration, gap formation, triggered formation of more planets, and wholly unstable disks. For Type I migration of the planet due to secular interactions with the disk material, the migration rate depends linearly on the disk mass, and is independent of the initial planet mass. For more massive disks, the planet can disturb the disk strongly enough to trigger the collapse of gas into additional giant planets. When additional planets form, their interaction as point masses dominates the subsequent behavior of the system. This mechanism allows for the rapid formation of Jupiter-mass and higher planets. Migration due to interaction with the disk can significantly change the orbits of giant planets in gas disks.

INTRODUCTION

The formation and evolution of giant planets are two important, largely unsolved problems in astrophysics. Observations of extrasolar planets offer an intriguing puzzle of explaining the presence of many giant planets with very small orbits. Planet migration via interaction with the gas disk the planet is embedded in appears to have the potential to explain these observations. Such migration has been explored before, both analytically [1, 2] and numerically [3, 4]. Here we present the results of a new numerical study, using an order of magnitude increase in resolution with a particle-based method of simulation.

TECHNIQUE

We simulated a disk of gas around a 1 $M_\odot$ star, specifying a radial temperature profile ($T \propto r^{-1}$) and initial density profile ($\Sigma \propto r^{-3/2}$). This disk initially extends to a radius of 25 AU, and is free to grow. The total disk mass was one of the parameters we varied, from 0.01 to 0.2 $M_\odot$. The disk was composed of 10^5 gas particles that interact via gravity and Smoothed Particle Hydrodynamics (SPH). All simulations included the self-gravity of the disk, were three-dimensional, and had free boundary conditions. A planet particle, interacting via gravity only, was inserted in the disk on a circular orbit. The initial mass and semi-major axis of the planet were the other parameters we varied, from 0.25 to 2.0 M_{Jup} and 5 to 12.5 AU, respectively. The disk-planet system was evolved for at least 326 years. Our simulations were performed with *Gasoline*, a tree-based N-body code for gravity and gas dynamics [5].

CP713, The Search for Other Worlds: Fourteenth Astrophysics Conference,
edited by S. S. Holt and D. Deming
© 2004 American Institute of Physics 0-7354-0190-X/04/$22.00

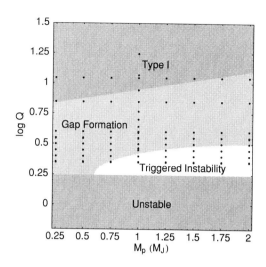

FIGURE 1. A parameter space plot showing the different behaviors observed in a disk-planet system. The points represent the initial conditions for each simulation. The labeled regions demarcate the different outcomes we observed after evolving each system for a fixed period of time. Q is the value of the Toomre stability criteria of the disk at the initial orbit of the planet.

RESULTS

Our parameter study of giant planet migration yielded a classification of behaviors, illustrated in Fig. 1. For each point in parameter space (a specific disk mass, initial planet mass, and initial planet orbital radius) we evolved the system for 326 years. We then classified the state of the system into one of four possible regions: Type I migration, gap formation, triggered formation of additional planets, and wholly unstable disks.

Parameter Map. In stable disks (the upper region of Fig. 1), planets of all masses migrate inward at a roughly constant rate. In more massive disks, planets can open a gap, with a bias toward more massive planets. To differentiate between the regions labeled 'Type I' and 'Gap Formation' we look for the halting of the migration of the embedded planet within the time we simulated. Thus our definition of a gap is operational; a gap has formed when the density depression along the orbit of the planet has caused the planet to stop migrating. In yet more massive disks, the density perturbations raised by the inserted planet are large enough to collapse gravitationally, leading to the formation of additional planets. In this scenario, a disk that is stable in the absence of a "seed" planet can fragment to form several additional planets via gravitational instability. Finally, even more massive disks are unstable even without a seed planet.

The demarcation between regions is based on the amount of time the system has evolved for. For example, some of the planetary systems in the region labeled 'Type I' will in fact open a gap eventually. So there will be a different parameter space diagram for every age of the systems. In addition, while we varied three parameters, the diagram

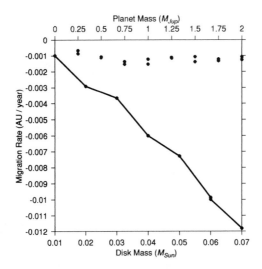

FIGURE 2. The migration rate of a planet embedded in a gas disk, as the planet mass (points) and disk mass (line) is varied. The migration rate appears independent of the planet mass, and linearly dependent on the disk mass. The migration was roughly constant over the duration of the simulation (several hundred years).

is two-dimensional. We found the disk stability to be the strongest predictor of the eventual outcome. To accommodate the third parameter, initial orbit size, our diagram displays disk stability at the initial orbit of the planet. While useful, this contraction of a parameter obscures interesting behavior in the triggered formation of additional planets region. In this region, close-in planets embedded in less-stable disks did not trigger the formation of new planets, while farther-out planets in more-stable disks did.

Migration Rates. We measured Type I migration rates for several disk-planet systems. Our findings, shown in Fig. 2, reveal that the migration rate appears to be independent of initial planet mass, and linearly dependent on the disk mass. This does not agree with the analytic results of Tanaka et al. [2], but is consistent with another numerical investigation, Nelson and Benz [6].

Forming Additional Planets. The gravitational instability mode of planet formation requires a cool, massive disk [7]. These conditions can be met locally given a strong density perturbation in the cooler regions of the disk. The spiral density waves excited by an embedded planet can thus become unstable and collapse into an additional planet. Our simulations have a fixed temperature profile that falls off with distance from the central star. Thus we expect that exterior spiral arms would fragment more easily than interior arms. Fig. 3 is a snapshot of a simulation where this scenario has occurred. This scenario increases the range of parameter space over which the gravitational instability mode of planet formation is applicable.

Once additional planets have formed, the interaction between the bound objects dominates the orbital evolution of the system. Therefore, the concept of migration via in-

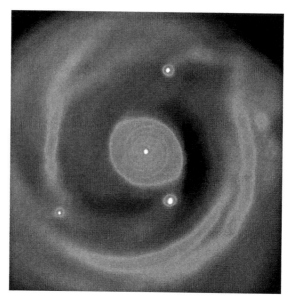

FIGURE 3. A 1.25 M_{Jup} planet has triggered the collapse of two additional planets in a 0.1 $M_\odot$ gas disk. In the absence of the initial planet, this disk will not fragment. The gray scale represents the logarithmic density of the gas.

teraction with the disk is not applicable in this scenario. Migration via the stochastic process of many-body interaction is still possible.

CONCLUSION

This work presents some initial findings on giant planet migration. A more detailed study of the migration process is necessary to truly understand the interaction between a gas disk and an embedded planet. Future work will carry out this study, exploring a wider range of parameter space and looking at migration over longer periods of time. In addition, we have found another region of parameter space where the migration scheme is not applicable, that of triggered formation of additional planets.

REFERENCES

1. Ward, W. R., *Icarus*, **126**, 261–281 (1997).
2. Tanaka, H., Takeuchi, T., and Ward, W. R., *ApJ*, **565**, 1257–1274 (2002).
3. Nelson, R. P., Papaloizou, J. C. B., Masset, F., and Kley, W., *MNRAS*, **318**, 18–36 (2000).
4. D'Angelo, G., Henning, T., and Kley, W., *A&A*, **385**, 647–670 (2002).
5. Wadsley, J., Stadel, J., and Quinn, T., *New Astronomy*, **9**, 137–158 (2004).
6. Nelson, A. F., and Benz, W., *ApJ*, **589**, 556–577 (2003).
7. Mayer, L., Quinn, T., Wadsley, J., and Stadel, J., *Science*, **298**, 1756–1759 (2002).

Propagation of Eccentricity Disturbances in Planetary Systems

Nadia L. Zakamska* and Scott Tremaine*

*Department of Astrophysical Sciences, Princeton University, Princeton NJ 08544

Abstract. One of the remarkable features of the ~ 120 known extrasolar planets is their relatively high eccentricities. We suggest that these eccentricities arise because of secular interactions in an extended planetary disk: eccentricities are excited in the outer part of the disk, for example by a passing star, and then propagate inward through the disk, somewhat like a wave. We explore how the inward propagation of eccentricity in planetary disks depends on the number and masses of the planets and on the overall surface density distribution. We find that the main governing factor is the large-scale surface-density distribution. If the latter is approximated by a power-law $\Sigma(r) \propto r^{-q}$, then eccentricity disturbances propagate inward efficiently for flat density distributions with $q \lesssim 1$.

The eccentricities of extrasolar planets are much larger than eccentricities of solar-system planets. The origin of these large eccentricities remains largely unexplained: in most scenarios, planets form from a quiescent disk and thus should have nearly circular orbits. A wide variety of eccentricity excitation mechanisms have been suggested, as we describe elsewhere in this volume [1].

In this paper, we explore the possibility that eccentricities are excited by encounters with passing stars. For example, in the solar neighborhood, a stellar encounter within 400 AU is expected every 10^{10} years. Such encounters are too distant to have any significant direct impact on a planet with semimajor axis of a few AU or less. However, there is growing evidence that most forming stars are surrounded by extended gas, dust, and debris disks, with radii of hundreds of AU [2, 3]. Much of the material in these disks may survive to the present in the form of planetesimals or planets. Outward migration can also drive planets to larger radii, up to tens of AU, where formation by core accretion would not normally be possible [4]. Planets at large radii can be excited to high eccentricities during an encounter, and then excite the eccentricities of the inner planets.

To study the effect of an external perturber on an extended planetary system, we first calculate the response of the planets to the direct interaction with the perturber, neglecting interactions between the planets since the encounter time is short. Then we use Laplace-Lagrange secular perturbation theory to study how the eccentricities are redistributed among the planets on much longer timescales.

In our simplified model, there are N planets on coplanar, circular orbits around the central star M_c, with masses $m_i \ll M_c$ and semimajor axes a_i. The perturbing star has a mass M_e and moves in the same plane (either prograde or retrograde), with velocity v_e and impact parameter p relative to the central star. We consider an example encounter with $v_e = 40\,\mathrm{km\,s^{-1}}$ (a typical encounter velocity in the solar neighborhood) and $p = 500$ AU (an encounter with this impact parameter is expected about once during the main-

CP713, *The Search for Other Worlds: Fourteenth Astrophysics Conference,*
edited by S. S. Holt and D. Deming

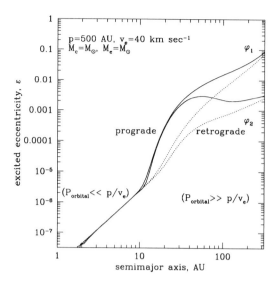

FIGURE 1. Excited eccentricity as a function of the semimajor axis of the planet for a perturber with $p = 500$ AU, $v_e = 40$ km sec^{-1}, $M_e = M_\odot$. We show results of the direct integration of the equation of motion (1) for two values of the orbital phase φ at the moment of closest approach of the perturber, and for two directions of the motion of the planet relative to the motion of the perturber, prograde (solid) and retrograde (dotted). Prograde perturbers are more efficient in exciting the eccentricity than retrograde ones. At small semimajor axes the excited eccentricity varies as $a^{5/2}$.

sequence lifetime of a solar-type star).

During the encounter, which has a duration $\tau_e \equiv p/v_e = 60$ yr, the interactions between the planets can be neglected, since they occur on much longer timescales $P_{\text{secular},i} \sim P_{\text{orbital},i} M_c/m_i$. Thus the planets move as test particles in the potential field of the central star and the external perturber. In the frame fixed on the central star,

$$\ddot{\mathbf{r}}_i = -G\frac{M_c\mathbf{r}_i}{r_i^3} - G\frac{M_e(\mathbf{r}_i - \mathbf{r}_e)}{|\mathbf{r}_i - \mathbf{r}_e|^3} - G\frac{M_e\mathbf{r}_e}{r_e^3}. \tag{1}$$

The result of the numerical integration of this equation is shown in Figure 1, as a function of the semimajor axis of the planet. For short-period planets ($a < 10$ AU) the perturbation is completely negligible, whereas planets at semimajor axes of a few tens of AU or more can be excited to substantial eccentricities.

After the perturber is gone, the excited eccentricities are redistributed among the planets by secular interactions. If mean-motion resonances between planets can be neglected and if the eccentricities are small, we can use Laplace-Lagrange secular theory [5]. The eccentricity components $h_i = \varepsilon_i \sin \varpi_i$ and $k_i = \varepsilon_i \cos \varpi_i$ (where $\varpi_i(t)$ are the longitudes of pericenters) evolve according to

$$\dot{h}_i = \sum_{j=1}^N A_{ij} k_j; \quad \dot{k}_i = -\sum_{j=1}^N A_{ij} h_j. \tag{2}$$

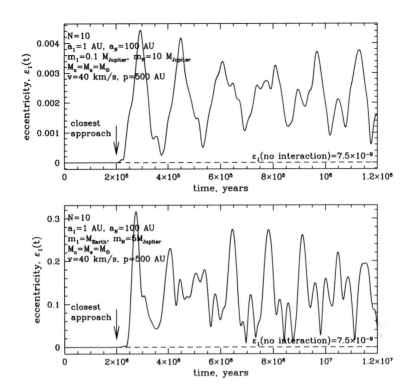

FIGURE 2. Evolution of the eccentricity of the innermost planet ($a_1 = 1$ AU) before and after the example perturbation, in two 10-planet systems with different mass distributions. In the top panel, masses of the planets range from 0.1 $M_{Jupiter}$ to 10 $M_{Jupiter}$, in the bottom from M_{Earth} to 5 $M_{Jupiter}$.

The matrix **A** is determined by the masses and semimajor axes of the planets and the mass of the central star. The initial conditions for these equations are determined by the eccentricities excited as a result of the interaction with the perturber.

In Figure 2 we show the solution of the equations of motion (2) for the innermost planet in two planetary systems. Each system contains ten planets that follow a geometric progression in semimajor axis from 1 AU (innermost planet) to 100 AU (outermost planet). The masses also follow a geometric progression, with a total range of a factor of 100 (top) or 1600 (bottom); the evolution is independent of the normalization of the mass distribution. The perturber passes at $p = 500$ AU at $v_e = 40$ km s^{-1}. Clearly, the presence of the outer planets made the perturbation much more efficient in exciting the inner planet: if there were no outer planets, the eccentricity of the innermost planet would be just 7.5×10^{-9} as a result of the direct interaction with the perturber.

The two systems differ only in the mass distribution, and the excited eccentricities are very different. In Figure 3 we show the dependence of the excitation efficiency on

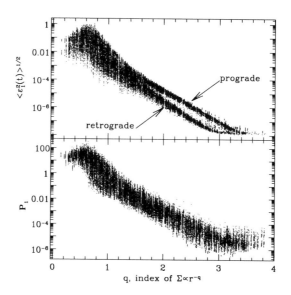

FIGURE 3. The rms eccentricity (top) and the propagation efficiency (bottom) for the innermost planet, as a function of the mass distribution, for a large ensemble of simulated systems. The perturber passes in the plane of the planetary orbits at $p = 500$ AU and $v_e = 40$ km s^{-1} in all cases; the innermost planet has semimajor axis 1 AU and the outermost planet is between 50 and 300 AU.

the mass distribution, which we parametrized by the power index q of the smeared-out surface density: $\Sigma(r) \propto r^{-q}$. The two parameters describing the excitation efficiency are the rms eccentricity of the innermost planet, $\langle \varepsilon_1^2(t) \rangle^{1/2}$, and the propagation efficiency $P_1 = \mathrm{rms}(\varepsilon_1)/\mathrm{rms}(\varepsilon_N)$. If $P_1 \ll 1$, the eccentricity of the innermost planet cannot be excited significantly without ejecting the outer planet. We find that systems with relatively flat mass distributions ($q \lesssim 1$) are very efficient in transporting the eccentricity excitation from the outer parts of the planetary system to the innermost planet.

We conclude that eccentricities of short-period planets could be excited in encounters with passing stars, in an environment like the solar neighborhood, as long as these planets are members of planetary systems extending to at least a few tens of AU, with a flat mass distribution ($q \lesssim 1$).

REFERENCES

1. Tremaine, S. & Zakamska, N.L. 2003, these proceedings
2. Koerner, D.W. 2001, in Tetons 4: Galactic Structure, Stars, and the Interstellar Medium, eds. C.E. Woodward, M.D. Bicay, & J.M. Shull (San Francisco: ASP), 563
3. Zuckerman, B. 2001, ARA&A, 39, 549
4. Veras, D. & Armitage, P.J. 2003, MNRAS, in press (astro-ph/0310161)
5. Murray, C.D., & Dermott, S.F. 1999, Solar System Dynamics (Cambridge: Cambridge University Press)

New TVD Hydro Code for Modeling Disk-Planet Interactions

Lawrence Mudryk* and Norman Murray[†]

*Department of Astronomy and Astrophysics, 60 St. George Street, Toronto, ON M5S 3H8,
Canada; mudryk@astro.utoronto.ca
[†]Canadian Institute for Theoretical Astrophysics, 60 St. George Street, Toronto, ON M5S 3H8,
Canada; murray@cita.utoronto.ca

Abstract. We present test simulations of a TVD hydrodynamical code designed with very few calculations per time step. The code is to be used to preform simulations of proto-planet interactions within gas disks in early solar systems.

HYDRODYNAMICAL CODE BASICS

The Euler Equations provide a description of fluid flow, prescribing conservation of mass, momentum and energy on a fixed grid. Hydrodynamical codes standardly use the weak (differential) form of the equations which are discontinuous across shocks. While these shock boundaries are discontinuous analytically, computationally they must be resolved over a finite number of grid cells. An effective hydro code has the ability to capture strong shocks over a minimum of grid cells.

Problems arise because the Euler equations are most easily discretized for a given grid cell in terms of the mass, momenta and energy densities at the cell *center* as functions of the flux through the cell *boundaries*. The spatial separation introduced by discretization complicates the relation between the densities and fluxes. Most of the computational effort of a hydrodynamical algorithm is spent trying to accurately calculate one in terms of the other.

We have developed a TVD code based on an algorithm by Trac and Pen [1] which was first designed for use in cosmological simulations. The algorithm's primary advantage is that it uses very few calculations per time step while still accurately capturing shocks, thus, enabling very high resolution simulations to be performed. We expect to be able to routinely run simulations on grid resolutions of at least $N_r \times N_\theta \times N_z = 1000 \times 1000 \times 100$.

TEST PROBLEMS

We show several results of the TVD code from a suite of test problems. Results are compared either to theoretical values for the problem or to the results of the well-tested hydrodynamic code, VH-1 [2] based on the piecewise-parabolic method (PPM) of Woodward and Collela [3]. In comparison to the TVD code, the PPM code has many

CP713, *The Search for Other Worlds: Fourteenth Astrophysics Conference,*
edited by S. S. Holt and D. Deming
© 2004 American Institute of Physics 0-7354-0190-X/04/$22.00

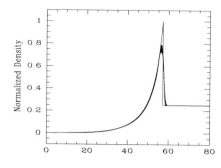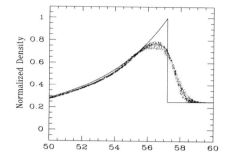

FIGURE 1. TVD results for 3D blast wave with analytical curves. Grid Resolution: 128x128x128.

more calculations per time step, but formally yields a higher order solution than the TVD. We expect our code to make up for this deficit with higher possible resolution.

3D Sedov Taylor Blast Wave. The center of an otherwise isotropic medium is initially seeded with a large thermal energy. This point "explosion" produces an expanding radial blast wave which travels spherically through the medium. Simulation results for the density are plotted in Figure 1 as points against analytical curves, normalized to the post-shock density. Notice in particular the accurate capture of the post-shock values (within 20%) and the shock resolution of 2 - 3 grid cells. The PPM code yielded very similar results to the TVD code (only TVD results are plotted).

2D Sod Shock Tube. Density and pressure discontinuities are set up at an angle across one corner of the grid. Propagation of the resulting shock wave, contact discontinuity and the expansion region are perpendicular to the initial shock angle. The solutions on the x and y boundaries are seen to reduce analytical results for the 1D Sod Shock Tube. Figure 2 illustrates that at the same resolution the TVD code is only slightly more diffusive even though the PPM code formally yields a higher order solution. In the more difficult problem where the simulation is run with another order of magnitude difference in the pressure and density discontinuities, the TVD code also compares well with the PPM code and the 1D analytical results.

Kelvin-Helmholtz Instability. Two fluids of slightly differing densities lie over top of one another, moving horizontally in opposite directions. The instability is excited with an initial sine wave perturbation. The PPM code is not shown for comparison; the results were similar.

Supersonic Flow Around Cylinder. We simulate supersonic flow around a half-cylinder (the flow is symmetric to either side of the cylinder). Both codes capture the strong bow shock (Figure 4) seen in experimental images and had very similar velocity profiles (not plotted). Experimental images also show a second weaker shock leading off the sphere, about 90 degree from the front. While both codes show indications of a weaker second shock, the locations and strengths differ slightly.

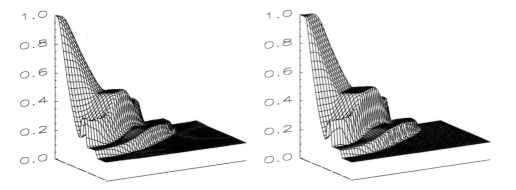

FIGURE 2. Simulation of a 2D Sod Shock Tube using the TVD code (left) and PPM code (right). Grid Resolution: 100x100.

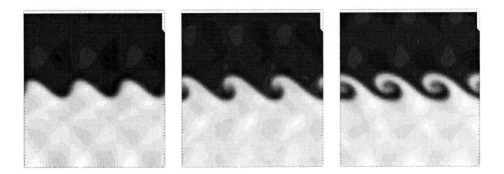

FIGURE 3. TVD simulation of Kelvin Helmholtz instability shown in density contours at three equally spaced time intervals. Grid Resolution: 256x256.

FUTURE PROJECTS

We intend to use the code described above to further examine planet-disk interactions in early solar systems.

The interactions between proto-planets and the disks in which they are embedded is a leading candidate to explain the small period orbits observed in extra-solar systems. Simulations of a single proto-planet embedded in a gas disk [5] show the evolution of spiral density waves as predicted by theory [4]. These deposit momentum at the planet's orbital radii which clears the surrounding gas. The back-reaction of the spiral waves in the disk on the planet causes inward migration.

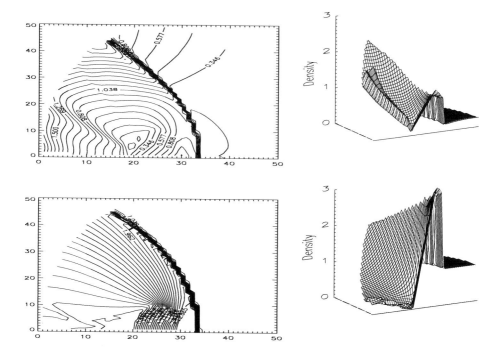

FIGURE 4. Density contours (left) and density surface plots (right) for the TVD (top) and PPM codes (bottom). The flow is from right to left in the contour plots and the half-cylinder is centered at (25,0) with a circular radius of 10.

Preliminary simulations with multiple proto-planets [6] have shown potential to increase the eccentricity of the proto-planets (as observationally observed). We hope to further examine this scenario with a wide selection of planet numbers, masses and initial radii which requires the high resolution, efficient algorithm which has been designed.

A more extensive test suite of simulations is maintained at:

http://www.astro.utoronto.ca/~mudryk/professional/test_suite.html

REFERENCES

1. Trac, H., and Pen, U., *PASP*, **115**, 303–321 (2003).
2. Blondin, J., See URL http://wonka.physics.ncsu.edu/pub/VH-1/ for VH-1 code and contact information.
3. Collela, P., and Woodward, P., *J. Comp. Phys.*, **54**, 174–201 (1984).
4. Goldreich, P. and Tremaine, S., *ApJ*, **233**, 857–871 (1979).
5. Nelson, R. P. et al., *MNRAS*, **318**, 18–36 (2000).
6. Kley, W., *MNRAS*, **313L**, L47–L51 (2000).

On the Stability of a Planetary System Embedded in the β Pictoris Debris Disk

Jared H. Crossley* and Nader Haghighipour†

*New Mexico Institute of Mining and Technology, 801 Leroy Place, Socorro, NM 87801
†Department of Terrestrial Magnetism and NASA Astrobiology Institute,
Carnegie Institution of Washington, 5241 Broad Branch Road, Washington, DC 20015

Abstract. It has recently been stated that the warp structure observed around the star β Pictoris may be due to four planets embedded in its debris disk [1]. It, therefore, becomes important to investigate for what range of parameters, and for how long such a multibody system will be dynamically stable. We present the results of the numerical integration of the suggested planetary system for different values of the mass and radii of the planets, and their relative positions and velocities. We also present a statistical analysis of the results in an effort to understand the relation between the different regions of the parameter-space of the system and the duration of the orbital stability of the embedded planets.

INTRODUCTION

Beta Pictoris, a type A5 IV star at a distance of ∼19 pc from the Earth, is one of the youngest close stars with an approximate age of 12^{+8}_{-4} Myr [2]. Observational evidence indicates that this star is surrounded by a planetary debris disk [3, 4, 5]. The close proximity of such a young circumstellar disk has made β Pictoris an ideal candidate for the study of the evolution of protoplanetary disks and planetary system formation.

In the past several years, there have been a number of reports of the detection of symmetric warps in the β Pictoris debris disk [1, 5, 6, 7, 8]. Models have best explained these warps as gravitational perturbations caused by planetary companions [6, 7, 8]. Among these reports, the disk warps discovered by Wahhaj et al. [1] were explained as the edge-on projection of debris rings orbiting the central star. Models of the flux density allow some parameters of these rings to be determined via χ^2 fitting. Wahhaj et al. [1] proposed a multiple planet system to account for ring formation, noting that all adjacent rings are in mean-motion resonances.

In consideration of these findings, we undertake here a study of the dynamical evolution of a multibody system similar to that proposed by Wahhaj et al. [1]. We perform a statistical analysis of the stability of randomly generated systems within a portion of the total available parameter-space and briefly analyze the relationship between the parameters of the system and its orbital stability.

CP713, *The Search for Other Worlds: Fourteenth Astrophysics Conference*,
edited by S. S. Holt and D. Deming
© 2004 American Institute of Physics 0-7354-0190-X/04/$22.00

NUMERICAL ANALYSIS

The planetary system proposed by Wahhaj et al. [1] consists of four planets with radial distances and orbital inclinations equal to those of their corresponding warps [see 1, Table 1]. To study the dynamics of this planetary system, we explore a parameter-space which includes the mass number of the four planets, their radii[1], and their positions and velocities. The mass and radius of β Pictoris are taken to be $2.0M_\odot$ and $1.9R_\odot$, respectively [9].

We consider planets with masses randomly chosen between one to three Jupiter-masses. For each value of the mass of a planet, we calculate its radius assuming an average density equal to that of Jupiter (1.33 g cm^{-3}). We also assume that all planets are initially on direct Keplerian circular orbits and their orbital phases are chosen randomly from the range $0° \leq \phi \leq 360°$.

To explore the orbital stability of this planetary system, we integrated the system for 50 Myr using Mercury Integrator Package VI [10]. We considered the system to be stable if no planet came closer than three Hill's radii or obtained a radial distance larger than 1000 AU.

We ran a total of 20457 simulations using randomly generated values of planet masses and initial orbital phases. Of this total, 14409 simulations used unique parameter sets. The remaining 6048 systems were exact replications of the systems in the unique set—a consequence of the random number generation routine. The duplicate systems have been kept for the sake of statistical analysis, since they are randomly distributed throughout the phase space.

Table 1 shows the statistical data for all simulations grouped in five 10 Myr intervals. The middle two columns show the number and percentage of simulations that became unstable within their corresponding time intervals. The majority of the randomly generated systems became unstable at early stages of the integration, with their number decreasing as time increases. The rightmost column shows the percentage of the systems remaining stable beyond their respective time interval. As shown here, approximately 41% of the systems remained stable after the first 10 Myrs. From these systems, 6.7% were still stable after 20 Myr from the beginning of the integrations–the upper estimate of the lifetime of β Pictoris (20 Myr). During the last 10 Myr, only 0.1% of all systems were still stbale.

Figure 1 shows how stability lifetimes are distributed across a random sampling of the parameter-space. This histogram shows the number of systems that became unstable in 10^5 year intervals. It can be seen from this figure that for a system chosen randomly within our assumed parameter-space, it is most probable that the stability lifetime of the system is approximately 7 Myr. It is also seen that no systems became unstable at an age of less than 1.0 Myr. We note that 15 systems remained stable for the entire 50 Myr and are therefore not accounted for in the histogram.

In a preliminary attempt to determine the relationship between initial conditions and stability lifetime, we have plotted the mass of each planet versus its initial phase-angle for those systems that remained stable for 50 Myr (Fig. 2). The initial phase of planet

---

[1] Planets' radii were needed to calculate their Hill's radii.

TABLE 1. Statistical data on five-body system stability.

Time (Myr)	Number of Unstable Systems	Percentage of Unstable Systems	Percentage of Remaining Stable Systems
0–10	12100	59.1	40.9
10–20	6980	34.1	6.7
20–30	1152	5.6	1.1
30–40	194	0.9	0.2
40–50	16	0.1	0.1

1 was set to $0°$ for all simulations. It is interesting to note that none of these systems contain a planet with a mass greater than $2.4 M_J$.

CONCLUSIONS

We have analyzed the stability of the proposed five-body planetary system embedded in the β Pictoris debris disk [1] for over 14000 initial conditions. Our results indicate that the majority of systems became unstable between a time of 1 to 10 million years. There were only 8 unique simulations that remained stable for the entire 50 Myr integration time. These systems contained planets with masses less than 2.4 times the mass of Jupiter.

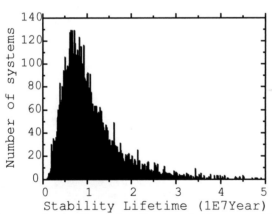

FIGURE 1. Histogram showing the number of systems that became unstable in 10^5 yr intervals. A randomly chosen system is most likely to become unstable near 7 Myr. No system in this sample became unstable in less than 1 Myr. The 15 systems that remained stable for the entire 50 Myr integration are not shown here.

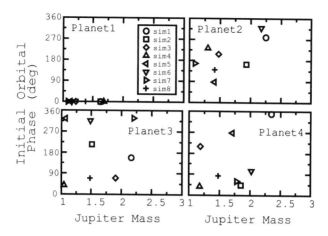

FIGURE 2. Mass versus initial phase for planets in the 8 unique systems that remained stable for the entire 50 Myr integration. Note that none of these planets have masses greater than 2.4 Jupiter mass.

ACKNOWLEDGMENTS

This work is partially supported by the Carnegie Institution of Washington Internship Program, and also an REU Site for Undergraduate Research Training in Geoscience, NSFEAR-0097569 for JHC, and NASA Origins of the Solar System Program under grant NAG5-11569, and also the NASA Astrobiology Institute under Cooperative Agreement NCC2-1056 for NH.

REFERENCES

1. Wahhaj, Z., Koerner, D. W., Ressler, M. E., Werner, M. W., Backman, D. E., and Sargent, A. I., *Astroph. J.*, **584**, L27–L31 (2003).
2. Zuckerman, B., Song, I., Bessel, M. S., and Webb, R. A., *Astroph. J.*, **562**, L87–L90 (2001).
3. Smith, B. A., and Terrile, R. J., *Science*, **226**, 1421–1424 (1984).
4. Hobbs, L. M., Vidal-Madjar, A., Ferlet, R., Albsert, C. E., and Gry, C., *Astroph. J.*, **293**, L29–L33 (1985).
5. Weinberger, A. J., Becklin, E. E., and Zuckerman, B., *Astroph. J.*, **584**, L33–L37 (2003).
6. Burrows, C. J., Krist, J. E., and Stapelfeldt, K. R., *Bul. Am. Astro. Soc.*, **27**, 1329 (1995).
7. Mouillet, D., Larwood, J. D., Papaloizou, J. C. B., and Lagrange A. M., *Month. Not. Roy. Astron. Soc.*, **292**, 896–904 (1997).
8. Heap, S. R., Lindler, D. J., Lanz, T. M., Cornett, R. H., Hubeny, I., Maran, S. P., and Woodgate, B., *Astroph. J.*, **539**, 435–444 (2000).
9. Carroll, B. W., and Ostlie, D. A., *An Introduction to Modern Astrophysics*, Addison-Wesley, New York, pp. A–13 (1996).
10. Chambers, J. E., *Month. Not. Roy. Astron. Soc.*, **304**, 793–799 (1999).

On the Dynamical Stability of γ Cephei, an S-Type Binary Planetary System

Nader Haghighipour

Department of Terrestrial Magnetism and NASA Astrobiology Institute,
Carnegie Institution of Washington, 5241 Broad Branch Road, Washington, DC 20015

Abstract. Precision radial velocity measurements of the γ Cephei (HR 8974) binary system suggest the existence of a planetary companion with a minimum mass of 1.7 Jupiter-mass on an elliptical orbit with a $\sim$2.14 AU semimajor axis and 0.12 eccentricity [1]. I present in this paper a summary of the results of an extensive numerical study of the orbital stability of this three-body system for different values of the semimajor axis and orbital eccentricity of the binary, and also the orbital inclination of the planet. Numerical integrations indicate that the system is stable for the planet's orbital inclination ranging from 0 to 60 degrees, and for the binary's orbital eccentricity less than 0.5. The results also indicate that for large values of the inclination, the system may be locked in a Kozai resonance.

INTRODUCTION

The existence of planets in binary star systems is no longer a mere idea. Approximately 35% of extrasolar planets discovered till 2002 exist in dual star systems [2]. These systems are mostly wide binaries with separations between 500 and 750 AU, and with planets revolving around one of the stars. At such large distances, the perturbative effect of one star on the formation and dynamical evolution of planets around the other star is entirely negligible. A recently discovered Jupiter-like planet around the primary of the γ Cephei binary system is, however, an exception to this rule.

Gamma Cephei is a close spectroscopic binary composed of a 1.59 solar-mass K1 IV subgiant and a 0.4 solar-mass red M dwarf [3]. The semimajor axis of this system has been reported to have a lower value of 18.5 AU [1] and an upper value of 21 AU [4]. Precision radial velocity measurements suggest that a planet with a minimum mass of 1.7 Jupiter-mass revolves around the primary of this system on an elliptical orbit with an eccentricity of 0.12 ± 0.05 and a semimajor axis of approximately 2.14 AU. Being the first discovered S-type binary-planetary system[1], it is quite valuable to investigate whether this system is dynamically stable, and for what values of its orbital parameters its stability will remain. In this paper I present a summary of the results of an extensive numerical study of the dynamical stability of this system for different values of the orbital parameters of the binary and also the orbital inclination of the planet. A more comprehensive study of the dynamics of this system are to be published elsewhere.

---

[1] As classified by Rabl & Dvorak (1988), a binary-planetary system is called S-type when the planet revolves around one of the stars, and P-type when the planet revolves around the entire binary.

CP713, *The Search for Other Worlds: Fourteenth Astrophysics Conference,*
edited by S. S. Holt and D. Deming
© 2004 American Institute of Physics 0-7354-0190-X/04/$22.00

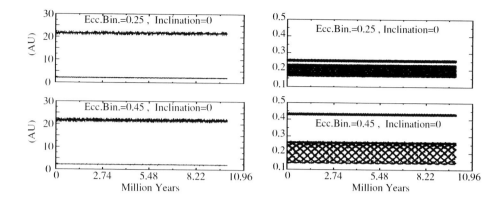

FIGURE 1. Semimajor axes (left) and eccentricities (right) of the binary and planet. The system was considered to be coplanar. At the beginning of the integration, the initial value of the semimajor axis of the binary was 21 AU and that of the planet was taken to be equal to 2.14 AU.

NUMERICAL ANALYSIS

An important quantity in determining the stability of a planet's orbit in a binary system is its semimajor axis. Rabl & Dvorak [5] and Holman & Wiegert [6] have presented an empirical formula for the maximum value of the semimajor axis of a stable planetary orbit in S-type binary-planetary systems. The value of the *critical semimajor axis*, a_c is given by

$$a_c/a_b = (0.464 \pm 0.006) + (-0.380 \pm 0.010)\mu + (-0.631 \pm 0.034)e_b$$
$$+ (0.586 \pm 0.061)\mu e_b + (0.150 \pm 0.041)e_b^2 + (-0.198 \pm 0.047)\mu e_b^2 \qquad (1)$$

where a_b is the semimajor axis of the binary, and $\mu = m_2/(m_1 + m_2)$ and e_b represent the mass-ratio and orbital eccentricity of the binary, respectively. In the definition of the mass-ratio μ, m_1 and m_2 are the masses of the primary and secondary stars. To study to what extent equation (1) can be applied to the orbital stability of γ Cephei, this system was numerically integrated using a conventional Bulirsch-Stoer integrator. Numerical integrations were carried out for values of e_b ranging from 0.25 to 0.65 with steps of 0.05. When considered coplanar, the system was stable for all the values of $e_b < 0.5$ at all times. However, for the value of the binary eccentricity larger than 0.5, the system became unstable in less than 1000 years. Figure 1 shows the semimajor axes and eccentricities of the system for two cases of $e_b = 0.25$ and 0.45.

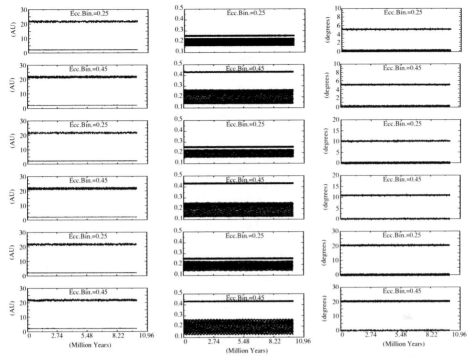

FIGURE 2. Graphs of semimajor axes (left), eccentricities (middle), and inclinations (right) of the binary and planet for three different values of the planet's orbital inclination (5, 10, 20 degrees).

The stability of the system was also studied for different values of the orbital inclination of the planet. For each above-mentioned value of the binary eccentricity, the initial inclination of the orbit of the planet with respect to the plane of the binary was chosen from the values of 1, 5, 10, 20, 40, and 60 degrees. For values of $e_b < 0.5$, the system was stable for all inclinations less than 60 degrees. Figure 3 shows the results of sample runs for the values of the planet's orbital inclination equal to 5, 10, and 20 degrees. Also, as expected, in some cases of large inclinations and for eccentric binaries, the planet was locked in a Kozai resonance (Fig. 3).

ACKNOWLEDGMENTS

This work is partially supported by the NASA Origins of the Solar System Program under grant NAG5-11569, and also the NASA Astrobiology Institute under Cooperative Agreement NCC2-1056.

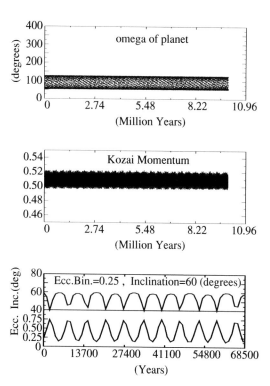

FIGURE 3. From top to bottom, ω of the planet, the value of the third Delaunay momentum $H_K = \cos I (1 - e^2)^{1/2}$, and the graph of e, the eccentricity of the planet and I, its inclination, in terms of time. As shown here, ω and H_K are constant, and the inclination and eccentricity of the planet have opposite extrema.

REFERENCES

1. Hatzes, A. P., Cochran, W. D., Endl, M., McArthur, B., Paulson, D. B., Walker, G. A. H., Campbell, B., and Yang, S., *Astrophys. J.*, 599, 1383-1394 (2003).
2. Kortenkamp, S. J., Wetherill, G. W., and Inaba, S., *Science*, **293**, 1127–1129 (2001).
3. Fuhrmann, K., *Astron.Nachr.*, in press (2003).
4. Griffin, R. F., Carquillat, J. M, and Ginestet, N., *The Observatory*, **122**, 90–109 (2002).
5. Rabl, G., and Dvorak, R., *Astron. Astrophys.*, **191**, 385–391 (1988).
6. Holman, M. J., and Wiegert, P. A., *Astron. J.*, **117**, 621–628 (1999).

Long-Term Stability of the Ups And Planetary System

Verene Lystad* and Frederic Rasio*

*Northwestern University, Dept. of Physics & Astronomy, 2145 Sheridan Rd., Evanston, IL
60208-0834

Abstract. Since the announcement of the triple planet system orbiting Upsilon Andromedae by Butler et al. in 1999 [1] the best-fit orbital parameters of the system have varied significantly with additional observations over the years. We have performed long-term numerical integrations and a new stability analysis using the most recent best-fit orbital parameters. All of our integrations run for a minimum of 10^8 years, unless a close encounter between two planets or the collision of one planet with the central star occurs first. We vary systematically the value of the unknown inclination angle with respect to the plane of the sky as well as the relative inclinations. Based on these numerical results we are able to provide improved constraints on angles of inclination and the masses of the planets. We also find evidence against the claim that the middle and outer planets exhibit a secular resonance.

1. MOTIVATION

The planetary system around Ups And was the first of only two known triple planet systems to be discovered (the other is 55 Cnc). It is one of only thirteen known multiple planet systems, to date. Few thorough studies with long-term numerical integrations for two or three giant planets have been done ([2], [3], [4]). Previous studies concerning the stability of the Ups And system have addressed only the short-term stability, and they were carried out several years ago with values of the orbital parameters that differ noticeably from the most up-to-date best-fit values (e.g., Stepinski et al. 2000 [5], Chiang et al. 2001 [8]). By studying the long-term stability of the system we can in principle better constrain the masses and orbital parameters of the planets as well as assess the role played by possible secular resonances ([8]).

2. METHOD

We used the most recent orbital parameters available (Table 1.) from directly measurable quantities [1].

Keeping the measurable quantities constant, we varied the overall inclination, i, and the relative inclinations (with respect to the overall inclination), e.g., $\cos i_r =$

---

[1] Taken from "A Triple Planet System Orbiting Upsilon Andromedae", California and Carnegie Planet Search Team, http://www.exoplanets.org/esp/upsandb/upsandb.shtml, 03 July 2003

CP713, *The Search for Other Worlds: Fourteenth Astrophysics Conference,*
edited by S. S. Holt and D. Deming
© 2004 American Institute of Physics 0-7354-0190-X/04/$22.00

TABLE 1. Orbital elements for the Ups And planetary system.

	Period (days)	e	$m \sin i$ (M_J)	a (AU)	ϖ (degrees)	T_{peri} (JD)	K (m/s)
b	4.6171	0.012	0.69	0.059	73	2450002.093	70.2
c	241.5	0.28	1.89	0.829	250	2450160.5	53.9
d	1284	0.27	3.75	2.53	260	2450064.0	61.1

$\cos i_c \cos i_d + \sin i_d \sin i_c \cos(\Omega_d - \Omega_c)$ for planets c and d. For the integrations reported here we kept $\Delta\Omega = 0$ so the equation for relative inclination reduces to the difference between the angles of inclination of two orbits. Unless otherwise noted, the inner planet is taken to be at $i + i_r$ relative to the middle planet, and the outer planet is taken to be $i - i_r$ relative to the middle planet. The total mass is $m_{\mathrm{tot}}(i_{\mathrm{tot}}) = m_{\mathrm{obs}}/\sin i$, where $i_{\mathrm{tot}} \equiv i \pm i_r$.

The code used to do our integrations is the multivariable symplectic (MVS) integrator from Mercury [6]. The symplectic integrator is particularly helpful for our integrations since it is fast for relatively large time steps while maintaining good energy conservation (to $\sim 10^{-8}$) for essentially regular orbits. Our integrations were stopped at the time these orbits first became irregular (i.e., at the onset of instability), before the fixed time step of the MVS integrator would cause energy errors to grow unacceptably large. For all integrations presented here, time steps were kept constant at 0.5 d, or about 10% of the period of the inner planet.

We integrated all three planets in the Ups And system either until a close encounter had occurred, defined as the approach of two bodies in the system to within 3 times the Hill radius, or until a maximum time of 10^8 yr had been reached (for most cases, but see below). Even though the integration time is roughly an order of magnitude smaller than the estimated age of the central star ($\simeq 1 - 3 \times 10^9$ yr [7]), this was the longest reasonable integration time considering the available computational resources.

Two integrations that seemed to border on stability limits for the given system were extended to beyond the standard integration time of 10^8 yr used for all other integrations. By this we attempted to approach the age of the central star, since we know the planetary system must be near the same age, and we tried to elucidate the drastic jump between times to first close encounter produced by two inclinations about one degree apart (see below).

3. RESULTS

Integrations for small relative inclination angles are shown in Fig. 1. When $i_r = \pm 1°$ the system becomes unstable at an overall inclination of $i = 31°$ after about 10^4 yr, while the integration with $i = 32°$ was extended to more than 9.7×10^8 yr without producing a close encounter. The same was true for the system with $i_r = \pm 3.5°$ from the middle planet's orbital plane, between $34°$ and $35°$, except that, to date, this integration has only reached $\sim 5 \times 10^8$ yr. Integrations for systems with the inner and outer planets' relative inclinations $i_r = \pm 5°, \pm 10°$, and $\pm 15°$ were stable for at least 10^8 yr when the middle

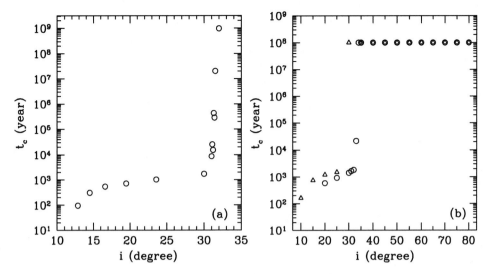

FIGURE 1. For small angles of relative inclination, the time to first close encounter, t_c, is shown as a function of overall inclination, i. (a) Relative inclinations $i_r = \pm 1°$. The circles indicate the time at which the first close encounter occurred in the system, except the last circle at an overall inclination of $32°$, which represents an integration that is still presently stable and running. (b) Relative inclinations $i_r = \pm 3.5°$. These relative inclinations are more representative of the range found in our Solar System. The circles indicate integrations that had relative inclinations $i_r = +3.5°$ between the inner and middle planets, and $i_r = -3.5°$ between the middle and outer planets. The triangles represent integrations with relative inclinations in the opposite configuration, with $i_r = -3.5°$ betwent the inner and middle planets and $i_r = +3.5°$ between the middle and outer planets.

planet had an overall inclination $i \geq 35°, 40°$, and $45°$, respectively.

For systems with relative inclinations $i_r = \pm 3.5°$ between each pair of planets, Fig. 1b shows that configurations where the inner planet has a negative relative inclination and the outer planet has the positive relative inclination retained stability when the middle planet had an overall inclination $i \geq 30°$, whereas with the inclinations of the inner and outer planets reversed, stability was retained only when the overall inclination of the middle planet is $i \geq 34°$.

Integrations where either the inner or middle planet was given a relative inclination $i_r = +30°$ and the remaining two planets had equal overall inclinations $i \geq 35°$ produced stable systems (Fig. 2a). However, the systems where the outer planet had a relative inclination $i_r = +30°$ with respect to the inner two planets were stable when the overall inclination was only $i \geq 15°$.

In the past, others to study this system (e.g. Lissauer and Rivera [2], Stepinski et al. [5], and Chiang et al. [8]) have found secular resonances between the middle and outer planets, where $|\Delta \omega| = |\omega_d - \omega_c| \ll 1$ for all time. Secular resonance in these cases has been used to explain the unusual proximity of ω_c to ω_d, and also to distinguish likely orbital configurations. In particular, approximating the orbit of the inner planet as circular, Chiang et al. [8] found that relative inclinations $i_r \leq 20°$, where $i_r = i_d - i_c = i_b - i_c$ (so

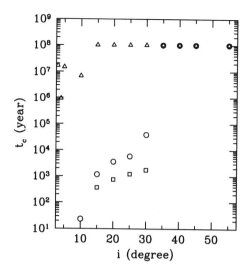

FIGURE 2. Same as Fig. 1, but with only one planet inclined with $i_r = +30°$ with respect to the overall inclination. The squares correspond to systems where the inner planet was inclined relative to the outer two planets, the circles correspond to systems where the middle planet was inclined, and the triangles correspond to systems where the outer planet was inclined.

that the inner and outer planets are in the same plane), limit variations in eccentricity, thus conserving the stability of the system. Using the same initial conditions cited by Chiang et al. we were able to reproduce their results and find secular resonance in the same orbital configurations. However, in our integrations with up-to-date orbital parameters, no secular resonances were detected, suggesting that the proximity of the observed longitudes of pericenter is coincidental and that there is therefore no apparent correlation between secular resonance and long-term stability for any inclinations considered here.

REFERENCES

1. Butler, P.; Marcy, G.; Fischer, D.; Brown, T.; Contos, A.; Korzennik, S.; Nisenson, P.; and Noyes, R.; *ApJ*, **526**, 916 (1999).
2. Lissauer, Jack J. & Rivera, Eugenio J., *ApJ*, **554**, 1141 (2001).
3. Marzari, F. & Weidenschilling, S.J., *Icarus*, Vol. 156, Issue 2, 570 (2002).
4. Ford, Eric B.; Havlickova, Marketa; Rasio, Frederic A.; *Icarus*, Vol. 150, Issue 2, 303 (2001).
5. Stepinski, Thomas F.; Malhotra, Renu; Black, David C.; *ApJ*, **545**, 1044 (2000).
6. Chambers, J.E., "A Hybrid Symplectic Integrator that Permits Close Encounters between Massive Bodies", *Monthly Notices of the Royal Astronomical Society*, **304**, 793, (1999).
7. Ford, E.B.; Rasio, F.A.; and Sills, A.; *ApJ*, **514**, 411 (1999).
8. Chiang, E.I.; Tabachnik, S.; Tremaine, S.; *ApJ*, **122**, 1607 (2001).

Formation and Migration
of Trans-Neptunian Objects

S. I. Ipatov

George Mason University, VA, USA; Institute of Applied Mathematics, Moscow, Russia

Abstract. Trans-Neptunian objects (TNOs) with diameter $d>100$ km moving now in not very eccentric orbits could be formed directly by the compression of large rarefied dust condensations (with semi-major axes $a>30$ AU), but not by the accretion of smaller solid planetesimals. A considerable portion of TNO binaries could be formed at the stage of compression of condensations. Five years before the first TNO was observed, we supposed that, besides TNOs formed beyond 30 AU and moving in low eccentric orbits, there were former planetesimals from the zone of the giant planets in highly eccentric orbits beyond Neptune. For the present mass of the trans-Neptunian belt, the collisional lifetime of 1-km TNO is about the age of the solar system. At the present time TNOs can supply a large amount of matter to the near-Earth space.

FORMATION OF TRANS-NEPTUNIAN OBJECTS

The total mass of the present Edgeworth–Kuiper belt (EKB) for objects with $30 \leq a \leq 50$ AU is estimated [1, 2] to be about $(0.06-0.25)m_\oplus$, where $m_\oplus$ is the mass of the Earth. Objects moving in highly eccentric orbits (mainly with $a>50$ AU) are called "scattered disk objects" (SDOs). The total mass of SDOs in eccentric orbits between 40 and 200 AU has been estimated to be about $0.05m_\oplus$ [3] or $0.5m_\oplus$ [4].

It was considered by many authors that a dust disk around the forming Sun became thinner until its density reached a critical value about equal to the Roche density. At this density, the disk became unstable to perturbations by its own self-gravity and developed dust condensations. These initial condensations coagulated under collisions and formed larger condensations, which compressed and formed solid planetesimals. In [5] it was considered that initial dimensions of planetesimals in the zone of Neptune were about 100 km, and in the terrestrial feeding zone they were about 1 km. According to [6], the mass of the largest condensation in the region of Neptune could exceed $2m_\oplus$. Formation and collisional evolution of the EKB were investigated in [7–10]. In these models, the process of accumulation of trans-Neptunian objects (TNOs) took place at small ($\sim$0.001) eccentricities and a massive belt. More references on the above problems are presented in [11, 12].

Our runs showed [11, 13] that maximum eccentricities of TNOs always exceed 0.05 during 20 Myr under the gravitational influence of the giant planets. Gas drag could decrease eccentricities of planetesimals, and the gravitational influence of the forming giant planets could be less than that of the present planets. Nevertheless, in our opinion, it is probable that, due to the gravitational influence of the forming giant planets, migrating planetesimals, and other TNOs, small eccentricities of TNOs could not exist during all

CP713, The Search for Other Worlds: Fourteenth Astrophysics Conference,
edited by S. S. Holt and D. Deming

the time needed for the accumulation of TNOs with diameter $d>100$ km.

Eneev [14] supposed that large TNOs and all planets were formed by compression of large rarefied dust–gas condensations. We do not think that planets could be formed in such a way, but we consider [12] that TNOs with $d\geq100$ km moving now in not very eccentric orbits could be formed directly by the compression of large rarefied dust condensations (with $a>30$ AU), but not by the accretion of smaller solid planetesimals. The role of turbulence could decrease with an increase of distance from the Sun, so, probably, condensations could be formed at least beyond Saturn's orbit.

Probably, some planetesimals with $d\sim100$–1000 km in the feeding zone of the giant planets and even some large main-belt asteroids also could be formed directly by the compression of rarefied dust condensations. Some smaller objects (TNOs, planetesimals, asteroids) could be debris of larger objects, and other such objects could be formed directly by compression of condensations. Even if at some instant of time at approximately the same distance from the Sun, the masses of initial condensations, which had been formed from the dust layer due to gravitational instability, had been almost identical, there was a distribution in masses of final condensations, which compressed into the planetesimals. As in the case of accumulation of planetesimals, there could be a "run–away" accretion of condensations. It may be possible that, during the time needed for compression of condensations into planetesimals, some largest final condensations could reach such masses that they formed planetesimals with diameter equal to several hundreds kilometers.

It is considered that TNO binaries can be produced due to the gravitational interactions or collisions of future binaries with an object (or objects) that entered their Hill sphere. In our opinion, binary TNOs (including Pluto–Charon) were probably formed at that time when heliocentric orbits of TNOs were almost circular. For such orbits, two TNOs entering inside their Hill sphere could move there for a long time (e.g., greater than half an orbital period [15]). We suppose that a considerable portion of TNO binaries could be formed at the stage of compression of condensations. At this stage, the diameters of condensations, and so probabilities of their mutual collisions and probabilities of formation of binaries were much greater than those for solid TNOs. The stage of condensations was longer for TNOs than that for asteroids, and therefore binary asteroids (which could be mainly formed after formation of solid objects) are less frequent and more differ in mass than binary TNOs. Besides, at the initial stage of solar system formation, eccentricities of asteroids could be mainly greater (due to the influence of the forming Jupiter and planetesimals from its feeding zone) than those for TNOs.

Five years before the first TNO was discovered in 1992, based on our runs of the formation of the giant planets we supposed [15] that there were two groups of TNOs and, besides TNOs formed beyond 30 AU and moving in low eccentric orbits, there were former planetesimals from the zone of the giant planets in highly eccentric orbits beyond Neptune. During accumulation of the giant planets, planetesimals with a total mass equal to several tens $m_\oplus$ could enter from the feeding zone of the giant planets into the trans-Neptunian region, increased eccentricities and inclinations of 'local' TNOs, which initial mass could exceed $10m_\oplus$, and swept most of them [15, 16]. A very small fraction of such planetesimals could be left in eccentrical orbits beyond Neptune and became SDOs.

The total mass of planetesimals in the feeding zones of the giant planets, probably, didn't exceed $300m_\oplus$, and only a smaller part of them could get into the Oort and Hills clouds and into the region between 50 and 1000 AU. So it seems more probable that the total mass of the objects located beyond Neptune's orbit doesn't exceed several tens $m_\oplus$.

Our computer runs [16–18], in which gravitational interactions of bodies were taken into account with the use of the spheres method, showed that the embryos of Uranus and Neptune could increase their semi-major axes from <10 AU to their present values, moving permanently in orbits with small eccentricities, due to gravitational interactions with the migrating planetesimals. Later on, Thommes et al. [19, 20] obtained similar results using direct numerical integration. The comparison of the results presented in [16–20] shows that the method of spheres can provide statistically reliable results for many bodies moving in eccentric orbits.

COLLISIONAL EVOLUTION OF TRANS-NEPTUNIAN OBJECTS

Our estimates [11, 21] of the frequency of collisions of bodies in the present EKB and in the main asteroid belt (MAB) are of the same order of magnitude as the estimates obtained by other scientists (e.g., in [10]). For the EKB with a total mass $M_{EKB} \sim 0.1 m_\oplus$ and the ratio s of masses of two colliding bodies, for which a collisional destruction of a larger body usually takes place, equal to 10^3, (s depends on composition and diameters of objects, a collisional specific energy, and collisional velocity) a collisional lifetime T_c of a body with d=100 km is about 30 Gyr [13]. For 10^{12} 100-m TNOs, 1-km TNO collides with one of 100-m TNOs on average ones in 3 Gyr. At the same s, the values of T_c for 1-km TNOs are of the same order of magnitude as those for main-belt asteroids.

The mean energy of a collision is proportional to v_c^2, where v_c is the relative velocity of a collision. For the EKB the mean energy of a collision and, for the same composition of two colliding bodies, also the ratio s needed for destruction of a larger colliding body in the EKB are smaller by about a factor of $k \approx 20$ than those for the MAB. However, as it can be more easy to destruct icy TNOs than rocky bodies in the MAB, then s can be much larger for the EKB, and collisional lifetimes of small bodies in the EKB can be of the same order as those in the MAB. If some TNOs are porous, then it may be more difficult to destroy them than icy and even rocky bodies and their collisional lifetimes can be larger than those for main-belt asteroids of the same sizes.

The total mass of SDOs moving in highly eccentric orbits between 40 and 200 AU is considered to be of the same order or greater than M_{EKB}. The mean energy of a collision of a SDO with a TNO is greater than that for two colliding TNOs of the same masses. Therefore, though SDOs spend a smaller part of their lifetimes at a distance $R<50$ AU, the probability of a destruction of a TNO (with $30<a<50$ AU) by SDOs can be of the same order of magnitude as that by TNOs. In [11, 21] we also investigated the orbital evolution of TNOs under the mutual gravitational influence.

The total mass of planetesimals that entered the trans-Neptunian region during the formation of the giant planets could be equal to several tens $m_\oplus$, and this time interval could be about several tens Myr. Besides, the initial mass of the EKB can be much larger ($\geq 10 m_\oplus$) than its present mass. Therefore, TNOs could be more often destroyed during planet formation than during the last 4 Gyr.

MIGRATION OF TRANS-NEPTUNIAN OBJECTS

As migration of TNOs to Jupiter's orbit was investigated by several authors, we [22–24] have made a series of simulations of the orbital evolution of 25,000 Jupiter-crossing objects (JCOs) under the gravitational influence of planets. Our runs showed that if one observes former comets in near-Earth object (NEO) orbits, then most of them could have already moved in such orbits for millions of years. Results of our runs testify in favor of at least one of these conclusions: 1) the portion of 1-km former TNOs among NEOs can exceed several tens of percents, 2) the number of TNOs migrating inside the solar system can be smaller by a factor of several than it was earlier considered, 3) most of 1-km former TNOs that had got NEO orbits disintegrated into mini-comets and dust during a smaller part of their dynamical lifetimes if these lifetimes are not small.

ACKNOWLEDGMENTS

This work was supported by NASA (NAG5-10776), INTAS (00-240) and RFBR (01-02-17540).

REFERENCES

1. Jewitt, D., Luu, J., and Trujillo, C., *Astron. J.*, **115**, 2125–2135 (1998).
2. Anderson, J. D., Lau, E. L., Scherer, K., et al., *Icarus*, **131**, 167–170 (1998).
3. Trujillo, C. A., Jewitt, D. C., and Luu, J. X., *Astrophys. J. Letters*, **529**, L103–L106 (2000).
4. Luu, J., Marsden, B. G., Jewitt, D., et al., *Nature*, **387**, 573-575 (1997).
5. Goldreich, P. and Ward, W. R., *Astrophys. J.*, **183**, 1051–1061 (1973).
6. Safronov, V. S. and Vityazev, A. V., *Sov. Sci. Reviews*, Sec. E., *Astrophys. and Space Physics Reviews*, Harwood Academic Publishers, **4**, 1–98 (1985).
7. Davis, D. R. and Farinella P., *Icarus*, **125**, 50–60 (1997).
8. Durda D. D. and Stern S. A., *Icarus*, **145**, 220–229 (2000).
9. Kenyon, S. J. and Luu, J. X., *Astron. J.*, **118**, 1101–1119 (1999).
10. Stern, S. A., *Astron. J.*, **110**, 856–868 (1995).
11. Ipatov, S. I., *Migration of Celestial Bodies in the Solar System*, Moscow, Editorial URSS, 320 pp., in Russian (2000).
12. Ipatov, S. I., *32nd LPSC*, N 1165 (2001).
13. Ipatov S. I., "Migration of Kuiper–Belt Objects Inside the Solar System", in *Planetary Systems — the Long View*, edited by L. M. Celnikier and Tran Thanh Van, Editions Frontieres, 157–160 (1998).
14. Eneev, T. M., *Sov. Astron. Letters*, **6**, 163–166 (1980).
15. Ipatov, S. I., *Earth, Moon, and Planets*, **39**, 101–128 (1987).
16. Ipatov, S. I., *Solar Syst. Res.*, **27**, 65–79 (1993).
17. Ipatov, S. I., *Sov. Astron. Letters*, **17**, 113–119 (1991).
18. Ipatov, S. I., *22nd LPSC*, 607–608 (1991).
19. Thommes, E. W., Duncan, M. J., and Levison, H. F., *Nature*, **402**, 635–638 (1999).
20. Thommes, E. W., Duncan, M. J., and Levison, H. F., *Astron. J.*, **123**, 2862–2883 (2002).
21. Ipatov, S. I., *Solar Syst. Res.*, **29**, 261–286 (1995).
22. Ipatov, S. I., "Formation and Migration of Trans-Neptunian Objects and Asteroids", in *Asteroids, Comets, Meteors, 2002*, edited by Barbara Warmbein, 371–374 (2002).
23. Ipatov, S. I. and Mather, J. C., *Advances in Space Research*, in press (2003)
24. Ipatov, S. I. and Mather, J. C., *Earth, Moon, and Planets*, in press (2004) (http://arXiv.org/format/astro-ph/0305519).

Programs and Techniques for Detection and Characterization of Extrasolar Planets

Detection and Characterization of Extra-Solar Planets: Future Space Missions

M. A. C. Perryman

European Space Agency, ESTEC, Noordwijk 2200AG, The Netherlands

Abstract. Various techniques are being used to search for extra-solar planetary signatures, including accurate measurement of positional (astrometric) displacements, gravitational microlensing, and photometric transits. Planned space experiments promise a huge increase in the detections and statistical knowledge arising from transit and astrometric measurements. In contrast, imaging of even nearby Earth-mass planets in the habitable zone and the measurement of their spectral characteristics, typified by the TPF and Darwin missions, represents an enormous challenge. A number of proposed precursors aimed at exploiting coronagraphy or occultations are being studied. Beyond TPF/Darwin, Life Finder would aim to produce confirmatory evidence of the presence of life, while an Earth 'imager', some massive interferometric array providing resolved images of a distant Earth, appears only as a distant vision. A 10 nas astrometric mission would detect 'Earths' systematically out to 100 pc.

INTRODUCTION

Many new results in the field of exoplanet detection and characterization from ground-based telescopes are expected over the coming years. Space experiments offer significant observational advantages which will be exploited by a series of planned and future missions, whose objectives focus on increasing our statistical knowledge in order to understand better the formation process, and detecting and characterizing habitable systems [1]. First on the horizon of approved missions comes Kepler (NASA), due for launch around 2007, and which should provide improved prospects for photometric transits following on from MOST (launched on 30 June 2003) and Corot (launch planned for June 2006). It focuses on the detection of Earth-size planets in the habitable zone, monitoring some 10^5 main-sequence stars. Detection of some 50–640 terrestrial inner-orbit planet transits are predicted, depending on whether their typical radii lie in the range $R \sim 1.0 - 2.2R_E$. Eddington (ESA) was designed to monitor some 5×10^5 stars, and proposed for launch around 2008, but at the time of writing appears in ESA's science program only as a 'reserve' mission. Predictions are for some 20 000 planets with $R < 15R_E$, some 2000 terrestrial planets, and perhaps some dozens of Earth-like planets in the habitable zone. An advance in transit statistics may also come from observations of the Galaxy bulge with HST in February 2004 (Sahu, private communication).

Gaia (ESA) and SIM (NASA) are two very different approaches to space astrometry, both currently being developed for launch around 2010–12. Gaia is a scanning, survey-type instrument [2], contributing to the large-scale systematic detection of Jupiter-mass planets (or above) in Jupiter-period orbits (or smaller); some 10–20 000 detections out to 150–200 pc are expected ([3], [4]), including most of the (longer-period) radial velocity

CP713, *The Search for Other Worlds: Fourteenth Astrophysics Conference,*
edited by S. S. Holt and D. Deming
© 2004 American Institute of Physics 0-7354-0190-X/04/$22.00

detections known to date. Planetary masses, M, rather than $M\sin i$, will be obtained, orbital parameters for some 5000 systems, and relative inclinations for multiple systems. Some 4–5000 transit systems, of the hot Jupiter type, might also be detected [5]. Gaia might also detect a handful of protoplanetary collisions photometrically [6]. The NASA SMEX proposal AMEX aims at 150 μas accuracy at 9 mag and 3 mas at 15 mag and, with a proposed launch in 2007–08, provides limited prospects for planet detection through the comparison of proper motions with Hipparcos (Johnston et al. 2003, these proceedings), including some 600 detections to 30 pc down to K5V stars, and transits to V =11 mag.

SIM is a pointed interferometer with a launch around 2009 [7]: accuracies of a few microarcsec down to 20 mag are projected, but such faint observations will be expensive in observing time, and brighter target stars are likely to be the rule. In the field of exoplanets, SIM will excel in any accurate and frequent astrometric follow-up of previously-detected systems. The baselined planet detection program includes 50 separate epoch 1-hour observations of selected fields, with some 250 stars measured at 1 μas accuracies, and some 2000 stars at accuracies of about 4 μas.

While these transit and astrometric discovery predictions must be taken with certain caveats, most notably as a result of our uncertainty of the exoplanet population properties, they promise a major advance in the detection and knowledge of the statistical properties of a wide range of exoplanets: ranging from heavy (Jupiter-mass) planets in long-period (Jupiter-type) orbits via astrometry, through to Earth-mass planets in the habitable zone, and the occurrence and properties of multiple systems.

The rest of this paper examines ideas extending beyond these accepted space missions. I have found no reference to transit missions beyond the planned Kepler and Eddington missions, and look first at a space-based lensing proposal, GEST, which expands on the parameter space for which statistical information on planet formation would be provided. Then I look at the various imaging missions. TPF/Darwin, SIRTF and JWST are covered in an accompanying contribution [8], and only the former are mentioned briefly here. The literature includes many ideas which fall under the category of scientific or technical precursors to TPF/Darwin, as well as some pointers to the missions which will follow a successful TPF/Darwin: 'life finders' and 'planet imagers'. I conclude with a look at nanoarcsec astrometry, which would have considerable discovery potential as well as demanding technology very similar to that of a 'mini life finder' which has already been examined under NASA contract. A convenient starting point for more information about each is contained in the ongoing programmes and future projects section of Jean Schneider's www page (http://www.obspm.fr/encycl/searches.html).

LENSING (AND TRANSITS)

Although primarily devoted to lensing searches, GEST (Galactic Exoplanet Survey Telescope [9]) is also sensitive to transits. GEST was proposed for a NASA mission in 2001–02 (a Survey for Terrestrial ExoPlanets (STEP) was also submitted to NASA's Extrasolar Planets Advanced Concepts Program at the same time). A 1.2-m aperture telescope with a 2 deg^2 field of view continuously monitors 10^8 Galactic bulge main-sequence stars. Sources in the bulge are lensed by foreground (bulge or disk) stars which are accompanied by the planets being sought. GEST was not selected in 2002, but is expected to be re-submitted for the 2004 Discovery round.

The sensitivity of such measurements is highest at orbital separations of 0.7–10 AU, but it will also detect systems with larger separations, masses as low as that of Mars, large moons of terrestrial planets, and some 50 000 giant planets via transits with orbital separations of up to 20 AU (the prime sensitivity of a transit survey extends inward from 1 AU, while the sensitivity of microlensing extends outwards). There are theoretical reasons to believe that free-floating planets may be abundant as a by-product of the planetary formation process and, uniquely, GEST will also be able to detect these.

The planetary lensing events have a typical duration of 2–20 hr (compared to the typical 1–2 month duration for lensing events due to stars), and must be sampled by photometry of ∼ 1% accuracy several times per hour over a period of several days, and with high angular resolution because of the high density of bright main-sequence stars in the central bulge. The proposed polar orbit is oriented to keep the Galactic bulge in the continuous viewing zone.

Most of the multiple-planet detections in the simulations of [9] are systems in which both 'Jupiter' and 'Saturn' planets are detected. Since multiple orbits are generally stable only if they are close to circular, a microlensing survey will be able to provide information on the abundance of giant planets with nearly circular orbits by measuring the frequency of double-planet detections and the ratios of their separations.

Just over 100 Earths would be detected if each lens star has one in a 1 AU orbit. The peak sensitivity is at an orbital distance of 2.5 AU, with 230 expected detections if each lens star had a planet in such an orbit. Based on certain observational conditions and physical arguments, the data can provide the mass of the host star, the planetary mass, the distance to the host star, and the planet-star separation in the plane of the sky.

One of the disadvantages of lensing experiments is that a planet event, once observed, can never (in practice) be seen again – follow-up observations for further characterizations are not feasible (unlike the case for any of the other principal detection methods). Nevertheless, a mission like GEST will provide important observational and statistical data on the occurrence of low-mass planets, low-mass planets at larger orbital radii, multiple systems and, significantly, free-floating planets formed as a by-product of the system formation. Lensing experiments, like transit experiments, can never realize their full potential unless placed in space, so that a mission designed for both lensing and transit measurements should have a place in the armoury of space missions.

IMAGING

The promised science and enabling technologies for high-contrast space imaging have been explored in the contexts of JWST, TPF/Darwin, and various NASA Explorer or Discovery class missions [8]. The many possible solutions involve combinations of active wavefront correction, coronagraphs, apodization, interferometers, and large free-flying occulters.

TPF will take the form of either a coronagraph operating at visible wavelengths or a large-baseline interferometer operating in the infrared. There are two aspects of this choice which should be distinguished: (a) the scientific aspect: is reflected (visible and near IR) light or thermal emission (mid-IR) the best regime to characterize planets (albedo, temperature, colour, etc; see [10] for a recent discussion); (b) the instrumental aspects: is an interferometer or a coronagraph the best? Here, the NASA Technology Plan for TPF states that 'Technology readiness, rather than a scientific preference for any wavelength region, will probably be the determining factor in the selection of a final architecture'. In May 2002, two architectural concepts were selected for further evaluation: an infrared interferometer (multiple small telescopes on a fixed structure or on separated spacecraft flying in precision formation and utilizing nulling), and a visible light coronagraph (utilizing a large optical telescope, with a mirror three to four times bigger and at least 10 times more precise in WFE than the Hubble Space Telescope). NASA and JPL will issue a series of calls for proposals seeking input on the development and demonstration of technologies to implement the two architectures (one has already been issued and awarded: http://acquisition.jpl.nasa.gov/rfp/TPF-HCIT-TFE/). It is anticipated that one of the two architectures will be selected in about 2006.

A visible light system can be smaller (some 10 m aperture) than a comparable interferometer, however advances in mirror technology are required: mirrors must be ultra-smooth ($\sim \lambda/15\,000$) to minimize scattered light, and in addition active optics would be needed to maintain low and mid-spatial frequency mirror structure at acceptable levels. IR interferometry would require either large boom technology or formation flying, typically with separation accuracies at the cm-level with short internal delay lines. For the detection of ozone at distances of 15 pc and S/N$\sim$25, apertures of about 40 m^2, and observing times of 2–8 weeks per object, are indicated.

The ESA effort is focussed on an interferometer: Darwin is a presently considered as a flotilla of eight spacecraft (6 telescopes, one beam combination unit, and one communication unit) that will survey 1000 of the closest stars in the infrared, searching for Earth-like planets and analyzing their atmospheres for the chemical signature of life [11], scientific objects in common with those of TPF. The system is presently planned for an L2 orbit, and a single Ariane launch. Specific precursor efforts include GENIE, a nulling interferometer prototype under development by ESA and ESO for the VLTI, and the space mission Smart-3, not yet approved, but included within the Darwin concept plans to demonstrate the concept of formation flying for two or three satellites.

Precursors: Interferometers, Coronagraphs and Apodizers

The McKee-Taylor Decadal Survey Committee [12] qualified its endorsement of the TPF mission with the condition that the abundance of Earth-size planets be determined prior to the start of the TPF mission. In these sections we look at ideas which have not (yet) been approved, and which may fall somewhere between scientific and technological precursors for TPF. Some are concepts, while some are specific mission proposals.

Eclipse (coronagraphy) is a proposed NASA Discovery-class mission to perform a direct imaging survey of nearby planetary systems, including a complete survey for Jovian-sized planets orbiting 5 AU from all stars of spectral types A–K within 15 pc of the Sun [13]. Its optical design incorporates a telescope with an unobscured aperture of 1.8 m, a coronagraphic camera for suppression of diffracted light, and precision active optical correction for suppression of scattered light, and imaging/spectroscopy. A three-year science mission would provide a survey of the nearby stars accessible to TPF. Eclipse will probably be resubmitted for NASA's next Discovery round in 2004.

Jovian Planet Finder (JPF) was a MIDEX proposal to directly image Jupiter-like planets around some 40 nearby stars using a 1.5-m optical imaging telescope and coronagraphic system, originally on the International Space Station (ISS) [14]. Its sensitivity results from super-smooth optical polishing, and should be sensitive to Jovian planets at typical distances of 2–20 AU from the parent star, and imaging of their dusty disks – potentially solar system analogues. A 3-yr mission lifetime is proposed. Some successor to JPF will probably be resubmitted for NASA's next Discovery round in 2004, probably through a merging with ESPI ('EPIC', Clampin, private communication).

Extra-Solar Planet Imager (ESPI) is another proposed precursor to TPF [15]. Originally proposed as a NASA Midex mission as a 1.5×1.5 m^2 apodized square aperture telescope, reducing the diffracted light from a bright central source, and making possible observations down to 0.3 arcsec from the central star. Jupiter-like planets could be detected around 160–175 stars out to 16 pc, with S/N > 5 in observations lasting up to 100 hours. Spectroscopic follow-up of the brightest discoveries would be made. The Extrasolar Planet Observatory (ExPO) is a similar concept proposed as a Discovery-class mission [16].

Self-luminous Planet Finder (SPF) is a further TPF precursor under study by N. Woolf and colleagues, aiming at the search for younger or more massive giant planets in Jupiter/Saturn like orbits, where they will be highly self-luminous and bright at wavelengths of 5–10 μm, where neither local nor solar system zodiacal glow will limit observations. SPF will demonstrate the key technologies of passive cooling associated with interferometric nulling and truss operation that are required for a TPF mission. SPF targets young Jupiter-like planets both around nearby stars such as ε Eri, and around A and early F stars.

The Fourier-Kelvin Stellar Interferometer (FKSI) is a concept under study at NASA GSFC [17]. It is a space-based mid-infrared imaging interferometer mission concept being developed as a precursor for TPF. It aims to provide 3 times the angular resolution of JWST and to demonstrate the principles of interferometry in space. In its minimum configuration, it uses two 0.5-m apertures on a 12.5-m baseline, and predicts that some 7 known exoplanets will be directly detectable in this configuration, with low-resolution spectroscopy ($R \sim 20$) being possible in the most favorable cases.

Optical Planet Discoverer (OPD) is a concept midway between coronagraphy and Bracewell nulling [18].

Phase-Induced Amplitude Apodization (PIAA, [19]) is an alternative to classical pupil apodization techniques (using an amplitude pupil mask). An achromatic apodized pupil is obtained by reflection of an unapodized flat wavefront on two mirrors. By carefully choosing the shape of these two mirrors, it is possible to obtain a contrast better than 10^9 at a distance smaller than $2\lambda/d$ from the optical axis. The technique preserves both the angular resolution and light-gathering capabilities of the unapodized pupil, and claims to allow efficient detection of terrestrial planets with a 1.5-m telescope in the visible.

Free-Flying Occulters

Occulting masks are another approach to tackle in a conceptually simple manner the basic problem of how to separate dim sources from bright ones, although interest in this approach at NASA level currently appears limited. The history of free-flying occulters to search for extra-solar planets goes back to at least the early 1960s, with ideas by Robert Danielson, Lyman Spitzer, Gordon Woodcock, and Christian Marchal (who found that the screen's efficiency can be enhanced by choosing complex shapes). Occulters have also been considered as precursor missions to TPF/Darwin.

UMBRAS (Umbral Missions Blocking Radiating Astronomical Sources) refers to a class of missions [20], Kochte et al. 2003, these proceedings), currently designed around a 4-m telescope and a 10-m occulter, with earlier concepts including a 5–8 m screen (CORVET), or as NOME (Nexus Occulting Mission Extension) a modification to Nexus, itself foreseen as an engineering test of key technologies for JWST, cancelled in 2000.

BOSS (Big Occulting Steerable Satellite [21]) consists of a large occulting mask, typically a 70×70 m^2 transparent square with a 35 m radius, and a radially-dependent, circular transmission function inscribed, supported by a framework of inflatable or deployable struts. The mask is used by appropriately aligning it with a ground- or space-based observing telescope. In combination with JWST, for example, both would be in a Lissajous-type orbit around the Sun-Earth Lagrange point L2, with the mask steered to observe a selected object using a combination of solar sailing and ion or chemical propulsion. Thermal emissivity, reflection, and scattering properties of the screen have also been considered. All but about 4×10^{-5} of the light at 1 μm would be blocked in the region of interest around a star selected for exoplanet observations (since occultation occurs outside the telescope, scattering inside the telescope does not degrade this performance). Their predictions suggest that planets separated by as little as 0.1–0.2 arcsec from their parent star could be seen down to a relative intensity of 1×10^{-9} for a magnitude 8 star. Their simulations indicate that for systems mimicking our solar system, Earth and Venus would be visible for stars out to 5 pc, with Jupiter and Saturn remaining visible out to about 20 pc.

BEYOND TPF/DARWIN

Within NASA's Origins Program HST, SIRTF and others are referred to as 'precursor missions', with SIM and JWST as 'First Generation Missions' leading to the 'Second Generation Mission' TPF which will begin to examine the existence of life beyond our Solar System. Once habitable planets are identified, a 'Life Finder' type of mission would expand on the TPF principles to detect the chemicals that reveal biological activities. And once a planet with life is found, 'Planet Imager' would be needed to observe it. These 'Third Generation Missions', Life Finder and Planet Imager, are currently just visions because the required technology is not on the immediate horizon.

Life Finder. Taking pictures of the nearest planetary system (TPF/Darwin) is considered to be a reasonable goal on a 10-year timescale, with low-quality spectra a realistic by-product. Life Finder, which would only be considered after TPF/Darwin results are available, and once oxygen or ozone has been discovered in the atmosphere, would aim to produce confirmatory evidence of the presence of life, searching for an atmosphere significantly out of chemical equilibrium, for example through its oxygen (20% abundance on Earth) and methane (10^{-6} abundance on Earth). Some pointers to the technology requirements and complexity of Life Finder have been described in the 'Path to Life Finder' [22]. Given that TPF/Darwin will take low-resolution low-S/N spectra, a large area high angular resolution telescope will be needed for detailed spectral study in order to confirm the presence of life. Recalling that the target objects will be as faint as the Hubble Deep Field galaxies, buried in the glare of their parent star some 0.05–0.1 arcsec away, the light collecting area of Life Finder will have to be substantially larger than TPF's 50 m^2: a useful target is 500–5000 m^2. One of the primary technical challenges will be to produce such a collecting area at affordable cost and mass.

The required development of new low mass and better wavefront optics, coronagraphy versus nulling, pointing control by solar radiation pressure, sunshield, vibration damping, and space assembly, were addressed by [22]. According to their study, a 'mini-Life Finder' might be a 50×10 m^2 telescope, made with 12 segments of 8.3×5 m^2, made of 5 kg m^{-2} glass, piezo-electric controlled adaptive optics, and a total mass (optics and structure) of about 10 tons. Cooling would be by an attached sunshade also used for solar pressure pointing, in a 'sun orbiting fall away' orbit to avoid the generation of thruster heat needed to maintain the L2-type orbit. There are still unsolved complexities underlying the actual science case for Life Finder: *if* the goal is to detect the 7.6 μm methane feature — which is not definitively the relevant goal; see, for example discussions of the use of the 'vegetation signature' in [23] — the required collecting area accelerates from a plausible 220 m^2 (four or five 8-m telescopes) for a planet at 3.5 pc, to a mighty 4000 m^2 (eighty 8-m telescopes) even at only 15 pc. A new proposal to study Life Finder has recently been submitted to NASA by Shao, Traub, Danchi & Woolf (N. Woolf, private communication). Various reports on related studies can be found under NASA's Institute for Advanced Concepts (NIAC) www pages (http://www.niac.usra.edu/) including 'Very large optics for the study of Extrasolar Terrestrial Planets' (N. Woolf); and 'A structureless extremely large yet very lightweight swarm array space telescope' (I. Bekey). The former includes an outline technology development plan for Life Finder, with costs simply stated as ≫$2 billion.

Planet Imager. TPF aims to image a reflected point-source image of a planet. Resolving the surface of a planet is, at best, a far future goal requiring huge technology development that is not yet even in planning. Much longer baselines will be required, from tens to hundreds of km in extent. Formation flying of these systems will require technology development well beyond even the daunting technologies of TPF/Darwin – complex control systems, ranging and metrology, wavefront sensing, optical control and on-board computing. Having accepted that we are now peering into a much more distant and uncertain future, we can examine some of the ideas which are being discussed.

Life Finder studies [22] have been used to evaluate the requirements for Planet Imager which, they consider, would require some 50–100 Life Finder telescopes used together in an interferometric array. Their conclusions were that *'the scientific benefit from this monstrously difficult task does not seem commensurate with the difficulty'*. This echoes the conclusions of [24] who undertook a partial design of a separated spacecraft interferometer which could achieve visible light images with 10×10 resolution elements across an Earth-like planet at 10 pc. This called for 15–25 telescopes of 10-m aperture, spread over 200 km baselines. Reaching 100×100 resolution elements would require 150–200 spacecraft distributed over 2000 km baselines, and an observation time of 10 years per planet. These authors noted that the resources they identified would dwarf those of the Apollo Program or the Space Station, concluding that it was *'difficult to see how such a program could be justified'*. The effects of planetary rotation on the time variability of the spectral features observed by an imager, complicates the imaging task although may be tractable, while more erratic time variability (climatic, cloud coverage, etc.) will greatly exacerbate any imaging attempts.

OVLA. Parallel to the Planet Imager studies in the US, in Europe the LISE group (Laboratoire d'Interférométrie Stellaire et Exo-planétaire) carries out research in the area of high-resolution astronomical imaging, including imaging extra-solar planets. The group is studying several complementary projects for 'hypertelescopes' on Earth and in space ([25], [26]). The steps needed to reach this goal are set out as requiring: (1) a hypertelescope on Earth – the OVLA (Optical Very Large Array); (2) a 100-m precursor geostationary version in space; (3) a km-scale version in a higher orbital location; (4) a 100 km version, including dozens of mirrors of typically 3 m aperture. Labeyrie et al. proposed the mission 'Epicurus', an extrasolar earth imager, to ESA in 1999 in response to the F2/F3 call for mission proposals.

Their basic 'hypertelescope' design involves a dilute array of smaller apertures (an imaging interferometer) having a 'densified' exit pupil, meaning that the exit pupil has sub-pupils having a larger relative size than the corresponding sub-apertures in the entrance pupil (see Fig. 1 of [27]). Their applicability extends to observing methods highly sensitive to the exit pupil shape, such as phase-mask coronagraphy.

In the most recent published studies [25] the hypertelescope is combined with such a coronagraph to yield attenuations at levels of 10^{-8}. Simulations of 37 telescopes of 60 cm aperture distributed over a baseline of 80 m in the IR, observing the 389 Hipparcos M5–F0 stars out to 25 pc (with simulated contributions from zodiacal and exo-zodiacal background) yields 10-hour snapshot images in which an Earth-like planet is potentially detectable around 73% of the stars. Gains of a factor 20–30 with respect to a simple Bracewell nulling interferometer are reported.

In space, the plans call for a flotilla of dozens or hundreds of small elements, deployed in the form of a large dilute mosaic mirror. Pointing is achievable by globally rotating the array, which is slowly steerable with small solar sails attached to each element. A 'moth-eye' version allows full sky coverage with fixed elements, using several moving focal stations ([28], [29]). The geostationary precursor hypertelescope could be a possible version of Terrestrial Planet Finder (TPF). An exo-Earth discoverer would require a 100–1000 m hypertelescope with coronagraph, while an exo-Earth imager would require a 150 km hypertelescope with coronagraph. In the approach of [28] a 30-min exposure using a hypertelescope comprising 150 3-m diameter mirrors in space with separations up to 150 km, would be sufficient to detect 'green' spots similar to the Earth's Amazon basin on a planet at a distance of 10 light-years (although these vegetation features are more prominent in the infrared, e.g. [23]).

NANOARCSEC ASTROMETRY

The ambitious technological nature of the imaging exoplanet missions being considered, as well as the subject of this review, offers an opportunity for some speculation about astrometry at the nas (nanoarcsec) levels. Earth-mass perturbations around a solar-mass star are 300 nas at 10 pc, or 30 nas at 100 pc, the latter requiring an instantaneous measurement accuracy a factor 3 better, i.e. 10 nas at, say, 12 mag. This is a factor of some 1000 improvement with respect to Gaia. Keeping all other mission parameters (efficiency, transverse field of view, mission duration, total observing time per star, and image pixel sampling, etc.) unchanged, we can consider reaching this accuracy simply through a scaling up of the primary mirror size. The Gaia primary mirror has an along-scan dimension $D = 1.4$ m and a transverse dimension $H = 0.5$ m; the final accuracy scales as $\sigma \propto D^{-3/2} H^{-1/2}$. These desired accuracies would therefore require a primary mirror size of order 50×12 m^2, and a focal length (scaling with D) of about 1600 m. Interestingly, this is precisely the scale of the optics derived for the mini-version of Life Finder. Such a system would provide sub-microarcsec accuracies on all objects to 20 mag out to 100 pc, independent of any *a priori* knowledge.

These accuracy levels of ~ 10 nas are still above the noise floors due to interplanetary and interstellar scintillation in the optical, or stochastic gravitational wave noise. Aside from many other scientific domains of interest at 1–10 nas, geometric cosmology becomes accessible. A study of the ultimate limits in astrometry is underway in Lund (Lindegren, private communication).

DISCUSSION

Kepler and Gaia will provide an improvement in the statistical knowledge of exoplanet formation and distributions through photometric transits and astrometry respectively. The proposed mission GEST would provide statistical information through microlensing, including the distribution of free-floating systems. TPF/Darwin aim at the first detection through imaging of specific systems which are potentially habitable. Various scientific or technical precursors for TPF/Darwin are being discussed in the literature. The investigations already carried out into Life Finder and Planet Imager provide an insight into the technology required to meet these ambitious goals, but again they will have to be underpinned by a clear and robust scientific judgment on the observational evidence that is needed to qualify the existence of life beyond our Solar System.

I have not considered missions in the 'Go There' class of traveling to the stars, but recent discussions are given in Alan Penny's Darwin page at http://ast.star.rl.ac.uk/darwin, and the NASA page at http://www.lerc.nasa.gov/WWW/PAO/warp.htm. NASA's Gossamer Spacecraft Initiative, started in 1999, opened further discussions about the technology to travel to the nearest stars: solar sailing (powered far out by solar-powered lasers or microwave beams), the interstellar probe, or missions to the solar gravitational lens point at 550 AU. Their Breakthrough Propulsion Physics Project (http://www.grc.nasa.gov/WWW/bpp/) studied more exotic propulsion systems, but since 2002 is no longer funded.

ACKNOWLEDGMENTS

I thank the following for valuable discussions: Jos de Bruijne, Malcolm Fridlund, Philippe Gondoin, and Jean Schneider.

REFERENCES

1. Perryman, M. A. C., *Rep. Prog. Phys.*, **63**, 1209–1272 (2000).
2. Perryman, M. A. C., de Boer, K. S., Gilmore, G., et al., *AA*, **369**, 339–363 (2001).
3. Lattanzi, M. G., Spagna, A., Sozzetti, A., and Casertano, S., *MNRAS*, **317**, 211–224 (2000).
4. Sozzetti, A., Casertano, S., Lattanzi, M. G., and Spagna, A., *AA*, **373**, L21 (2001).
5. Robichon, N., in *Proc. Les Houches Summer School on Gaia*, edited by O. Bienaymé and C. Turon, EDP, 2002, pp. 215–221.
6. Zhang, B., and Sigurdsson, S., *ApJ*, **596**, L95–L98 (2003).
7. Danner, R., and Unwin, S., *SIM: Taking the Measure of the Universe*, NASA/JPL, 1999.
8. Beichman, C., these proceedings (2003).
9. Bennett, D. P., and Rhie, S. H., *ApJ*, **574**, 985–1003 (2002).
10. Schneider, J., in *Proc. Heidelberg Conference*, edited by M. Fridlund, ESA SP-339, 2003, in press.
11. Fridlund, C. V. M., in *Darwin and Astronomy*, edited by B. Schürmann, ESA SP–451, Noordwijk, 2000, pp. 11–18.
12. McKee, C. F., and Taylor, J. H., Astronomy & Astrophysics in the New Millennium, Washington DC, National Academy Press (2000).
13. Trauger, J., Hull, A., Backman, D., et al., *35th meeting, DPS, September 2003*, in press (2003).
14. Clampin, M., Ford, H. C., Illingworth, G., et al., *AAS 199th meeting, Washington*, in press (2002).
15. Lyon, R. G., Gezari, D. Y., Melnick, G. J., et al., *Proc. SPIE*, **4860**, 45–53 (2003).
16. Gezari, D. Y., Nisenson, P., D., P. C., et al., *Proc. SPIEE*, **4860**, 302–310 (2003).
17. Danchi, W. C., Deming, D., Kuchner, M. J., et al., *ApJ*, **in press** (2003).
18. Mennesson, B., Shao, M., Serabyn, E., et al., *Proc. SPIE*, **4860** (2003).
19. Guyon, O., *AA*, **404**, 379–387 (2003).
20. Schultz, A. B., Lyon, R. G., Kochte, M., et al., *Proc. SPIE*, in press (2003).
21. Copi, C. J., and Starkman, G. D., *ApJ*, **532**, 581–592 (2000).
22. Woolf, N., et al., Path to life finder, Tech. Rep., NASA Institute for Advanced Concepts: presentation at http://peaches.niac.usra.edu/files/library/fellows_mtg/jun01_mtg/pdf/374Woolf.pdf (2001).
23. Arnold, L., Gillet, S., Lardière, O., et al., *AA*, **392**, 231–237 (2002).
24. Bender, P. L., and Stebbins, R. T., *JGR*, **101(E4)**, 9309–9312 (1996).
25. Riaud, P., Boccaletti, A., Gillet, S., et al., *AA*, **396**, 345 (2002).
26. Gillet, S., Riaud, P., Lardière, O., et al., *AA*, **400**, 393–396 (2003).
27. Pedretti, E., Labeyrie, A., Arnold, L., et al., *AA*, **147**, 285–290 (2000).
28. Labeyrie, A., *Science*, **285**, 1864–1865 (1999).
29. Labeyrie, A., in *Working on the Fringe: Optical and IR Interferometry from Ground and Space*, edited by S. Unwin and R. Stachnik, ASP Conf. Ser. 194, San Francisco, 1999, p. 7.

UKIRT's Wide Field Camera and the Detection of 10 M$_{\text{Jupiter}}$ Objects

Sandy Leggett*, the WFCAM Team† and the UKIDSS Team**

*Joint Astronomy Centre Hawaii
†www.roe.ac.uk/atc/projects/wfcam/
**www.ukidss.org

Abstract. In mid–2004 a near–infrared wide field camera will be commissioned on UKIRT. About 40% of all UKIRT time will go into sky surveys and one of these, the Large Area Survey using *YJHK* filters, will extend the field brown dwarf population to temperatures and masses significantly lower than those of the T dwarf population discovered by the Sloan and 2MASS surveys. The LAS should find objects as cool as 450 K and as low mass as 10 M$_{\text{Jupiter}}$ at 10 pc. These planetary–mass objects will possibly require a new spectral type designation.

THE WIDE FIELD CAMERA

The Wide Field Camera (WFCAM) is being built at the Astronomy Technology Centre in Edinburgh ([1]) and will be delivered to the Joint Astronomy Centre in Hawaii in mid–2004 for commissioning on the UK Infrared Telescope (UKIRT) on Mauna Kea. WFCAM will use about 60% of UKIRT time and 70% of this time will be used for the UKIRT Infrared Deep Sky Surveys (UKIDSS). Survey data is immediately available to the UK and ESO communities and to designated Japanese UKIDSS consortium members, and becomes public after 18 months. The pipeline data processing and science archive are being handled by the Cambridge (UK) Astronomy Survey Unit and the Edinburgh Wide Field Astronomy Unit. The goals of the surveys are outlined by [2] and more information can be found at *www.ukidss.org*.

The instrument has four Rockwell Hawaii–II 2048^2 arrays with a pixel scale of 0.4". SDSU controllers will be used and a data rate of 120 Gb a night is expected. The arrays are not buttable and are spaced at 94% of the detector width. After 2×2 steps the width of a single tile is 0.9 degrees giving a solid angle of 0.78 deg^2. To improve image sampling, 2×2 microstepping will also be employed. *ZYJHK* filters have been designed to optimize selection of cool brown dwarfs, high redshift starburst galaxies, elliptical galaxies and galaxy clusters at $1 < z < 2$, as well as the highest redshift quasars at $z = 7$.

THE SURVEYS

WFCAM will survey 7500 deg^2 of the Northern sky to $K = 18.5$; limited areas will be deeper. UKIDSS actually consists of five surveys which will use about 1000 nights of UKIRT time over a seven year period. The Large Area Survey (LAS) will survey 4000

CP713, *The Search for Other Worlds: Fourteenth Astrophysics Conference*,
edited by S. S. Holt and D. Deming
© 2004 American Institute of Physics 0-7354-0190-X/04/$22.00

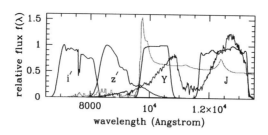

FIGURE 1. Spectrum of a 1000 K brown dwarf (dark grey or red) and a $z = 7$ quasar (light grey or green) with Sloan and WFCAM filter profiles overlaid.

deg^2 to $K = 18.4$, targeting Sloan survey regions. The Galactic Plane Survey (GPS) will survey 1800 deg^2 to $K = 19.0$ and the Galactic Clusters Survey (GCS) will survey 1400 deg^2 to $K = 18.7$. The Deep Extragalactic Survey (DXS) will survey 35 deg^2 to $K = 21.0$ and the Ultra Deep Survey (UDS) will survey 0.77 deg^2 to $K = 23.0$. Cool brown dwarfs will primarily be found by the LAS which in the first two years is expected to receive 105 nights surveying 2000 deg^2 to $Y = 20.5$, $J = 19.7$, $H = 18.8$ and $K = 18.4$.

Cool brown dwarfs are extremely red in the optical but have blue near–infrared colors due to the presence of strong molecular absorption bands. As shown by e.g. [3], brown dwarfs and high redshift quasars can both be found as extremely red objects, often detected only in the reddest, z, band of the Sloan survey. Figure 1 shows a far–red spectrum of a 1000 K brown dwarf and a $z = 7$ quasar, with the Sloan iz and WFCAM YJ filter bandpasses overlaid. It can be seen that the Y filter allows the detection of even higher redshift quasars than those found in the Sloan survey and also that, while both quasars and brown dwarfs will be selected as optically red objects, they can be distinguished by their $Y - J$ colors.

Figure 2 shows a color:magnitude diagram with theoretical isochrones for 1—25 M_{Jupiter} objects. The models indicate that the near–infrared colors of brown dwarfs remain blue for effective temperatures as cool as 400 K. At lower temperatures, water clouds form in the photosphere and $J - K$ becomes redder. Separating these cooler and redder brown dwarfs from hotter stars, using near–infrared colors only, will be difficult.

Figure 3 shows the calculated flux distributions of 10 and 25 M_{Jupiter} objects at 10pc. The WFCAM LAS sensitivity limits are indicated, as well as those for Sloan z, *SIRTF* and *JWST*. It can be seen that the LAS will be able to detect brown dwarfs as cool as 450 K at *YJH*. The models predict that NH$_3$ will be apparent in near–infrared spectra at $T_{\text{eff}} < 500$ K and this may signal the next spectral type after T (Y has been suggested). However these models assume chemical equilibrium which appears not to be valid (see Saumon et al. [5]). In the case of disequilibrium the abundance of NH$_3$ is reduced, so a spectral change may not occur until even cooler temperatures. Note also that the new chemistry strengthens CO absorption at 5 μm which slightly compromises the *SIRTF* detection limits.

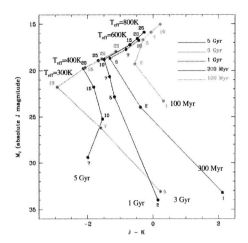

FIGURE 2. Isochrones of absolute J magnitude against $J - K$ color from [4]. Note increasingly blue $J - K$ color as T_{eff} decreases, for $T_{eff} > 400$ K.

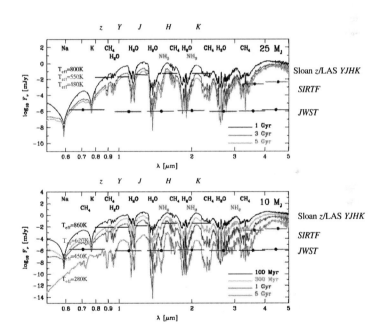

FIGURE 3. Spectral energy distributions of low mass brown dwarfs from [4]. Detection limits are indicated for Sloan z, WFCAM LAS $YJHK$, $SIRTF$ and $JWST$.

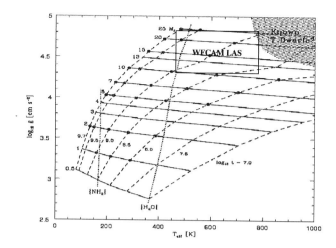

FIGURE 4. Gravity against effective temperature for low mass brown dwarfs from [4]. Solid lines show evolutionary tracks for masses indicated at left, dased lines are isochrones. The coolest dwarf currently known is 2MASS 0415 − 09 with $T_{eff} \sim 700$ K ([6, 7]).

THE WFCAM FIELD BROWN DWARF POPULATION

The Large Area Survey will detect brown dwarfs with T_{eff} as cool as 450 K as objects red in $Y - J$ and blue in $J - H$ and $J - K$. For field dwarfs aged $\sim$1–5 Gyr, evolutionary models indicate that this temperature limit corresponds to a mass limit of $\sim$10 $M_{Jupiter}$ (see Figure 4). While the Sloan and 2MASS surveys discovered the field brown dwarf population with $T_{eff} \sim$700–1400 K and mass $\sim$25–75 $M_{Jupiter}$ (e.g. [8, 9]), the LAS will detect the population with $T_{eff} \sim$450–700 K and mass $\sim$10–40 $M_{Jupiter}$. That is, while Sloan and 2MASS probed the high–mass brown dwarf field population, WFCAM will detect the low–mass population, objects with more planetary–like masses. If the mass function is approximately flat, extrapolating the Sloan discovery rate implies that we will find about 10 T+ or Y dwarfs in the first two years of the survey (2004 to 2006).

REFERENCES

1. Henry, D.M. et al. 2003, SPIE 4841, 63. Instrument Design and Performance for Optical/Infrared Ground–based Telescopes, ed. Iye & Moorwood.
2. Warren, S. 2002, SPIE 4836, 313. Survey and Other Telescope Technologies, ed. Tyson & Wolff.
3. Fan, X. et al. 2001, AJ 122, 2833.
4. Burrows, A., Sudarsky, D., & Lunine, J.I. 2003, ApJ 596, 587.
5. Saumon, D., Marley, M. S., Lodders, K., & Freedman, R. S. 2003, Proc. IAU Symposium 211, 'Brown Dwarfs', eds. E. Martín, ASP Conf. Series Vol 211, p. 345.
6. Golimowski, D. et al. 2004, AJ submitted.
7. Vrba, F. et al. 2004, AJ submitted.
8. Burgasser, A. et al. 1999, ApJ 522, L65.
9. Strauss, M. et al. 1999, ApJ 522, L61.

The Fourier-Kelvin Stellar Interferometer: A Concept for a Practical Interferometric Mission for Discovering and Investigating Extrasolar Giant Planets

D. J. Benford*, W. C. Danchi*, R. J. Allen†, D. Deming*, D. Y. Gezari*,
M. Kuchner**‡, D. T. Leisawitz*, R. Linfield§, R. Millan-Gabet¶,
J. D. Monnier‖, M. Mumma*, L. G. Mundy††, C. Noecker§, J. Rajagopal*,
S. A. Rinehart‡‡, S. Seager§§ and W. A. Traub**

*NASA / Goddard Space Flight Center
†Space Telescope Science Institute
**Harvard-Smithsonian Center for Astrophysics
‡Russell Fellow
§Ball Aerospace
¶California Institute of Technology
‖University of Michigan
††University of Maryland
‡‡NRC / Goddard Space Flight Center
§§Carnegie Institution of Washington

Abstract. The Fourier-Kelvin Stellar Interferometer (FKSI) is a mission concept for a nulling interferometer for the near-to-mid-infrared spectral region ($3 - 8\,\mu m$). FKSI is conceived as a scientific and technological precursor to TPF. The scientific emphasis of the mission is on the evolution of protostellar systems, from just after the collapse of the precursor molecular cloud core, through the formation of the disk surrounding the protostar, the formation of planets in the disk, and eventual dispersal of the disk material. FKSI will answer key questions about extrasolar planets:

- What are the characteristics of the known extrasolar giant planets?
- What are the characteristics of the extrasolar zodiacal clouds around nearby stars?
- Are there giant planets around classes of stars other than those already studied?

We present preliminary results of a detailed design study of the FKSI. Using a nulling interferometer configuration, the optical system consists of two 0.5 m telescopes on a 12.5 m boom feeding a Mach-Zender beam combiner with a fiber wavefront error reducer to produce a 0.01% null of the central starlight. With this system, planets around nearby stars can be detected and characterized using a combination of spectral and spatial resolution.

THE ROLE OF FKSI

The Fourier-Kelvin Stellar Interferometer (FKSI) is envisioned as a practical interferometric system for discovering and investigating extrasolar giant planets. Its capability is designed to enable a significant near term step in planet finding and characterization as compared to other present planet studies. FKSI supports key NASA Strategic Plan

CP713, *The Search for Other Worlds: Fourteenth Astrophysics Conference,*
edited by S. S. Holt and D. Deming
© 2004 American Institute of Physics 0-7354-0190-X/04/$22.00

objectives in the area of expolanet research, and will serve as a pathfinder for the Terrestrial Planet Finder (TPF) and Darwin mission. The key questions FKSI seeks to find answers to are:

1. What are the characteristics of the known extrasolar giant planets?
2. What are the characteristics of the extrasolar zodiacal clouds around nearby stars?
3. Are there giant planets around classes of stars other than those already studied?

To answer these questions, the FKSI mission will complete four science projects:

- Detect >25 extrasolar giant plants (based on conservative predictions)
  - Characterize their atmospheres with $\lambda / \delta \lambda = 20$ spectroscopy
  - Observe secular changes in spectrum
  - Observe orbit of the planet
  - Estimate density of planet; to determine if rocky or gaseous
  - Determine main constituents of atmospheres from broad spectral features
- Search for exo-planets around nearby stars (not planned for observations with TPF)
  - Nearby M dwarfs (>30 such stars) at distances within 10 pc
  - Nearby F, G, K giants and subgiants, luminosity classes III and IV (>50 stars) within 30 pc
- Study circumstellar material
  - Exozodiacal dust cloud measurements around nearby stars
  - Search for companion stars
  - Resolve debris disks, looking for clumpiness due to planets
- See star formation at high angular resolution
  - Taxonomy of the evolution of circumstellar disks
  - Characterize morphology, gaps, rings, etc.

There are currently more than 100 known extrasolar giant planets [1]. Because the discovery method uses only the light of the parent star, almost nothing is known observationally about the planets' atmospheres. The biggest surprise in this list is that there are many giant planets very close to their stars. Is this a bias of the radial velocity technique, or due to migration of the planet inward? FKSI will provide an important test of the bias of this observing technique.

More than 25 extrasolar giant planets are detectable with FKSI, and spectroscopy is possible for about half of these. A broad survey of nearby (K, M, class IV and III) stars is also possible. These studies will provide a firm observational foundation for the new field of exoplanetary science.

FKSI MISSION CONCEPT

FKSI is a two-element nulling interferometer (figure 1). The two telescopes, separated by 12.5 m, are precisely targeted (by small steering mirrors) on the target start. The two path lengths are accurately controlled to be precisely the same. A phase shifter /

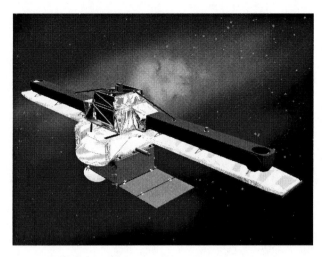

FIGURE 1. Artist's rendition of FKSI in operation.

beam combiner (via Mach-Zender interferometer) produces an output beam consisting of the nulled sum of the planet's light and the star's light. When properly oriented, the starlight is nulled by a factor of 10^{-4}, and the planet light is undimmed. Accurate stellar spectroscopy is used to subtract this residual starlight, permitting the detection of planets much fainter than the host star, and at distances less than the $\sim 0.005''$ required for a similar imaging interferometer.

FINDING PLANETS WITH FKSI

FKSI's sensitivity is sufficient to detect and characterize extrasolar giant planets within $\sim 10\,\mathrm{pc}$. FKSI is comparable (to within an order of magnitude) in sensitivity to other missions in the $\simeq 5\,\mu\mathrm{m}$ range (SIRTF, JWST), and superior to ground-based facilities. This is illustrated in the figure below, adapted from [2]. However, without a nuller, SIRTF and JWST will be hard pressed to extract the faint planet signal from the overwhelming star signal. As a result, FKSI will be a more capable planet finding instrument than other $\simeq 5\,\mu\mathrm{m}$ systems.

We have simulated the response of FKSI to a planet and nulled star as the spacecraft rotates about its axis [3]. At a single wavelength, the intensity varies as a function of this rotation angle, as shown in figure 3. Ratios of various wavelengths produce the unmodulated spectrum of the star, and the modulated spectrum of the planet. After rotating FKSI several times, a high signal-to-noise spectrum of the planet and star are obtained. The spectrum of the planet tells us about the atmospheric composition, mass, and age of the planet. The nature of the modulation pattern of the planet spectrum reveals the orbit of the planet, resolving the $M \sin i$ ambiguity present in radial velocity measurements. Watching this orbit varying over time will permit the study of the atmosphere of the planet as a function of its distance to its sun.

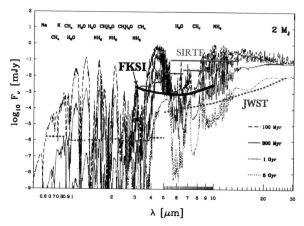

FIGURE 2. Predicted flux density of a $2M_J$ planet as a function of age, at visible through mid-IR wavelengths, at a distance of 10 pc. FKSI can detect young giant planets of this sort with ease.

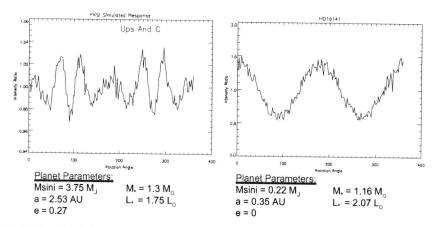

FIGURE 3. Simulation of FKSI response at a single wavelength to two simple planetary systems.

ACKNOWLEDGMENTS

We thank the FKSI engineering team at NASA/GSFC for their excellent work in producing the detailed design of the FKSI mission.

REFERENCES

1. Marcy, G.W., Butler, R.P., Fischer, D.A. and Vogt, S.S., in *Scientific Frontiers in Research on Extrasolar Planets*, ASP Conference Series, V.294, edited by D. Deming and S. Seager, pp.1-16 (2003)
2. Burrows, A., Sudarsky, D. and Lunine, J.I., *ApJ*, **596**, pp.587-596 (2003)
3. Danchi, W.C., Deming, D., Kuchner, M.J. and Seager, S., *ApJL*, **597**, pp. 57-60 (2003)

A Proposed Search for Planets in Binaries with the Navy Prototype Optical Interferometer

J. T. Armstrong*, G. C. Gilbreath*, T. A. Pauls*, R. B. Hindsley*,
D. Mozurkewich† and D. J. Hutter**

*Naval Research Laboratory, Code 7215, 4555 Overlook Ave. SW, Washington, DC 20375
†Seabrook Engineering, 9310 Dubarry Ave., Lanham, MD 20706
**US Naval Observatory, Flagstaff Station, P.O. Box 1149, Flagstaff, AZ 86002

Abstract. We are developing a search for planets via differential astrometry in binary systems, using the high-precision capabilities of the Navy Prototype Optical Interferometer. The targets include not only the FGKM dwarfs observed with radial-velocity techniques, but also stars of earlier types.

Differential astrometry in binary systems addresses three important questions: Do planets form around stars outside the FGKM spectral types surveyed by radial velocity techniques? How frequent are planets in orbits larger than those found by these techniques, that is, larger than roughly 4 AU? Can planets form in relatively close binary systems, those in with separations less than a few hundred AU? Exploring these questions will open up areas of parameter space that have barely been examined.

The basic outline is simple: use the binary components as positional references for one another and then track their relative motion. Deviations from Keplerian motion suggest the presence of a companion orbiting one of the components—although they don't tell us which. Knowing the orbit of the binary is not critical: the period of a stable planetary orbit around one component is much shorter than that of the binary orbit.

Detecting these deviations calls for a high degree of astrometric precision. Optical stellar interferometry, which has already produced dozens of spectroscopic binary orbits, offers that precision, e.g., the orbit of Mizar A with a median residual of 40 μas given by Hummel et al. [1] from Navy Prototype Optical Interferometer (NPOI)[1] data. See Armstrong et al. [2] for a description of the NPOI.

An interferometer is a spatial filter, measuring the Fourier components of an image at the spatial frequencies $\mathbf{B}/\lambda$, where $\mathbf{B}$ is the baseline vector and λ is the wavelength. The Fourier transform[2] of the image of a binary is a set of parallel stripes of maximum V^2 in the spatial frequency (u, v) domain. A baseline sampling a range of wavelength channels measures along a ray in the (u, v) plane, as sketched in Fig. 1.

---

[1] The NPOI, a collaboration between the Naval Research Laboratory (NRL) and the US Naval Observatory (USNO), in cooperation with the Lowell Observatory, with funding from the Oceanographer of the Navy and the Office of Naval Research, is located on the Lowell Observatory site on Anderson Mesa, AZ.

[2] It is convenient for dealing with photon-counting bias effects to use the squared modulus of the Fourier transform; when normalized to unity, it becomes the squared fringe contrast V^2.

CP713, *The Search for Other Worlds: Fourteenth Astrophysics Conference*,
edited by S. S. Holt and D. Deming
© 2004 American Institute of Physics 0-7354-0190-X/04/$22.00

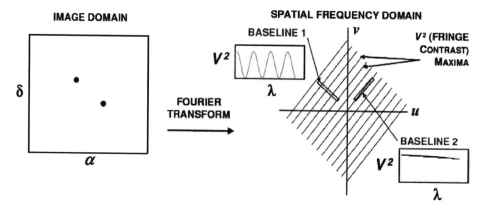

FIGURE 1. Schematic view of an interferometer's response to a binary. The Fourier transform of the binary image is a set of parallel maxima of V^2 in the spatial frequency domain, with the spacing of the maxima inversely proportional to the separation of the binary components. An interferometer baseline that samples a range of wavelength channels sees varying numbers of these maxima according to its orientation. The two example baselines shown here correspond roughly to the test data shown in Fig. 3.

The current configuration of the NPOI, which we will use for this program, includes two moveable imaging siderostats and four fixed astrometric siderostats in an array with $B_{max} = 37$ m. The siderostats are 50 cm in diameter, but the feed system reduces the usable aperture to 12 cm. The wavelength coverage is $\lambda\lambda 570 - 850$ nm in 16 channels. Future improvements include 35 cm beam compressing telescopes for the astrometric array (now nearing deployment), four additional imaging siderostats in an array with $B_{max} = 437$ m, and wavelength coverage to $\lambda 450$ nm in 32 channels.

In order to determine the available parameter space, a number of factors must be considered. The first is the key equation in astrometric searches for planets:

$$a_{reflex} = 20 \ \mu as \left(\frac{D}{50 \ pc}\right)^{-1} \left(\frac{m_{pl}}{m_J}\right) \left(\frac{M_*}{M_\odot}\right)^{-2/3} \left(\frac{P_{pl}}{1 \ yr}\right)^{2/3} , \tag{1}$$

where a_{reflex} is the semimajor axis of the reflex motion of the parent star, D is the distance, m_{pl} and m_J are the masses of the planet and of Jupiter, M_* and $M_\odot$ are the masses of the star and of the sun, and P_{pl} is the period of the planetary orbit.

Two important NPOI system constraints are the sensitivity limit (currently $\approx 6^m$; at this flux, we see ~ 10 photons per 2 ms sample per channel at $\lambda 800$ nm, where our sensitivity peaks) and the system visibility V^2_{sys}, the visibility of an unresolved source, which is $\approx 0.1 - 0.2$ at $\lambda 800$ nm. For a perfect interferometer, $V^2_{sys} = 1$.

The precision σ_ρ of the separation measurement is given by

$$\sigma_\rho \approx core \ of \ PSF \ \frac{F_2}{F_1} \frac{1}{SNR_{V^2}} \frac{1}{\sqrt{M}} \approx \frac{\lambda}{B} \frac{F_2}{F_1} \frac{4}{NV^2} \frac{1}{\sqrt{M}} \approx 3 \ \mu as , \tag{2}$$

TABLE 1. Rough constraints on an NPOI search for planets in binary systems

Constraint	Value	Comment
Magnitude	$V < 6$	Current NPOI sensitivity
Magnitude difference	$\Delta m < 3$	Keeps V^2 variations detectable
Spectral type	No constraint	No dependence on spectral lines
Maximum separation	$\rho_{bin} < 500$ mas	Component images overlap
Separation precision	$\sigma_{\rho_{bin}} > 3\ \mu$as	Photon noise limit for 1000 s
Planetary orbital size	$a_{pl} < 7a_{bin}$	Stable orbit
Maximum reflex motion	$a_{reflex} < 400\ \mu$as	Keeps a_{bin} within limits

where $F_{1,2}$ are the fluxes from the primary and secondary, N is the number of photons per sample per spectrometer channel, and M is the number of samples. The 3 μas estimate is for a 5^m star with a $F_2/F_1 = 0.5$ and 1000 s of observation with one spectral channel of the current NPOI array (PSF $\approx$ 1.5 mas). When the flux ratio falls below ≈ 15 ($\Delta m \approx 3$), the V^2 variations due to the binary become too small to detect reliably.

A less familiar constraint is that the binary separation ρ_{bin} is limited by the aperture size A, in the sense that larger apertures lead to smaller maximum separations. The images of the components must overlap to produce fringes, so ρ_{bin} must be significantly less than $1.22\lambda/A$, or ≈ 1.3 arcsec for the NPOI. We will take $\rho_{bin} < 500$ mas.

We must also take into account the stability of the planetary orbit. The criterion we have chosen here is an extension of a recent analysis by David et al. [3]: for stability on Gyr time-scales, the periastron of the binary must be at least 7 times the apastron of the planet, i.e., $a_{bin}(1 - e_{bin}) > 7a_{pl}(1 + e_{pl})$, where a_{pl} and a_{bin} are the semimajor axes of the planetary and binary orbits, and e_{pl} and e_{bin} are the orbital eccentricities.

The size of the *maximum* reflex motion we can observe is given very roughly by the maximum binary separation, the stability criterion, and the mass ratio $m_{pl}/M*$. With $a_{pl,max} < a_{bin,max}/7$, and assuming for the moment that $a_{bin,max} \approx \rho_{bin,max}$, we have $a_{reflex,max} \approx 140\ \mu$as for $m_{pl} = 2m_J$ and $M* = M_\odot$. This constraint is very fuzzy because many binaries have orbital orientations that make ρ_{bin} significantly smaller than a_{bin} over considerable parts of their orbits. We will assume that we gain about a factor of three for a significant fraction of our targets, and set $a_{reflex,max} = 400\ \mu$as.

As a result of these considerations, the binary systems accessible to this technique are a complicated function of the stellar and planetary masses, their orbital periods, and the distance. See Fig. 2 for one illustration of the constraint, and Table 1 for a summary.

We have taken trial data to start verifying these ideas. Figure 3 shows data taken during one scan with the NPOI on 4 June 2003. We observed β CrB, a 10-year spectroscopic and speckle binary (Sp = F0p, $m_I = 3.45, \Delta m = 1.5$) with a separation of 194 mas on that date. The left panel is from a baseline roughly perpendicular to the V^2 maxima, while the right panel shows data from a baseline nearly parallel to the maxima (compare with Fig. 1). The formal precision from 36 thirty-second scans over 3.7 hours, without taking systematics into account, is $\sim 1\ \mu$as.

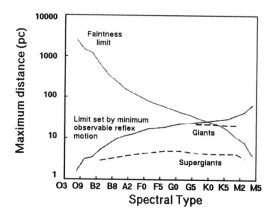

FIGURE 2. Maximum distance for observing planets in binaries with the NPOI as a function of spectral type. The planetary mass and orbital period are $2m_J$ and 5 yr; the assumed minimum observable a_{reflex} is 5 μas; the luminosity limit is 5^m at I. The solid curves are for main-sequence stars. Reflex motions below both these lines are observable, while motions below only one of the lines are unobservable either because the parent star is so massive that the reflex motion is too small (left) or because the parent star is too faint (right). The dashed curves shows the limit set by reflex motion for giants and supergiants.

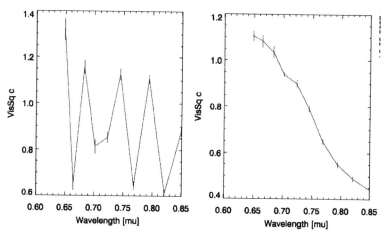

FIGURE 3. Test observations of β CrB taken with the NPOI on 4 June 2003. β CrB is a 10-year binary for which $(\rho, \theta) \sim (194\ \text{mas}, 165.2°)$ on that date. The two panels correspond roughly with the two panels in Fig. 1.

REFERENCES

1. Hummel, C. A., et al., *Astron. J.*, **116**, pp. 2536–2548 (1998).
2. Armstrong, J. T., et al., *Astrophys. J.*, **496**, pp. 550–571 (1998).
3. David, E.-M., Quintana, E. V., Fatuzzo, M., & Adams, F. C., *Pub. Astron. Soc. Pacific*, **115**, pp. 825–836 (2003).

Wide-Field Interferometric Imaging: A Powerful Tool for the Terrestrial Planet Finder

Stephen A. Rinehart*[†], David T. Leisawitz*, Bradley J. Frey*,
Marc Kuchner**, Douglas B. Leviton*, Luke W. Lobsinger*,
Anthony J. Martino* and Lee G. Mundy[‡]

*NASA's GSFC, Greenbelt MD 20771
[†]National Research Council Research Associate
**Princeton University, Peyton Hall, Princeton NJ 08544
[‡]University of Maryland, College Park, MD 20742

Abstract. We present a discussion of the design and construction of the Wide-Field Imaging Interferometry Testbed (WIIT) at the Goddard Space Flight Center. The testbed is motivated by a need to develop techniques for wide-field interferometric imaging in the optical/IR for future space missions such as TPF. Theoretically, using multipixel detectors for spatial multiplexing/mosaicing allows this, but there are a number of practical questions which WIIT will address. We present some initial data acquired with this system and demonstrate its capability to produce wide-field interferometry data.

INTRODUCTION

Quoting the Decadal Report, "To ensure a broad science return from TPF, the committee recommends that, in planning the mission, comparable weight be given to the two broad science goals: studying planetary systems and studying the structure of astronomical sources at infrared wavelengths." [1]. The Wide-Field Imaging Interferometry Testbed (WIIT) was originally conceived to develop techniques and tools necessary for wide field-of-view interferometric imaging with Far-IR/Sub-mm missions such as SPECS (Submillimeter Probe of the Evolution of Cosmic Structure). For extragalactic surveys and studies of star formation regions, small fields of view (FOVs) inherent to Michelson stellar interferometers are insufficient. Similarly, an interferometric TPF requires wide FOV techniques in order to image star formation regions, debris disks, and thus to achieve other NASA Origins objectives.

In Figure 1, we show a model image of ε Eridani. This mid-IR image show the debris disk around the star, and from images such as this the presence and location of a planet within the disk can be inferred. In order to image the disk using an interferometric TPF (with 4-m telescopes) would require over 400 separate pointings, while by using the techniques being explored with WIIT, the FOV could be enlarged by an areal factor of over 10000, allowing observation of such a debris disk in a single pointing. These techniques would have broad application to a variety of astrophysical themes [2].

CP713, The Search for Other Worlds: Fourteenth Astrophysics Conference,
edited by S. S. Holt and D. Deming
© 2004 American Institute of Physics 0-7354-0190-X/04/$22.00

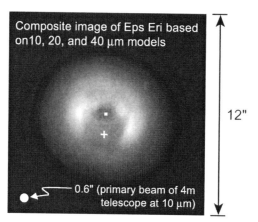

Composite image of Eps Eri based on 10, 20, and 40 μm models

12"

0.6" (primary beam of 4m telescope at 10 μm)

FIGURE 1. A model of the ε Eri debris disk. The star is marked with the white square, and the location of a planet within the disk is marked with the + sign.

TESTBED CHARACTERIZATION

Since the first acquisition of white light fringes with WIIT in August 2001, we have striven to improve the behavior of the testbed and the quality of the data [3][4]. We have compiled a complete error budget for WIIT, which allows us to identify the major sources of both visibility loss and uncertainty, and to recognize where further improvements could be made, if warranted [5]. WIIT works at optical wavelengths, and the dominant source of visibility loss occurs due to the imperfect optics used in the system. This leads to a 7% loss in visibility, and we can calibrate the data in post-processing to correct for the loss.

Likewise, we have identified the two largest sources of *uncertainty* in the observed visibility: the distortion of the wavefront due to turbulent effects (a roughly 5% effect) and the number-counting statistics intrinsic to the detector (a 3% effect). We plan to address the latter by acquisition of a camera with superior well-depth and noise characteristics. The former can be addressed through improved environmental isolation, including the use of hard plastic tubes surrounding the science beams and/or operation of WIIT within the Diffraction Grating Evaluation Facility (DGEF) at Goddard Space Flight Center [6].

EARLY RESULTS

In Figure 2, we show a raw interferogram obtained by WIIT of a pinhole source with a broadband spectrum. Adjacent to the interferogram is the spectral reconstruction from the interferogram. These early data show the production of a reasonably accurate spectrum, and this quality is improving as we finish system upgrades.

Also to verify system performance, we obtained interferograms at multiple baselines. We observed the visibility to decrease as the baseline length was increased, as expected

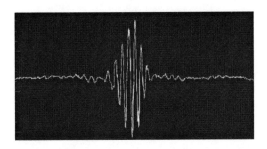
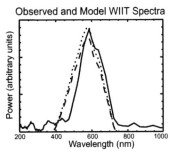

Observed and Model WIIT Spectra

Power (arbitrary units)

200   400   600   800   1000
Wavelength (nm)

FIGURE 2. On the left is a white light interferogram as captured by WIIT. On the right is the reconstructed white light spectrum. The solid line in this plot is the FFT of the interferogram. The dotted line is the independently measured source spectrum times the spectral response of WIIT. The dashed line is the predicted FFT, allowing for the noise and visibility loss characteristics of WIIT.

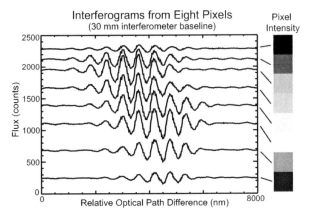
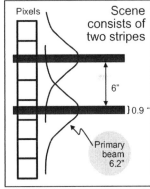

Interferograms from Eight Pixels
(30 mm interferometer baseline)

Pixel Intensity

Pixels

Scene consists of two stripes

6"

0.9"

Primary beam 6.2"

Flux (counts)

Relative Optical Path Difference (nm)

FIGURE 3. On the left is a set of 8 interferograms from 8 different pixels on the detector, corresponding roughly to the schematic diagram on the right. The separation of the interferograms formed from the two stripes seen by comparing the upper and lower interferograms, and in between the fringe patterns are blended.

for a resolved uniform disk. The observations were in agreement with theoretical predictions to within the estimated uncertainty. This figure is not shown here due to space limitations, but will be presented in an upcoming paper [7].

A major goal for WIIT is to demonstrate that spatial multiplexing using a detector array can be used to provide a wide field-of-view. Figure 3 shows the first demonstration of this technique, in a very basic form. The source consists of two stripes separated by a small (resolvable) distance. These two sources fall onto the detector 2 pixels apart. By looking at the interferograms across 8 pixels, we see two blended interferograms and the expected shift in relative zero path difference (ZPD) point corresponding to spatial information in the scene.

CONCLUSIONS AND FUTURE

WIIT has been well-characterized, and tests so far have indicated good overall performance. The data are consistent with models and predictions, and we have used the data to provide the first demonstration of spatial multiplexing using a detector array for wide field-of-view interferometry. For comparison, in conventional Michelson interferometry (pupil plane beam combination), a single pixel detector is used to measure an interferogram. Final improvements to WIIT are in progress at the moment. Over the next few months, we will be acquiring additional data to start the development and testing of 1-D image reconstruction algorithms for wide-field imaging. As these techniques develop, we will continue by working with 2 dimensional data using full u-v plane coverage, and we will experiment with spatially and spectrally complex scenes, not unlike the one shown in Figure 1.

ACKNOWLEDGMENTS

The authors would like to thank Sean Moran, whose thesis work, supervised by M. Kuchner, produced the models of ε Eri. S. Rinehart is supported by NASA as a National Research Council Resident Research Associate. The authors would also like to thank members of the WIIT Science and Technical Advisory Group (STAG) for continued advice and assistance. This material is based upon work supported by NASA under the Research Opportunities in Space Science program (APRA03-0000-0037).

REFERENCES

1. *Astronomy and Astrophysics in the New Millenium*, National Academies Press, Washington, DC, 2001.
2. Leisawitz, D. T., Frey, B. J., Leviton, D. B., Martino, A. J., Maynard, W. L., Mundy, L. G., Rinehart, S. A., Teng, S. H., and Zhang, X. "Wide-field imaging interferometry testbed I: purpose, testbed design, data, and synthesis algorithms," in *Interferometry in Space*, edited by M. Shao, Proc. SPIE 4852, SPIE, Bellingham WA, 2003, pp. 255.
3. Rinehart, S. A., Frey, B. J., Leisawitz, D. T., Leviton, D. B., Martino, A. J., Maynard, W. L., Mundy, L. G., Teng, S. H., and Zhang, X. "Wide-field imaging interferometry testbed II: Implementation, Performance, and Plans," in *Interferometry in Space*, edited by M. Shao, Proc. SPIE 4852, SPIE, Bellingham WA, 2003, pp. 674.
4. Leviton, D. B., Frey, B. J., Leisawitz, D. T., Martino, A. J., Maynard, W. L., Mundy, L. G., Rinehart, S. A., Teng, S. H., and Zhang, X. "Wide-field imaging interferometry testbed III: Metrology Subsystem," in *Interferometry in Space*, edited by M. Shao, Proc. SPIE 4852, SPIE, Bellingham WA, 2003, pp. 827.
5. Rinehart, S. A., Frey, B. J., Leisawitz, D. T., Leviton, D. B., Lobsinger, L. W., Martino, A. J., and Mundy, L. G. "WIIT: A Spatially Multiplexed Michelson Stellar Interferometer", In preparation, 2004.
6. Osantowski, J. F. and Leviton, D. B. "The GSFC Diffraction Grating Evaluation Facility – an Overview", in *X-Ray Instrumentation in Astronomy II*, edited by L. Golub, Proc. SPIE 982, 342.
7. Leisawitz, D. T., Frey, B. J., Leviton, D. B., Lobsinger, L. W., Martino, A. J., Mundy, L. G., and Rinehart, S. A. "WIIT: First Results and Demonstration of Wide Field-of-View Imaging", In preparation, 2004.

Electromagnetic Signals from Planetary Collisions

Bing Zhang and Steinn Sigurdsson

Department of Astronomy & Astrophysics, Penn State University, University Park, PA 16802

Abstract. We investigate the electromagnetic signals accompanied with planetary collisions and their event rate, and explore the possibility of directly detecting such events. A typical Earth–Jupiter collision would give rise to a prompt EUV-soft-X-ray flash lasting for hours and a bright IR afterglow lasting for thousands of years. With the current and forthcoming observational technology and facilities, some of these collisional flashes or the post-collision remnants could be discovered.

INTRODUCTION

More than 100 extra-solar planets have been detected. At the same time, our understanding of astrophysical phenomena has been greatly boosted through studying cataclysmic transient events such as supernovae, X-ray bursts, and gamma-ray bursts. Another type of electromagnetic transient events that arises from collisions of extra-solar planets has been long predicted in the context of planet formation[1]. Recently we discussed the possible electromagnetic signals accompanying such collisional events[2]. Discovering such events would undoubtedly have profound implications of understanding the formation rate, dynamical instability, as well as internal composition and structure of extrasolar planets.

ELECTROMAGNETIC SIGNALS: THREE STAGES

Generally one can categorize the collision-induced signals into three stages, defined by three characteristic time scales. Below we will take a head-on collision between a Jovian planet and an Earth-size planet as an example. A schematic lightcurve is presented in Figure 1.

Stage 1: Prompt EUV-Soft-X-ray Flash

Assuming zero velocity at infinity, the total energy of the collision is about 6×10^{40} erg. Upon impact, a reverse shock propagates into the impactor (i.e. the Earth-size planet), and the shock crossing time scale is ~ 10 minutes. This defines the rising timescale of the collision-induced electromagnetic flash. The energy deposition rate during this stage is $\sim 10^{38}$ erg s^{-1}, much greater than Jupiter's Eddington luminosity.

CP713, *The Search for Other Worlds: Fourteenth Astrophysics Conference*,
edited by S. S. Holt and D. Deming
© 2004 American Institute of Physics 0-7354-0190-X/04/$22.00

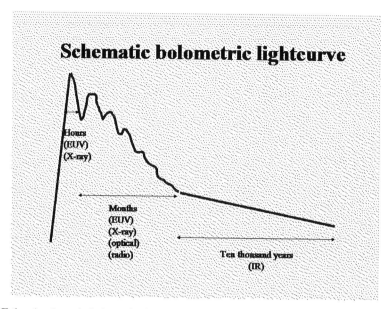

FIGURE 1. A schematic bolometric lightcurve for a Jupiter-Earth collision event. Three stages are highlighted: (1) a prompt EUV-soft-X-ray flash, characterized by a sharp rise and milder decay lasting for hours; (2) a spin-modulated decaying lightcurve lasting for about a month, which could be visible in the EUV, X-ray, optical and radio bands; (3) a long-term IR warm afterglow lasting for thousands of years.

A likely picture is that the prompt heat generated upon collision would dissociate the molecules and ionize the atoms within a short period of time, and the emission quickly becomes Eddington-limited. After the peak, the lightcurve decays mildly as the deceleration of the impactor is still going on inside the Jovian planet. The timescale of the prompt flash can be estimated to be of order of 10 times of the rising time, i.e.,

$$\tau_1 \sim 2 \text{ hr.} \tag{1}$$

The bolometric luminosity is near Eddington, i.e. $L_{pk} \sim 5 \times 10^{34}$ erg s^{-1}, with a thermal temperature $T_{pk} \sim 1.1 \times 10^5$ K emitting from a hot spot with a radius comparable of the Earth radius. The peak flux is $F_\nu(pk) \sim 60$ μJy $(D/10\text{kpc})^{-2}$, peaking in the EUV band. A non-thermal tail due to Comptonization would extend into the soft-X-ray band, detectable through out the Galaxy if the neutral hydrogen absorption is not important, and is even visible from some nearby galaxies. The prompt flash greatly increases the planet-to-star flux ratio, making them detectable in the optical band through photometry monitoring [$f(\text{U,pk}) \sim 0.2$, $f(\text{V,pk}) \sim 0.02$, and $f(\text{I,pk}) \sim 0.008$, where $f(\nu) \equiv F(\nu,\text{planet})/F(\nu,\text{star})$]. The total energy radiated during this prompt phase is a tiny fraction ($\sim 0.5\%$) of the total energy deposited. The majority of energy is stored as latent heat and radiated over a much longer time scale.

Stage 2: Spin-Modulated Decaying Phase

After the prompt phase ends (i.e. the impactor is stopped inside the giant planet), the luminosity steadily drops. The heat deposited deep inside the giant planet would excite a vigorous convective flow. The area of the hot spot gradually gets larger and larger until the whole surface reaches the same temperature. The time scale for retaining a hot spot could be estimated as

$$\tau_2 \sim 1 \text{ month}. \tag{2}$$

During this time, an observer would see some quasi-periodic signal due to the modulation of the planet spin (with a period $\lesssim 1$ day), as the hot spot enters and leaves the field of view. The modulation pattern could be visible in EUV-X-ray, or in optical through photometric monitoring, or in radio. A radio flare is expected during the vigorous convective epoch due to the enhanced dynamo activity inside the giant planet. The periodic pattern would be gradually smeared out as the hot spot boundary gradually increases.

Stage 3: Long-Term IR Afterglow

After the surface temperature becomes uniform, the giant planet keeps cooling, radiating away the majority of the energy deposited during the impact. This time scale is much longer, typically

$$\tau_3 \sim 10^3 - 10^4 \text{ yr}. \tag{3}$$

The channel of emission is mainly in IR[3]. During this epoch, the planet-to-star flux ratio in the IR band is very high. For a G2 host star, the typical I- and K-band flux contrasts are $f(\mathrm{I},\mathrm{ag}) \sim 2.6 \times 10^{-4}$ and $f(\mathrm{K},\mathrm{ag}) \sim 1.7 \times 10^{-3}$. This is favorable to be detected in nearby star forming regions[3].

DETECTABILITY

Numerical simulations[4, 5] suggest that collisions of the type we are discussing are plausible. The basic picture is that there are secular perturbations of the inner planets, which over time scales comparable to the age of the system lead to large changes in eccentricity and semi-major axis for one or more planets, leading to a large probability of collision.

The event rate can be roughly estimated. Assuming that on average a solar-system like our own has 5 collisions over its life time, and that essentially every star harbors a planetary system, the collision event rate in our galaxy would be about $5 \times 10^{11}/(10^{10} \text{ yr}) \sim 50/\text{yr}$.

Considering an ensemble of stars with an average age $\bar{t}$, in order to detect one event with duration τ after a continuous observation time of t_{obs}, the critical number of stars in this ensemble that have to be searched is

$$N_* = (f_p \bar{N}_c)^{-1} \frac{\bar{t}}{\max(\tau, t_{obs})}, \tag{4}$$

311

where $f_p \sim 100\%$ is the fraction of the stars in the ensemble that have planets, and $\bar{N}_c \sim 5$ is the average total number of collisions during the lifetime of a typical star in the ensemble. One can also define a characteristic flux $F_{v,c}$ of the collisional events. Given a number density n_* of the stars with the average age $\bar{t}$, one can estimate a critical distance one has to search in order to find one collisional event, i.e., $D_c \sim (N_*/n_*)^{1/3}$. The typical flux can be then estimated with D_c. Although N_* is sensitive to the average age $\bar{t}$ of the ensemble of stars being investigated, D_c and $F_{v,c}$ are essentially independent of the ensemble adopted, given a constant birth rate of stars. According to such an estimate, the *Extreme Ultraviolet Explorer (EUVE)* all-sky survey[6] may have recorded ~ 10 such events, with the brightest one having a count rate of about $(170 - 680)$ counts ks^{-1} at 100 angstrom. The predicted flux level is also well above the sensitivity of soft X-ray detectors, such as *ROSAT*, *Chandra*, and *XMM-Newton*, so that there might be such events recorded in their archival data as well.

SEARCHING STRATEGY

A future dedicated wide field detector sensitive to (50-200) angstrom would be able to detect 10's of EUV-soft-X-ray flares per year due to planetary collisions. Photometrically monitoring a huge number of stars over a long period of time by missions such as *GAIA*[7] (or other synoptic all-sky surveys) for several years should lead to detections of several collisional events. If an EUV-soft-X-ray flare is detected, a search of its optical and radio counterpart (like catching afterglows in gamma-ray burst study) is desirable. A planetary-rotation-period modulated fading signal would be an important clue. Doppler radial velocity measurements and IR monitoring may be performed later for the collision candidate to verify the existence of the planet(s). Each collisional event would leave a bright remnant glowing in IR. The duration of the afterglow is thousands of years. Such afterglows could be directly searched in the nearby star forming regions, and the planets could be directly imaged.

This work is supported by NASA NAG5-13286, the NSF grant PHY-0203046, the Center for Gravitational Wave Physics (under NSF PHY 01-14375), and the Penn State Astrobiology Research Center.

REFERENCES

1. Wetherill, G. W. Ann. Rev. Earth Planet. Sci., **18**, 205 (1990)
2. Zhang, B. & Sigurdsson, S. S. ApJ, **596**, L95 (2003)
3. Stern, S. A. AJ, **108**, 2312 (1994)
4. Chambers, J. E., Wetherill, G. W., & Boss, A. P. Icarus, **119**, 261 (1996)
5. Ford, E. B., Havlickova, M. & Rasio, F. A. Icarus, **150**, 303 (2001)
6. McDonald, K. et al. AJ, **108**, 1843 (1994)
7. Perryman, M. A. C., et al., A&A, **369**, 339 (2001)

LIST OF ATTENDEES

Name	Affiliation	Email
Appenzeller, Tim	National Geographic	Tappenzeller@usnews.com
Aspin, Colin	Gemini Observatory	caa@gemini.edu
Balbus, Steven	University of Virginia	sb@virginia.edu
Barnes, Jason W.	University of Arizona	jbarnes@c3po.barnesos.net
Barranco, Joseph	Univ. of California-Berkeley	jbarranco@astro.berkeley.edu
Beichman, Charles	Jet Propulsion Lab	Charles.A.Beichman@jpl.nasa.gov
Benford, Dominic	NASA/GSFC	Dominic.J.Benford@nasa.gov
Bennett, David	Univ. of Notre Dame	bennett@nd.edu
Bennett, Chuck	NASA/GSFC	charles.l.bennett@nasa.gov
Bhatia, Anand	NASA/GSFC	bhatia@stars.gsfc.nasa.gov
Bland, George		bland44@mchsi.com
Boldt, Elihu	NASA/GSFC	boldt@lheavx.gsfc.nasa.gov
Bond, Howard E.	STScI	bond@stsci.edu
Bowers, Charles W.	NASA/GSFC	bowers@band2.gsfc.nasa.gov
Brittain, Sean	Univ. of Notre Dame	sbrittai@nd.edu
Burrows, Adam	Steward Observatory	burrows@as.arizona.edu
Carpenter, Kenneth	NASA/GSFC	kgc@stargate.gsfc.nasa.gov
Chambers, John	NASA/Ames	john@mycenae.arc.nasa.gov
Charbonneau, Dave	Caltech	dc@caltech.edu
Cheung, Cynthia	NASA/GSFC	cynthia.y.cheung@nasa.gov

Cho, James	Carnegie Institution Of Washington	jcho@dtm.ciw.edu
Clampin, Mark	NASA/GSFC	mark.clampin@nasa.gov
Crannell, Carol Jo	NASA/GSFC	crannell@gsfc.nasa.gov
Crossley, Jared	New Mexico Tech & Carnegie Institution of Washington	jaredc@nmt.edu
Cuk, Matija	Cornell University	cuk@astro.cornell.edu
Deming, Drake	NASA/GSFC	drake.deming@nasa.gsfc.gov
Dent, Bill	UKATC	dent@roe.ac.uk
Douthitt, Bill	National Geographic	bdouthit@ngs.org
Drachman, Richard	NASA/GSFC	Richard.J.Drachman@nasa.gov
Dwek, Eliahu	NASA/GSFC	eli.dwek@gsfc.nasa.gov
Fahey, Richard	NASA/GSFC	Dick.Fahey@nasa.gov
Fischer, Debra	Univ. of California -Berkeley	fischer@astro.berkeley.edu
Ford, Eric	Princeton University	eford@astro.princeton.edu
Ford, Holland	The Johns Hopkins U.	ford@pha.jhu.edu
Fortney, Jonathan	University of Arizona	jfortney@lpl.arizona.edu
Freed, Melanie	Steward Observatory	freed@as.arizona.edu
Gaudi, Scott	Harvard-Smithsonian-CfA	sgaudi@cfa.harvard.edu
Gaume, Ralph	US Naval Observatory	rgaume@usno.navy.mil
Gehrels, Neil	NASA/GSFC	
Glenar, Dave	NASA/GSFC	dave.glenar@gsfc.nasa.gov
Goldreich, Peter	IAS & Caltech	pmg@sns.ias.edu

Golimowski, David	The Johns Hopkins U.	dag@pha.jhu.edu
Grady, Carol A.	Eureka Sci. & GSFC	cgrady@echelle.gsfc.nasa.gov
Greaves, Jane	UK Astronomy Technology Center	jsg@roe.ac.uk
Greenhouse, Matthew	NASA/GSFC	matt.greenhouse@nasa.gov
Haghighipour, Nadir	Carnegie Institution Of Washington	nader@dtm.ciw.edu
Haiman, Zoltan	Columbia University	zoltan@astro.columbia.edu
Hamilton, Doug	University of Maryland	Hamilton@astro.umd.edu
Harrington, Joe	Cornell University	jh@oobleck.astro.cornell.edu
Herbst, Tom	MPI fur Astronomie	herbst@mpia.de
Herbst, Bill	Wesleyan University	wherbst@wesleyan.edu
Holman, Matt	Harvard-Smithsonian-CfA	mholman@cfa.harvard.edu
Holt, Steve	Olin College	Stephen.Holt@olin.edu
Ipatov, Sergei	George Mason University	siipatov@hotmail.com
Jang-Condell, Hannah	Harvard-Smithsonian-CfA	hjang@cfa.harvard.edu
Jennings, Don	NASA/GSFC	
Johnston, Ken J.	US Naval Observatory	kjj@astro.usno.navy.mil
Jones, Frank	NASA/GSFC	
Jones, Hugh	Liverpool John Moores U.	hraj@astro.livjm.ac.uk
Kaplan, George	US Naval Observatory	gkaplan@usno.navy.mil
Kaye, Tom	Spectrashift.com	tom@airgun.com
Kazanas, Demos	NASA/GSFC	Demos.Kazanas-1@nasa.gov
Kerr, Dick	Science	rkerr@aaas.org

Kimble, Randy A.	NASA/GSFC	Randy.A.Kimble@nasa.gov
Kochte, Mark	STScI	kochte@stsci.edu
Koller, Josef	Rice University	josef@rice.edu
Kotredes, Lewis	Caltech	ltk@astro.caltech.edu
Kuchner, Marc	Princeton University	mkuchner@astro.princeton.edu
Leckrone, David	NASA/GSFC	dleckrone@hst.nasa.gov
Lee, Brian	University of Toronto	blee@astro.utoronto.ca
Leggett, Sandy	UKIRT JAC Hawaii	s.leggett@jach.hawaii.edu
Leinhardt, Zoe Malka	University of Maryland	zoe@astro.umd.edu
Leisawitz, David	NASA/GSFC	David.T.Leisawitz@nasa.gov
Lestrade, Jean-Francois	Observatoire de Paris/LERMA	jean-francois.lestrade@obspm.fr
Leventhal, Marv	University of Maryland	ml@astro.umd.edu
Lewin, Walter H. G.	MIT	lewin@space.mit.edu
Liu, Michael	University of Hawaii	mliu@ifa.hawaii.edu
Lufkin, Graeme	Univ. of Washington	gwl@u.washington.edu
Lystad, Verene	Northwestern University	vela@northwestern.edu
Maguire, William	NASA/GSFC	u3wcm@lepvax.gsfc.nasa.gov
Mallen-Ornelas, Gabriela	Harvard-Smithsonian-CfA	gmalleno@cfa.harvard.edu
Maran, Stephen P.	NASA/GSFC	stephen.p.maran@nasa.gov
Martin, Eduardo	Institute for Astronomy/ U. of Hawaii	ege@ifa.hawaii.edu
Mather, John C.	NASA/GSFC	john.c.mather@nasa.gov

Matsumura, Soko	McMaster University	soko@physics.mcmaster.ca
Matsuyama, Isamu	University of Toronto	isamu@astro.utoronto.ca
Mazzuca, Lisa	NASA/GSFC	lisa.m.mazzuca@nasa.gov
McCarthy, Chris	Carnegie Institution of Washington	chris@dtm.ciw.edu
McKee, Maggie	Astronomy Magazine	mmckee@astronomy.com
McNaughton, Rosemary	mcnaughr@dickinson.edu	Dickinson College
Menard, Francois	Lab. d'Astrophysique de Grenoble	menard@obs.ujf-grenoble.fr
Mitchell, John	NASA/GSFC	
Mudryk, Lawrence	University of Toronto	mudryk@astro.utoronto.ca
Mugrauer, Markus	AIU JENA	markus@astro.uni-jena.de
Mumma, Michael	NASA/GSFC	mmumma@lepvax.gsfc.nasa.gov
Mundy, Lee	University of Maryland	lgm@astro.umd.edu
Noll, Keith	STScI	noll@stsci.edu
Norman, Colin A.	The Johns Hopkins U.	norman@stsci.edu
O'Donovan, Francis T.	Caltech	ftod@astro.caltech.edu
Oegerle, William R.	NASA/GSFC	oegerle@s2.gsfc.nasa.gov
Oliversen, Ronald	NASA/GSFC	Ronald.J.Oliversen@nasa.gov
Olling, Rob P.	US Naval Observatory & USRA	olling@usno.navy.mil
Ormes, Jonathan	NASA/GSFC	Jonathan.F.Ormes.1@gsfc.nasa.gov
Pepper, Joshua	Ohio State University	pepper.16@osu.edu
Perryman, Michael	European Space Agency	mperryma@rssd.esa.int
Petre, Robert	NASA/GSFC	rob@hatrack.gsfc.nasa.gov

Porter, F. Scott	NASA/GSFC	
Rasio, Fred	Northwestern University	rasio@northwestern.edu
Rettig, Terrence	Univ. of Notre Dame	trettig@nd.edu
Reuter, Dennis	NASA/GSFC	
Richardson, Jeremy	USRA/NASA/GSFC	lee.richardson@colorado.edu
Richardson, Derek	University of Maryland	dcr@astro.umd.edu
Roberge, Aki	Carnegie Institution of Washington	akir@dtm.ciw.edu
Rojo, Patricio	Cornell University	pmr27@cornell.edu
Romanova, Marina	Cornell University	romanova@astro.cornell.edu
Rose, William K.	University of Maryland	wrose@astro.umd.edu
Rosenbaum, Doris		DRTeplitz@AOL.com
Sahu, Kailash	STScI	ksahu@stsci.edu
Schultz, Al	CSC/STScI	schultz@stsci.edu
Seager, Sara	Carnegie Institution of Washington	seager@dtm.ciw.edu
Serlemitsos, Peter	NASA/GSFC	pjs@astron.gsfc.nasa.gov
Shkolnik, Evgenya	U. of British Columbia	shkolnik@physics.ubc.ca
Silverberg, Robert F.	NASA/GSFC	Robert.Silverberg@nasa.gov
Simons, Bright	University of Durham	baronsimons@yahoo.co.uk
Sonneborn, George	NASA/GSFC	george.sonneborn-1@nasa.gov
Soummer, Remi	University of Nice	soummer@unice.fr
Spergel, David	Princeton University	dns@astro.princeton.edu
Stevenson, Dave	Caltech	djs@gps.caltech.edu

Streitmatter, Robert	NASA/GSFC	
Sudarsky, David	Steward Observatory	sudarsky@as.arizona.edu
Temkin, Aaron	NASA/GSFC	temkin@stars.gsfc.nasa.gov
Teplitz, Vigdor	NASA/GSFC	Teplitz@mail.physics.smu.edu
Terebey, Susan	Cal State-LA	sterebe@calstatela.edu
Terquem, Caroline	Inst. d'Astrophysique de Paris	terquem@iap.fr
Thiessen, Mark	National Geographic	mthiesse@ngs.org
Torres, Guillermo	Harvard-Smithsonian-CfA	gtorres@cfa.harvard.edu
Trasco, John	University of Maryland	jtrasco@astro.umd.edu
Tremaine, Scott	Princeton University	tremaine@astro.princeton.edu
Trimble, Virginia	University of Maryland & U. of Calif.-Davis	vtrimble@astro.umd.edu
Underwood, David R.	Open University (UK)	d.r.underwood@open.ac.uk
Volgenau, Nikolaus	University of Maryland	volgenau@astro.umd.edu
von Braun, Kaspar	Carnegie Institution of Washington	kaspar@dtm.ciw.edu
Vondrak, Richard	NASA/GSFC	vondrak@gsfc.nasa.gov
Wadsley, James	McMaster University	wadsley@mcmaster.ca
Weinberger, Alycia	Carnegie Institution of Washington	weinberger@dtm.ciw.edu
White, Nicholas	NASA/GSFC	nwhite@lheapop.gsfc.nasa.gov
Winn, Josh	Harvard-Smithsonian-CfA	jwinn@cfa.harvard.edu
Wollack, Edward J.	NASA/GSFC	Edward.J.Wollack@nasa.gov

Wood, John	Smithsonian Astrophysical Observatory	jwood@cfa.harvard.edu
Woodgate, Bruce E.	NASA/GSFC	Bruce.E.Woodgate@nasa.gov
Woodruff, Robert A.	Lockheed Martin Astronautics/Civil Space	robert.a.woodruff@lmco.com
Wyatt, Mark	UK Astronomy Technology Centre	wyatt@roe.ac.uk
Youdin, Andrew	Princeton University	youd@astro.princeton.edu
Zakamska, Nadia	Princeton University	nadia@astro.princeton.edu
Zehnle, Ron	AAS*TRA	rzquid@hotmail.com
Zhang, Ke	University of Maryland	kzh@astro.umd.edu
Zhang, William	NASA/GSFC	zhang@rosserv.gsfc.nasa.gov
Zhang, Bing	Penn State University	bzhang@astro.psu.edu

AUTHOR INDEX

321

Jones, H. R. A., 17
Jordan, I. J. E., 161, 223

K

Kinzel, W., 161
Kochte, M., 161, 223
Koller, J., 63
Konacki, M., 165
Kotredes, L., 169, 173
Kuchner, M., 297, 305
Kulesa, C., 107

L

Latham, D. W., 151
Lee, B. L., 177, 181
Leggett, S., 293
Leisawitz, D. T., 297, 305
Lestrade, J.-F., 115
Leviton, D. B., 305
Li, H., 63
Linfield, R., 297
Linz, H., 47
Lobsinger, L. W., 305
Looney, L. W., 59
Looper, D. L., 151, 173
Lufkin, G., 253
Lyon, R. G., 223
Lystad, V., 273

M

Mallén-Ornelas, G., 177, 181
Malumuth, E., 103
Mandushev, G., 151
Marcus, P. S., 67
Marcy, G., 13
Martino, A. J., 305
Mazeh, T., 31
McCarthy, C., 13
Ménard, F., 37, 123
Meyer, M. R., 193
Millan-Gabet, R., 297
Miskey, C., 223
Monnier, J. D., 297
Mozurkewich, D., 301

Mudryk, L., 261
Mugrauer, M., 31
Mumma, M., 297
Mundy, L. G., 59, 297, 305
Murray, N., 261

N

Newhäuser, R., 31
Noecker, C., 297

O

O'Donovan, F. T., 169, 173

P

Pauls, T. A., 301
Pepper, J., 185
Perryman, M. A. C., 283
Pinte, C., 123

Q

Quinn, T., 253

R

Rajagopal, J., 297
Rasio, F., 273
Rassuchine, J., 161
Rettig, T., 107
Richardson, L. J., 189
Rinehart, S. A., 297, 305
Roberge, A., 103
Rodmann, J., 47
Rodrigue, M., 161, 223
Rojo, P., 189
Rose, W. K., 55

S

Sasselov, D. D., 127, 165
Schultz, A. B., 161, 223

Seager, S., 177, 181, 213, 297
Sigurdsson, S., 309
Simon, T., 107
Skelton, D., 223
Sleep, P. N., 227
Stecklum, B., 47
Stevenson, D. J., 133
Steyert, D., 189
Storrs, A., 161
Sudarsky, D., 143

T

Taylor, D. C., 161
Terquem, C. E. J. M. L. J., 235
Torres, C. A. O., 47
Torres, G., 165
Traub, W. A., 297
Tremaine, S., 243, 257
Trimble, V., 3

U

Underwood, D. R., 227

V

Vogt, S., 13, 161
Volgenau, N. H., 59
von Braun, K., 177, 181

W

Wang, H., 47
Weinberger, A. J., 83, 103
Welch, W. J., 59
Welsh, W. F., 161
Wiedemann, G., 189
Wilkinson, E., 47
Williger, G. M., 47
Wood, J. A., 73
Woodgate, B., 47
Wyatt, M. C., 93

Y

Yee, H. K. C., 177, 181

Z

Zakamska, N. L., 243, 257
Zeehandelaar, D., 189
Zhang, B., 309

Subject Index

276,
Rho CrB, 228–229
rings, 6, 39, 44, 51–53, 86, 88, 94–98, 111–112, 215, 265, 298

Sherlock, 159, 172–173, 175–176
shocks, 48, 57, 63–66, 139, 198, 261–263, 309
siderophile, 79, 209
silicates, 48, 79–80, 88–90, 103, 143, 203–205, 209–210
SIM, 17, 30, 283–284, 288
Sleuth, 151, 169, 172–173, 175–176
Sloan survey, 169, 293–296
snow line, 130, 137
solar nebula, 41, 73, 75, 80, 88, 119, 137–139, 203, 237, 244, 250
sound speed, 61–65, 128, 237
spatial resolution, 47, 59, 61, 85, 89, 91, 104, 297
spectroscopic binaries, 244, 250, 269, 301
Spitzer Space Telescope (SIRTF), 17, 43, 45, 84–85, 114, 116, 118, 144–145, 284, 288, 294–295, 299
STARE, 151, 160, 169, 173
star formation, 48, 56, 59, 62, 83, 133-134, 298, 305
STIS, 47–50, 86–87, 103–104, 161–164
submillimeter observations, 111-112, 115-117
surface density, 40, 63–64, 94, 118, 128, 134, 137–138, 205–206, 209, 235–236, 239, 244, 250, 257, 260

Terrestrial Planet Finder (TPF), 17, 146, 213–214, 224, 283–291, 297–298, 305
terrestrial planets, 9, 25, 48, 73, 79, 83–85, 88, 134, 138, 140, 203–206, 208, 210–211, 213–214, 216, 219, 222–225, 227, 240, 277, 283, 285, 287
thermal emission, 38, 42–44, 86, 115, 146, 286
thermal equilibrium, 90, 111
transits, 7, 9, 14, 17, 20, 143–144, 151–193, 196, 213–216, 221, 283–285, 291
trans-Neptunian objects (TNOs), 277–280
triggered formation, 253–256
T Tauri stars, 37, 39–44, 47–48, 73, 84, 103–104, 123
turbulence, 59, 62, 67, 119, 204, 240, 278
TW Hya, 84, 89, 103–105

UKIRT, 32–33, 293
Ups And, 15, 228–229, 251, 273–274

Vega, 97–100, 115, 117
Vega-like stars, 43, 51, 89
Very Large Telescope (VLT), 38, 45, 47, 145, 189–192, 286
viscosity, 134, 240
volatiles, 4, 130, 136, 204, 209, 210, 213
vortices, 6, 63–64, 67, 69
vorticity, 63, 65, 68–70

water, 4, 88–89, 135–137, 140, 143,

CONSTANT	SYMBOL	MKS	CGS	OTHER
Permeability in vacuum	μ_0	$4\pi\cdot10^{-7}$ Hen/m		
Stefan-Boltzmann constant	σ	$5.67\cdot10^{-8}$ W/m²·K⁴	$5.67\cdot10^{-5}$ erg/s·cm²·K⁴	
Rydberg (=$m_e e^4/2\hbar^2$)	R_∞	$2.18\cdot10^{-18}$ J	$2.18\cdot10^{-11}$ erg	13.6 eV
1 amu		$1.66\cdot10^{-27}$ kg	$1.66\cdot10^{-24}$ gm	931.5 MeV
1 calorie		4.19 J	$4.19\cdot10^{7}$ erg	
1 year		$3.16\cdot10^{7}$ s	$3.16\cdot10^{7}$ s	
1 atmosphere		$1.01\cdot10^{5}$ N/m²	$1.01\cdot10^{6}$ dyne/cm²	14.2 lbs/in²
		$1.01\cdot10^{5}$ Pascal		760 Torr
1 eV		$1.6\cdot10^{-19}$ J	$1.6\cdot10^{-12}$ erg	
		$1.24\cdot10^{-6}$ m	$1.24\cdot10^{-4}$ cm	
		1 Tesla	10^{4} gauss	
				11,605 K

ASTROPHYSICAL CONSTANTS

CONSTANT	SYMBOL	MKS	CGS	OTHER
astronomical unit	AU	$1.50\cdot10^{11}$ m	$1.50\cdot10^{13}$ cm	
	AU/year			4.74 km/s
parsec	pc	$3.09\cdot10^{16}$ m	$3.09\cdot10^{18}$ cm	3.26 LY
solar mass	$M_\odot$	$1.99\cdot10^{24}$ kg	$1.99\cdot10^{33}$ gm	
solar luminosity	$L_\odot$	$3.90\cdot10^{26}$ J/s	$3.90\cdot10^{33}$ erg/s	
solar effective temperature	$T_{eff\odot}$	5780 K	5780 K	
solar radius	$R_\odot$	$6.96\cdot10^{8}$ m	$6.96\cdot10^{10}$ cm	
Earth radius	$R_\oplus$	$6.38\cdot10^{6}$ m	$6.38\cdot10^{8}$ cm	
Earth mass	$M_\oplus$	$5.98\cdot10^{24}$ kg	$5.98\cdot10^{27}$ gm	
Earth density	$\rho_\oplus$	5520 kg/m³	5.52 gm/cm³	
Jansky	Jy	$1.0\cdot10^{-26}$ W/m²·Hz	$1\cdot10^{-23}$ erg/s·cm²·Hz	
Hubble constant	H_0	$3.24h\cdot10^{-18}$ s⁻¹	$3.24h\cdot10^{-18}$ s⁻¹	$100h$ km/s·Mpc
critical density (=$3H_0^2/8\pi G$)	ρ_0	$1.88h^2\cdot10^{-26}$ kg/cm³	$1.88h^2\cdot10^{-29}$ gm/cm³	
plasma frequency				$8.98\sqrt{n_e(cm^{-3})}$ kHz/gauss
radian				$57.29578° = 206{,}265''$
CMB photon density	n_γ	$4.15\cdot10^{5}$ m⁻³	415 cm⁻³	